Applied Probability and Statistics

BAILEY · The Elements of Stochastic Processes with Applications to the Natural Sciences
BAILEY · Mathematics, Statistics and Systems for Health
BARTHOLOMEW · Stochastic Models for Social Processes, *Second Edition*
BECK and ARNOLD · Parameter Estimation in Engineering and Science
BENNETT and FRANKLIN · Statistical Analysis in Chemistry and the Chemical Industry
BHAT · Elements of Applied Stochastic Processes
BLOOMFIELD · Fourier Analysis of Time Series: An Introduction
BOX · R. A. Fisher, The Life of a Scientist
BOX and DRAPER · Evolutionary Operation: A Statistical Method for Process Improvement
BOX, HUNTER, and HUNTER · Statistics for Experimenters: An Introduction to Design, Data Analysis, and Model Building
BROWN and HOLLANDER · Statistics: A Biomedical Introduction
BROWNLEE · Statistical Theory and Methodology in Science and Engineering, *Second Edition*
BURY · Statistical Models in Applied Science
CHAMBERS · Computational Methods for Data Analysis
CHATTERJEE and PRICE · Regression Analysis by Example
CHERNOFF and MOSES · Elementary Decision Theory
CHOW · Analysis and Control of Dynamic Economic Systems
CLELLAND, deCANI, BROWN, BURSK, and MURRAY · Basic Statistics with Business Applications, *Second Edition*
COCHRAN · Sampling Techniques, *Third Edition*
COCHRAN and COX · Experimental Designs, *Second Edition*
COX · Planning of Experiments
COX and MILLER · The Theory of Stochastic Processes, *Second Edition*
DANIEL · Application of Statistics to Industrial Experimentation
DANIEL · Biostatistics: A Foundation for Analysis in the Health Sciences
DANIEL and WOOD · Fitting Equations to Data
DAVID · Order Statistics
DEMING · Sample Design in Business Research
DODGE and ROMIG · Sampling Inspection Tables,
DRAPER and SMITH · Applied Regression Analysis
DUNN · Basic Statistics: A Primer for the Biom *Second Edition*
DUNN and CLARK · Applied Statistics: Analysis of Variance and Regression
ELANDT-JOHNSON · Probability Models and Statistical Methods in Genetics
FLEISS · Statistical Methods for Rates and Proportions
GALAMBOS · The Asymptotic Theory of Extreme Order Statistics
GIBBONS, OLKIN, and SOBEL · Selecting and Ordering Populations: A New Statistical Methodology
GNANADESIKAN · Methods for Statistical Data Analysis of Multivariate Observations
GOLDBERGER · Econometric Theory
GOLDSTEIN and DILLON · Discrete Discriminant Analysis

continued on back

*The Asymptotic Theory
of Extreme Order Statistics*

The Asymptotic Theory of Extreme Order Statistics

JANOS GALAMBOS

Temple University
Philadelphia, Pennsylvania

John Wiley & Sons, New York • Chichester • Brisbane • Toronto

QA
274
.G34

Copyright © 1978 by John Wiley & Sons, Inc.

All rights reserved. Published simultaneously in Canada.

Reproduction or translation of any part of this work beyond that permitted by Sections 107 or 108 of the 1976 United States Copyright Act without the permission of the copyright owner is unlawful. Requests for permission or further information should be addressed to the Permissions Department, John Wiley & Sons, Inc.

Library of Congress Cataloging in Publication Data:

Galambos, János, 1940-
  The asymptotic theory of extreme order statistics.

(Wiley series in probability and mathematical statistics)
  Bibliography: p.
  Includes index.
  1. Stochastic processes.   2. Random variables.
3. Extreme value theory.   4. Distribution
(Probability theory)   I. Title.
QA274.G34   519.2   78-1916
ISBN 0-471-02148-2

Printed in the United States of America

10 9 8 7 6 5 4 3 2 1

Preface

The asymptotic theory of extreme order statistics provides in some cases exact but in most cases approximate probabilistic models for random quantities when the extremes govern the laws of interest (strength of materials, floods, droughts, air pollution, failure of equipment, effects of food additives, etc.). Therefore, a complicated situation can be replaced by a comparatively simple asymptotic model if the basic conditions of the actual situation are compatible with the assumptions of the model. In the present book I describe all known asymptotic models. In addition to finding the asymptotic distributions, both univariate and multivariate, I also include results on the almost sure behavior of the extremes. Finally, random sample sizes are treated and a special random size, the so-called record times, is discussed in more detail. A short section of the last chapter dealing with extremal processes is more mathematical than the rest of the book and intended for the specialist only.

Let me stress a few points about the asymptotic theory of extremes. I have mentioned that an asymptotic model may sometimes lead to the exact stochastic description of a random phenomenon. Such cases occur when a random quantity can be expressed as the minimum or maximum of the quantities associated with an arbitrarily large subdivision (for example, the strength of a sheet of a metal is the minimum of the strengths of the pieces of the sheet). But whether a model is used as an exact solution or as an approximation, its basic assumptions decide whether it is applicable in a given problem. Therefore, if the conclusions for several models are the same, each model contributes to the theory by showing that those conclusions are applicable under different circumstances. One of the central problems of the theory is whether the use of a classical extreme value distribution is justified—that is, a distribution which can be obtained as the limit distribution of a properly normalized extreme of independent and identically distributed random variables. Several of the models of the book give an affirmative answer to this question. In several other cases, however, limiting distributions are obtained that do not belong to the three classical types. This is what Bayesian statisticians and reliability scientists expected all along (and they actually used these distributions without appealing to extreme value theory). I sincerely hope that these distributions will be widely used in other fields as well.

One more point that is not encountered in most cases of applied statistics comes up in the theory of extremes. Even if one can accept that the basic random variables are independent and identically distributed, one cannot make a decision on the population distribution by standard statistical methods (goodness of fit tests). I give an example (Example 2.6.3) where, by usual statistical methods, both normality and lognormality are acceptable but the decision in terms of extremes is significantly different depending on the actual choice of one of the two mentioned distributions. It follows that this choice has to be subjective (this is the reason for two groups coming to opposite conclusions, even though they had the same information).

The book is mathematically rigorous, but I have kept the applied scientist in mind both in the selection of the material and in the explanations of the mathematical conclusions through examples and remarks. These remarks and examples should also make the book more attractive as a graduate text. I hope, further, that the book will promote the existing theory among applied scientists by giving them access to results that were scattered in the mathematical literature. An additional aim was to bring the theory together for the specialists in the field. The survey of the literature at the end of each chapter and the extensive bibliography are for these purposes.

The prerequisites for reading the book are minimal; they do not go beyond basic calculus and basic probability theory. Some theorems of probability theory (including expectations or integrals), which I did not expect to have been covered in a basic course, are collected in Appendix I. The only exception is the last section in Chapter 6, which is intended mainly for the specialists. By the nature of the subject matter, some familiarity with statistics and distribution functions is an advantage, although I introduce all distributions used in the text. The book can be used as a graduate text in any department where probability theory or mathematical statistics are taught. It may also serve as a basis for a nonmathematical course on the subject, in which case most proofs could be dropped but their ideas presented through special cases (e.g., starting with a simple class of population distributions). In a course, or at a first reading, Chapter 4 can be skipped; Chapters 5 and 6 are not dependent on it.

No books now in print cover the materials of any of Chapters 1 or 3–6. The only overlap with existing monographs is Chapter 2, which is partially contained in the book by Gumbel (1958) and in the monograph of de Haan (1970) (see the references). It should be added, however, that Gumbel's book has an applied rather than theoretical orientation. His methods are not applicable when the restrictive assumptions of Chapter 2 are not valid.

Although many proofs are new here, the credit for the theory, as it is known at present, is due to those scientists whose contributions raised it to its current level. It is easy to unify and simplify proofs when the whole material is collected at one place.

I did not have time to thank the many scientists individually who responded to my requests and questions. My heartiest thanks go to them all over the world. My particular thanks are due to those scientists who apply extreme value theory and who so patiently discussed the problems with me either in person or in letters.

I am indebted to Professor David G. Kendall, whom I proudly count among my friends, for presenting my plan to John Wiley & Sons, Inc. I should also like to thank Mrs. Mittie Davis for her skill and care in typing the manuscript.

JANOS GALAMBOS

Willow Grove, Pennsylvania
March 1978

Contents

Notations and conventions		xiii

	INTRODUCTION: ESTIMATES IN THE UNIVARIATE CASE	1
1.1	Problems leading to extreme values of random variables	1
1.2	The mathematical model	3
1.3	Preliminaries for the independent and identically distributed case	7
1.4	Bounds on the distribution of extremes	15
1.5	Illustrations	25
1.6	Special properties of the exponential distribution in the light of extremes	30
1.7	Survey of the literature	42
1.8	Exercises	44

2.	WEAK CONVERGENCE FOR INDEPENDENT AND IDENTICALLY DISTRIBUTED VARIABLES	49
2.1	Limit distributions for maxima and minima: sufficient conditions	51
2.2	Other possibilities for the normalizing constants in Theorems 2.1.1–2.1.6	57
2.3	The asymptotic distribution of the maximum and minimum for some special distributions	63
2.4	Necessary conditions for weak convergence	69
2.5	The proof of part (iii) of Theorem 2.4.3	79
2.6	Illustrations	88
2.7	Further results on domains of attraction	93

	2.8	Weak convergence for the kth extremes	102
	2.9	The range and midrange	105
	2.10	Speed of convergence	111
	2.11	Survey of the literature	117
	2.12	Exercises	120

3. WEAK CONVERGENCE OF EXTREMES IN THE GENERAL CASE — 124

3.1	A new look at failure models	124
3.2	The special role of exchangeable variables	127
3.3	Preparations for limit theorems	134
3.4	A limit theorem for mixtures	137
3.5	A theoretical model	143
3.6	Segments of infinite sequences of exchangeable variables	148
3.7	Stationary sequences	156
3.8	Stationary Gaussian sequences	163
3.9	Limiting forms of the inequalities of Section 1.4	174
3.10	Minimum and maximum of independent variables	180
3.11	The asymptotic distribution of the kth extremes	184
3.12	Some applied models	188

3.12.1.	Exact models via characterization theorems	188
3.12.2.	Exact distributions via asymptotic theory (strength of materials)	189
3.12.3.	Strength of bundles of threads	191
3.12.4.	Approximation by i.i.d. variables	192
3.12.5	Time to first failure of equipments with a large number of components	195

3.13	Survey of the literature	197
3.14	Exercises	201

4. DEGENERATE LIMIT LAWS; ALMOST SURE RESULTS — 205

4.1	Degenerate limit laws	205
4.2	Borel-Cantelli lemmas	210
4.3	Almost sure asymptotic properties of extremes of i.i.d. variables	213
4.4	Lim sup and lim inf of normalized extremes	225
4.5	The role of extremes in the theory of sums of random variables	232

	CONTENTS	
4.6	Survey of the literature	238
4.7	Exercises	240

5. MULTIVARIATE EXTREME VALUE DISTRIBUTIONS — 244

5.1	Basic properties of multivariate distributions	245
5.2	Weak convergence of extremes for i.i.d. random vectors: basic results	248
5.3	Further criteria for the i.i.d. case	254
5.4	On the properties of $H(\mathbf{x})$	258
5.5	Concomitants of order statistics	267
5.6	Survey of the literature	271
5.7	Exercises	274

6. MISCELLANEOUS RESULTS — 280

6.1	The maximum queue length in a stable queue	280
6.2	Extremes with random sample size	282
6.3	Record times	290
6.4	Records	299
6.5	Extremal processes	304
6.6	Survey of the literature	308
6.7	Exercises	311

APPENDIXES

I.	SOME BASIC FORMULAS FOR PROBABILITIES AND EXPECTATIONS	314
II.	THEOREMS FROM FUNCTIONAL ANALYSIS	322
III.	SLOWLY VARYING FUNCTIONS	329

REFERENCES — 331

INDEX — 349

Notations and Conventions

$X_1, X_2, \ldots, X_n$	basic random variables.
Z_n	maximum of $X_1, X_2, \ldots, X_n$.
W_n	minimum of $X_1, X_2, \ldots, X_n$.
$F(x) = P(X < x)$	distribution function of X.
$H_n(x)$	distribution function of Z_n.
$L_n(x)$	distribution function of W_n.
$\alpha(F)$	$\inf\{x : F(x) > 0\}$.
$\omega(F)$	$\sup\{x : F(x) < 1\}$.
$H(x)$	limit of $H_n(a_n + b_n x)$ with some constants a_n and $b_n > 0$.
$L(x)$	limit of $L_n(c_n + d_n x)$ with some constants c_n and $d_n > 0$.
$X_{r:n}$	the rth order statistic of $X_1, X_2, \ldots, X_n$. Thus $X_{1:n} \leq X_{2:n} \leq \ldots \leq X_{n:n}$ and $X_{1:n} = W_n$ and $X_{n:n} = Z_n$.
$\sum_{k=1}^{t}, \prod_{k=1}^{t}$	summation and product, respectively, from one to the integer part of t.
$H_{1,\gamma}(x)$	defined at (11) on p. 51.
$H_{2,\gamma}(x)$	defined at (13) on p. 51.
$H_{3,0}(x)$	defined at (18) on p. 52.
$L_{1,\gamma}(x)$	defined at (28) on p. 56.
$L_{2,\gamma}(x)$	defined at (30) on p. 56.
$L_{3,0}(x)$	defined at (35) on p. 57.
A' or A^c	the complement of the set A.
i.i.d.	independent and identically distributed.
i.o.	infinitely often.

*The Asymptotic Theory
of Extreme Order Statistics*

CHAPTER 1

Introduction: Estimates in the Univariate Case

We shall describe a number of situations where the extremes govern the laws that interest us. Both practical and theoretical problems will be listed, which will then be unified by a general mathematical model. Our aim is to investigate this mathematical model and to describe our present stage of knowledge about it under different sets of assumptions. The beauty of this subject matter is that it leads us to the understanding of regularities of extreme behavior—an expression that seems to contradict itself.

After the introduction of the mathematical model, the present chapter is devoted to inequalities which involve the distribution of extremes in a set of observations. Those inequalities serve two purposes. On one hand, they may provide good bounds on the distribution of extremes for a given number of observations without resorting to an asymptotic theory. On the other hand, they will constitute some of the basic tools of the asymptotic theory to be developed in later chapters. It should be emphasized that in several situations, contrary to general belief, an asymptotic theory may provide the exact model, while a fixed number of observations can be used only as an approximation. The reader is referred to Section 1.2 for a specific example.

1.1. PROBLEMS LEADING TO EXTREME VALUES OF RANDOM VARIABLES

We now list a number of cases when a mathematical solution to the problems involved is in terms of the largest or the smallest "measurements."

Natural disasters. Floods, heavy rains, extreme temperatures, extreme atmospheric pressures, winds and other phenomena can cause extensive human and material loss, if society is unprepared for them. While such disasters cannot be completely avoided, communities can take preventive action to minimize their effects. In dams, dikes, canals, and other structures the choice of building materials and methods of architecture can take some of these disasters into account. Engineering decisions that confront such problems should be based on a very accurate theory, because inaccuracies can be very expensive. For example, dams built at a huge expense may not last long before collapsing.

Failure of equipments. Assume that an equipment fails if one of its components fails. In other words, we consider only those of its components, the failure of any one of which leads to a halt in its operation. This is an extreme situation in the sense that the weakest component alone makes the equipment fail. While this assumption may seem a simplification, the most general failure model of a complicated equipment can be reduced to this model. As a matter of fact, if one first considers groups of components where the failure of a group results in the failure of the equipment, then the weakest group of components with the assumed property effects the first failure of the equipment.

Service time. Consider an equipment with large number of components and assume that components can be serviced concurrently. Then the time required for servicing the equipment is determined by that component which requires the longest service.

Corrosion. We say that a surface with a large number of small pits fails due to corrosion if any one of the pits penetrates through the thickness of the surface. Initially the pits are of random depths, which increase in time due to chemical corrosion. Again one extreme measurement, the deepest pit, causes failure.

Breaking strength. An absolutely homogeneous material would break under stress by a deterministic law. However, no material is absolutely homogeneous; indeed, engineering experience shows that the breaking strength of materials under identical production procedures varies widely. The explanation is that each point, or at least each small area, has a random strength, and thus varying amounts of force will be needed to break the material at different points. Evidently the weakest point will determine the strength of the whole material.

Air pollution. Air pollutant concentration is expressed in terms of proportion of a specific pollutant in the air. Concentrations are recorded at equal time intervals (present investigations are based on data obtained at five-minute intervals), and the aim of society is to keep the largest measurement below given limits.

Statistical samples. Observations are made on a given quantity; often one would like to know how large or how small a measurement can be expected.

Statistical estimators. After the collection of observations, the data are used to calculate estimators of certain characteristics of the quantity under observation. One would like to estimate these characteristics as accurately as possible, but over- or underestimation is unavoidable. Of considerable interest, therefore, is the investigation of the largest or the smallest estimator.

These problems, though they do not exhaust the possibilities, indicate that any successful theory of extremes unifies a great number of interesting topics. The theory to be developed can also show the beauty of mathematical abstraction: a single language will speak to the engineer, the physicist, the service person, the statistician, and others.

Further examples of fields for application of the theory will be spread in the text and among the problems for solution. Problems leading to multivariate extremes are postponed until Chapter 5.

1.2 THE MATHEMATICAL MODEL

In all the examples of the preceding section we were faced with a number n of random measurements $X_1, X_2, \ldots, X_n$, and the behavior of either

$$Z_n = \max(X_1, X_2, \ldots, X_n)$$

or

$$W_n = \min(X_1, X_2, \ldots, X_n)$$

was of interest.

As a matter of fact, in terms of floods, X_j may denote the water level of a given river on day j, "day 1" being, for example, the day of publication of this book. Since we do not know the water levels in advance, they are random to us. A question such as, "How likely is it that in this century the water level of our river remains below 230 cm?" is evidently asking the value $P(Z_n \leq 230)$, the probability of the event $\{Z_n \leq 230\}$. Here n is, of course, the number of days remaining in this century after the publication of this book. On the other hand, if we want to use the river as a source of energy, then our interest is in how far the water level can fall. This translates into $P(W_n \geq a)$, where we specify the value of a.

Similarly, if X_j is the amount of rainfall, the highest temperature, and so on, during the jth time unit, then whether a new agricultural product can be produced under certain climates is again dependent on the extreme rainfalls, temperatures, etc.—that is, Z_n or W_n. The present author lived through this argument in two different climates with two different products: the widespread production of rice in Hungary, for which a minimum amount of rainfall per week is needed (that is, decision is in terms of W_n with time unit a week), and the production of potatoes in West Africa, where high daily temperatures were the problem (that is, decision is in terms of Z_n with time unit one day).

The translation of the other examples to our mathematical model is evident. In case of failure of equipment, the fault-free operation of the equipment lasts for W_n time units, where X_j is the time to failure of the jth component (or jth group of components) and n is the number of components (or of groups of components). On the other hand, the service time is Z_n where X_j is the time needed for servicing the jth component and n is the number of components to be serviced. In the corrosion model, if the surface has n pits and X_j is the depth of the jth pit at a given time, then our interest is $P(Z_n < a)$, where a is the thickness of the surface. The reader is invited to complete this translation procedure for the remaining examples.

Since our measurements $X_1, X_2, \ldots$ are random, so are Z_n and W_n. Consequently, questions and solutions in regard to Z_n and W_n are not deterministic: we speak of the magnitude of Z_n and W_n only in probabilistic terms. In other words, questions and solutions in our model can, and will, concern only the magnitude of probabilities of Z_n, or W_n, falling into specified sets. In particular, we would like to evaluate or approximate the distributions

$$H_n(x) = P(Z_n < x) \tag{1}$$

and

$$L_n(x) = P(W_n < x). \tag{2}$$

Before we finally define our mathematical problem, let us go back to two of the examples of the previous section.

Let us first analyze the breaking strength S of a sheet of a metal with rectangular shape. The strength S is a random quantity and it thus has a distribution $L(x) = P(S < x)$. Let us now divide the sheet into n^2 equal parts by dividing each edge by n. Let the random strength of the jth part

1.2 THE MATHEMATICAL MODEL

be X_j. Since the whole sheet will break as soon as one part, the weakest one, breaks, evidently S equals the minimum W_n of X_j, $1 \leq j \leq n^2$. This means that $L_n(x) = L(x)$ for all n. In particular, $L(x)$ is the limit of $L_n(x)$ as $n \to +\infty$. While finding the distribution of X_j, and thus of W_n as well, is as difficult as finding $L(x)$ itself (the nature of the problem is evidently the same), a successful asymptotic theory for $L_n(x)$ may provide the answer for $L(x)$ without facing the problem of actually determining $L_n(x)$. This will indeed be the case for several models. The limiting form of $H_n(x)$ or $L_n(x)$ will be the same, or one of a small number of possibilities, regardless of the actual distribution of the X_j. Notice that here we seek the exact form of $L(x)$ through developing a theory which guarantees the existence of the limit of L_n (or H_n).

As a second case, let us choose the problem of flood on a river. Any practical-minded person would argue that if I have a method of predicting with high probability that the river will not need new dikes (for example, the water level remains below 230 cm) until, say, 1995, then I can use the same method of prediction until 1997 as well. In other words, translating into the notations of the present section, $H_n(x)$ does not change much with n after certain values. Since the distribution function $H_n(x)$ expresses the variation of Z_n with chance, the accurate mathematical equivalent to the above "naive" approach is that, after perhaps some normalization with constants a_n and $b_n > 0$, $(Z_n - a_n)/b_n$ becomes more and more independent of n. That is,

$$\lim_{n = +\infty} P\left(\frac{Z_n - a_n}{b_n} < z\right) = \lim_{n = +\infty} H_n(a_n + b_n z) = H(z)$$

exists for fixed z. Therefore if we could determine $H(z)$ itself, then the X_j, and their distributions, would not matter any more, provided a_n and b_n can somehow be calculated.

Notice that, in all of our examples, n is indeed large, and so our aim is to seek conditions under which the naive approach is mathematically justified. In addition, of course, we want to make our result meaningful by giving the actual limiting distribution and methods of calculating the constants a_n and $b_n > 0$.

Let us summarize this aim:

The main aim of the present book is to give conditions under which there are constants $a_n, c_n, b_n > 0$ and $d_n > 0$ such that, as $n \to +\infty$, $H_n(a_n + b_n z)$ and/or $L_n(c_n + d_n y)$ approach some distribution functions $H(z)$ and/or $L(y)$, respectively. In addition, we want to determine $H(z)$ and $L(y)$ and to give methods for the calculation of the constants a_n, b_n, c_n, and d_n.

For strictly mathematical reasons, the existence of the limits $H(z)$ and $L(y)$ will be required for continuity points of $H(z)$ and $L(y)$. As is well known, continuity points uniquely determine a distribution function.

Besides our main aim, we shall present several additional interesting mathematical results which are partly easy consequences of the methods developed for our main aim expressed above. Others will require some additional arguments or modifications. In particular, extensions will be made when n itself is a random quantity as well as to vector variables.

Throughout this book, we use the denotations

$$F_j(x) = P(X_j < x) \tag{3}$$

and

$$F_n^*(x_1, x_2, \ldots, x_n) = P(X_1 < x_1, X_2 < x_2, \ldots, X_n < x_n). \tag{4}$$

It is evident that

$$H_n(x) = F_n^*(x, x, \ldots, x). \tag{5}$$

Also, putting

$$G_j(x) = 1 - F_j(x) = P(X_j \geqslant x) \tag{6}$$

and

$$G_n^*(x_1, x_2, \ldots, x_n) = P(X_1 \geqslant x_1, X_2 \geqslant x_2, \ldots, X_n \geqslant x_n), \tag{7}$$

we get

$$1 - L_n(x) = G_n^*(x, x, \ldots, x). \tag{8}$$

Equations (5) and (8) show that a solution to our main aim means the knowledge of, or a good approximation to, the functions F_n^* and G_n^* when all of their n variables are identical. We shall use different methods of solution to this problem under different assumptions on the interrelation of $X_1, X_2, \ldots, X_n$. In several models of interdependence, we shall be able to find the asymptotic behavior not only of Z_n and W_n but also of the kth largest and the kth smallest among the X_j's (for the accurate mathematical definition, see Section 1.4).

Let us conclude this section with a simple but useful remark. Let $h(u)$ be a strictly decreasing function of u, defined for all possible values of $X_1, X_2, \ldots, X_n$. Putting $Y_j = h(X_j)$, we get

$$h\{\max(X_1, X_2, \ldots, X_n)\} = \min(Y_1, Y_2, \ldots, Y_n)$$

and

$$h\{\min(X_1, X_2, \ldots, X_n)\} = \max(Y_1, Y_2, \ldots, Y_n).$$

The most natural choice for $h(u)$ is $h(u) = -u$ for showing that the theory for Z_n is identical to that for W_n. For this reason, some statements will be made for either Z_n or W_n only. In such cases, however, the counterpart of the statement will be left as a problem for solution in order to record all statements for both maxima and minima.

1.3. PRELIMINARIES FOR THE INDEPENDENT AND IDENTICALLY DISTRIBUTED CASE

Looking at the mathematical model of Section 1.2 in the light of the practical problems of Section 1.1, one immediately sees that the variables $X_1, X_2, \ldots, X_n$ are dependent. The only exception can be the theoretical concept of statistical samples when one assumes that the experimenter is free to choose his data and therefore he does so independently at each observation. We would like to cover all practical situations, however, hence the following discussion is not representative of our aim and method, but will serve as an approximation and guide in several situations.

Let $X_1, X_2, \ldots, X_n$ be independent and identically distributed (i.i.d.) random variables. With the notations

$$Z_n = \max(X_1, X_2, \ldots, X_n),$$
$$W_n = \min(X_1, X_2, \ldots, X_n)$$

and

$$F(x) = P(X_j < x),$$

formulas (4)–(8) now reduce to

$$P(Z_n < x) = H_n(x) = F^n(x) \tag{9}$$

and

$$P(W_n \geqslant x) = 1 - L_n(x) = (1 - F(x))^n. \tag{10}$$

We give below two simple approximations to $H_n(x)$ and $L_n(x)$. These will serve as guides—one for obtaining limit theorems, the other for obtaining estimates for dependent systems. We start with two lemmas.

Lemma 1.3.1. *For any* $0 < z < \frac{1}{2}$,

$$e^{-nz} - (1-z)^n \{\exp(2nz^2) - 1\} < (1-z)^n \leq e^{-nz}.$$

The upper inequality remains to hold for $0 \leq z \leq 1$.

Proof. Let us first observe that, for any $0 < z < 1$,

$$(1-z)^n < e^{-nz}. \tag{11}$$

Indeed, since

$$(1-z)^n = \exp\{n \log(1-z)\},$$

one has to establish

$$\log(1-z) < -z.$$

But this last inequality immediately follows by comparing $\log(1-z)$ and its tangent at $z = 0$.

For reversing (11) with an error term, we show that, for $0 < z < \frac{1}{2}$,

$$(1-z)^{n-nz} > e^{-nz}. \tag{12}$$

As a matter of fact, by turning to the exponential form of the left hand side again, we see that (12) is equivalent to

$$(1-z) \log(1-z) > -z. \tag{12a}$$

Inequalities of this kind can easily be proved by integrating an appropriate inequality. Here we start with

$$\frac{1}{1-z} < 1 + 2z, \qquad 0 < z < \frac{1}{2},$$

which is evidently true. Integration yields

$$\log(1-z) > -z - z^2,$$

from which (12a), and thus (12) as well, now follows. Inequalities (11) and (12) imply that, for $0 < z < \frac{1}{2}$,

$$0 < e^{-nz} - (1-z)^n < (1-z)^n \{(1-z)^{-nz} - 1\}. \tag{13}$$

It now remains only to see that, for $0 < z < \frac{1}{2}$,

$$(1-z)^{-nz} = \exp\{-nz \log(1-z)\} < \exp(2nz^2),$$

1.3 PRELIMINARIES FOR THE I.I.D. CASE

the last step following from

$$-\log(1-z) < 2z, \quad 0 < z < \tfrac{1}{2}.$$

This completes the proof of the lemma. ▲

Lemma 1.3.2. *Let* $n \geq 1$ *be an integer. Then for any* $0 \leq z \leq 1$ *and for all integers* $s \geq 0$,

$$\sum_{k=0}^{2s+1} (-1)^k \binom{n}{k} z^k \leq (1-z)^n \leq \sum_{k=0}^{2s} (-1)^k \binom{n}{k} z^k.$$

Proof. We prove the lemma by induction over n. Put

$$f_{m,n}(z) = \sum_{k=0}^{m} (-1)^k \binom{n}{k} z^k$$

and

$$g_{m,n}(z) = f_{m,n}(z) - (1-z)^n.$$

Then evidently, for any $n \geq 1$ and $0 \leq z \leq 1$,

$$g_{0,n}(z) = 1 - (1-z)^n \geq 0. \tag{14}$$

It is also immediate that, for $m \geq 1$,

$$g_{m,1}(z) \equiv 0.$$

Therefore, in view of (14), the lemma is true for $n = 1$. Let us now fix $n > 1$ and assume that for $n-1$, the lemma has been proved. That is, for $0 \leq z \leq 1$ and for integers $s \geq 0$,

$$g_{2s, n-1}(z) \geq 0, \quad g_{2s+1, n-1}(z) \leq 0. \tag{15}$$

Let us consider $g_{m,n}(z)$. By comparing the function $(1-z)^n$ with its tangent $1 - nz$ at $z = 0$, we get

$$g_{1,n}(z) = 1 - nz - (1-z)^n \leq 0, \quad 0 \leq z \leq 1. \tag{16}$$

The inequalities (14) and (16) can be combined to state that, for $s = 0$,

$$g_{2s,m}(z) \geq 0, \quad g_{2s+1,n}(z) \leq 0, \quad 0 \leq z \leq 1. \tag{17}$$

For proving (17) for $s \geq 1$, we first observe that, for $m \geq 1$,

$$g'_{m,n}(z) = \sum_{k=1}^{m}(-1)^k \binom{n}{k} kz^{k-1} + n(1-z)^{n-1}$$

$$= n\sum_{k=1}^{m}(-1)^k \binom{n-1}{k-1} z^{k-1} + n(1-z)^{n-1} = -ng_{m-1,n-1}(z).$$

Second, for $m \geq 0, n \geq 1, g_{m,n}(0) = 0$. Thus, (15) implies (17) for $s \geq 1$, which completes the proof. ▲

An easy application of Lemma 1.3.1 leads to the following useful inequalities for the distribution of Z_n and of W_n.

Theorem 1.3.1. *Let $X_1, X_2, \ldots, X_n$ be i.i.d. random variables with common distribution function $F(x)$. Let x be such that*

$$1 - F(x) < \frac{1}{2\sqrt{n}}. \tag{18}$$

Then, for $n \geq 1$,

$$T(x) - 4n[1 - F(x)]^2 F^n(x) < P(Z_n < x) < T(x),$$

where

$$T(x) = \exp\{-n[1 - F(x)]\}.$$

Theorem 1.3.2. *Let $X_1, X_2, \ldots, X_n$ be i.i.d. with common distribution function $F(x)$. Let x be such that*

$$F(x) < \frac{1}{2\sqrt{n}}. \tag{19}$$

Then, for $n \geq 1$,

$$U(x) - 4nF^2(x)[1 - F(x)]^n < P(W_n \geq x) < U(x),$$

where

$$U(x) = \exp[-nF(x)].$$

Proof of Theorems 1.3.1 and 1.3.2. We first apply Lemma 1.3.1 with $z = 1 - F(x)$. Then, by (9), $(1-z)^n$ becomes $P(Z_n < x)$. The conclusion of

Theorem 1.3.1 follows by the elementary estimate

$$|e^w - 1| < 2w, \qquad 0 < w < \tfrac{1}{2},$$

which can be applied in Lemma 1.3.1 by assumption (18). The proof of Theorem 1.3.2 is identical to the one above with the choice of $z = F(x)$ and by an appeal to (10) and (19). Theorems 1.3.1 and 1.3.2 are thus established. ▲

Let us record the limiting forms of Theorems 1.3.1 and 1.3.2. In the statements below we permit infinity as limit, for which case we adopt the relation $\exp(-\infty) = 0$.

Corollary 1.3.1. *Let us use the notations of Theorem 1.3.1. Assume that there are sequences a_n and $b_n > 0$ of real numbers such that, for all y, as $n \to +\infty$*

$$\lim n[1 - F(a_n + b_n y)] = u(y) \qquad (20)$$

exists. Then, as $n \to +\infty$,

$$\lim P(Z_n < a_n + b_n y) = \exp[-u(y)]. \qquad (21)$$

Proof. Let us consider first the case of $u(y) = +\infty$. Then the upper inequality in Lemma 1.3.1 with $z = 1 - F(a_n + b_n y)$ immediately yields (21) (recall $e^{-\infty} = 0$). We can therefore assume that $u(y) < +\infty$. Now formula (20) implies (18) for all sufficiently large n. Thus, if we apply the estimates $F^n(x) < 1$, and

$$n[1 - F(a_n + b_n y)]^2 = \tfrac{1}{n}\{n[1 - F(a_n + b_n y)]\}^2 \to 0,$$

as $n \to +\infty$, on account of $u(y) < +\infty$ in (20), the conclusion of Theorem 1.3.1 leads to (21). This completes the proof. ▲

Appealing to Theorem 1.3.2 and arguing as above, we get the following limit theorem for W_n.

Corollary 1.3.2. *Let us use the notations of Theorem 1.3.2. Assume that there are sequences c_n and $d_n > 0$ of real numbers such that, for all y, as $n \to +\infty$,*

$$\lim n F(c_n + d_n y) = w(y) \qquad (22)$$

exists. Then, as $n \to +\infty$,

$$\lim P(W_n < c_n + d_n y) = 1 - \exp\{-w(y)\}. \qquad (23)$$

The difficulty in applying Corollaries 1.3.1 and 1.3.2 lies in the fact that they do not give conditions guaranteeing the validity of (20) and (22). Neither do they give methods of determining the sequences a_n, b_n, c_n, and d_n. We shall discuss these problems in Chapter 2. Here we limit ourselves to some examples.

Example 1.3.1 (The Exponential Distribution). Let $X_1, X_2, \ldots, X_n$ be i.i.d. random variables with common distribution function

$$F(x) = 1 - e^{-x}, \quad x \geq 0. \tag{24}$$

Then

$$1 - F(a_n + b_n z) = e^{-a_n} e^{-b_n z}.$$

In order to satisfy (20), one can choose $a_n = \log n$ and $b_n = 1$. Hence $u(z) = e^{-z}$ and thus, as $n \to +\infty$,

$$\lim P(Z_n < \log n + z) = \exp\{-e^{-z}\}. \tag{25}$$

On the other hand, Corollary 1.3.2 yields that one can choose $c_n = 0$ and $d_n = 1/n$ for (22) to hold. With this choice, (22) becomes

$$\lim n F\left(\frac{y}{n}\right) = y \quad (n \to +\infty).$$

Hence, by (23), as $n \to +\infty$,

$$\lim P\left(W_n < \frac{y}{n}\right) = 1 - e^{-y}. \tag{26}$$

In the special case of the exponential distribution (24), however, (26) can be obtained directly from (10) without referring to Corollary 1.3.2. Indeed, by (10),

$$P(W_n \geq x) = e^{-nx}$$

and thus

$$P\left(W_n < \frac{y}{n}\right) = 1 - e^{-y} \tag{26a}$$

for all n, not only in limit. Such property is, however, shared by a very few distributions only (see Section 2.4 of Chapter 2 for details).

As a comparison, let us calculate $P(Z_{50} < 5)$ by the exact formula (9) and let us also calculate its approximation by (25). We get from (9)

$$P(Z_{50} < 5) = (1 - e^{-5})^{50} = 0.71317,$$

while (25) yields

$$P(Z_{50}<5) \sim \exp\{-\exp(\log 50 - 5)\} = 0.71398.$$

On the other hand, Theorem 1.3.1 results in the following estimates. First, we have to check (18), which indeed holds. Then, replacing $F^n(x)$ by one on the left hand side of the estimate in Theorem 1.3.1, this error term becomes

$$200e^{-10} = 0.00908.$$

Thus, since

$$\exp(-50e^{-5}) = 0.71398,$$

we get

$$0.70490 \leq P(Z_{50}<5) \leq 0.71398.$$

▲

Example 1.3.2. Let $X_1, X_2, \ldots, X_n$ be i.i.d. with common distribution function

$$F(x) = 1 - \frac{1}{x}, \qquad x \geq 1.$$

For determining the limiting distribution of Z_n, let us observe that (20) is satisfied with $a_n = 0$ and $b_n = n$. Equation (20) becomes

$$\lim n \cdot (nz)^{-1} = \frac{1}{z}, \qquad z > 0,$$

and thus, by (21), as $n \to +\infty$,

$$\lim P(Z_n < nz) = \exp\left(-\frac{1}{z}\right), \qquad z > 0.$$

▲

Example 1.3.3. Let the common distribution function of the independent random variables $X_j, 1 \leq j \leq n$, be

$$F(x) = 1 - \frac{1}{\log x}, \qquad x \geq e.$$

In Chapter 2 (see Example 2.5.1) we shall see that (20) cannot be satisfied and thus Corollary 1.3.1 is not applicable. Theorem 1.3.1, however, may provide good estimates on the distribution $H_n(z)$ of Z_n. More importantly, it may help us in "guessing" how large Z_n can be. It is worthwhile going through the following shocking figures. Let us choose $n=4$. We want to choose x in $P(Z_n < x)$ so that the error term

$$4n(1-F(x))^2 = 16(\log x)^{-2}$$

be "small." If our aim is that it should not affect at least the first digit in the main estimate

$$\exp\{-n(1-F(x))\} = \exp\left(\frac{-4}{\log x}\right) \qquad (27)$$

of Theorem 1.3.1, then we should have

$$16(\log x)^{-2} < 0.05,$$

from which we get

$$x > 58{,}734{,}861.$$

Guided by this unexpectedly large figure, let us estimate

$$P(Z_4 < 60 \times 10^6).$$

It can easily be checked that (18) is satisfied, and thus (27) and our control of the error term yield

$$0.7498 < P(Z_4 < 60 \times 10^6) \leqslant 0.7998.$$

This implies that $P(Z_4 \geqslant 60 \times 10^6) > 0.2$. That is, in more than 20% of the cases, among 4 independent observations, the largest one will exceed 60 million. It is therefore of no surprise that, with increasing n, Z_n does not show stability in the sense of the existence of a limiting distribution. ▲

While Theorems 1.3.1 and 1.3.2 and Corollaries 1.3.1 and 1.3.2 made essential use of the independence of the X_j, Lemma 1.3.2 will lead us to another type of estimate on $H_n(x)$ and $L_n(x)$ which can be extended to dependent systems. For this aim, let us introduce the following notations. For $k \geqslant 1$, let

$$S_{k,n}(x) = \sum_{1 \leqslant i_1 < i_2 < \cdots < i_k \leqslant n} P(X_{i_1} \geqslant x, X_{i_2} \geqslant x, \ldots, X_{i_k} \geqslant x). \qquad (28)$$

If the X_j are i.i.d. with common distribution function $F(x)$, then

$$S_{k,n}(x) = \binom{n}{k}(1-F(x))^k.$$

Thus, writing $F(x) = 1 - (1-F(x)) = 1-z$ in (9), Lemma 1.3.2 leads to the following estimates.

Theorem 1.3.3. *Let $X_1, X_2, \ldots, X_n$ be i.i.d. random variables. Let $S_{0,n}(x) = 1$ and define $S_{k,n}(x)$ by (28) for $k \geq 1$. Thus, for any real number x and for any integer $s \geq 0$,*

$$\sum_{k=0}^{2s+1} (-1)^k S_{k,n}(x) \leq P(Z_n < x) \leq \sum_{k=0}^{2s} (-1)^k S_{k,n}(x). \tag{29}$$

Similar inequalities can be obtained for $L_n(x)$ after redefining $S_{k,n}(x)$ as the corresponding sum of probabilities of the joint occurrences of any k of the events $\{X_j < x\}$. Since, in the next section, we present more general inequalities, we do not restate Theorem 1.3.3 for W_n.

The importance of Theorem 1.3.3 is that it leads away from the restrictive assumption of independence. That is, the statement remains true with no assumption on the interdependence of the sequence X_j. (In fact, the validity of (29) for independent X_j's implies that it holds for arbitrary random variables; see Exercise 8.) This will be dealt with in the next section, where several extensions of (29) will also be proved.

1.4. BOUNDS ON THE DISTRIBUTION OF EXTREMES

Let us go back to the mathematical model of Section 1.2. We have a sequence $X_1, X_2, \ldots, X_n$ of random variables, which, in most applications, are dependent. Let the joint distribution of $X_{i_1}, X_{i_2}, \ldots, X_{i_k}$, $1 \leq i_1 < i_2 < \cdots < i_k \leq n$, be

$$F_{i_1, i_2, \ldots, i_k}(x_1, x_2, \ldots, x_k) = P(X_{i_1} < x_1, X_{i_2} < x_2, \ldots, X_{i_k} < x_k). \tag{30}$$

We also introduce the notations

$$G_{i_1, i_2, \ldots, i_k}(x_1, x_2, \ldots, x_k) = P(X_{i_1} \geq x_1, X_{i_2} \geq x_2, \ldots, X_{i_k} \geq x_k). \tag{31}$$

Notice that (30) and (31) reduce to (3) and (6), respectively, if $k=1$, while they become (4) and (7) if $k=n$ (and thus $i_j = j$). In this latter case we shall also use the shorter forms (4) and (7) for (30) and (31), respectively.

Our aim is to obtain bounds on, or exact expressions for, the distribution of the extremes of the $X_j, 1 \leq j \leq n$. Let us extend our concept of extremes to cover more than just the maximum and minimum.

Definition 1.4.1. Let us rearrange the random variables $X_1, X_2, \ldots, X_n$ into a nondecreasing sequence

$$X_{1:n} \leq X_{2:n} \leq \cdots \leq X_{n:n}. \tag{32}$$

When actual equalities apply, we do not make any requirement about which variable should precede the other one. The sequence (32) is called the order statistics of $X_1, X_2, \ldots, X_n$. $X_{r:n}$ is called the rth order statistic.

Notice that $X_{1:n} = W_n$ and $X_{n:n} = Z_n$. We shall keep using these previous notations as well as the new ones.

Definition 1.4.2. For fixed $k \geq 1$, as $n \to +\infty$, $X_{k:n}$ and $X_{n-k+1:n}$ will be called the kth extremes. We shall also use the term kth lower extreme for $X_{k:n}$ and kth upper extreme for $X_{n-k+1:n}$. The first extremes are the minimum and maximum and they will be called extremes regardless of n's being large or small.

The emphasis in the above definition is that we speak of the kth extremes, $k > 1$, in a limiting sense: k is fixed in advance and n increases indefinitely. When n is also fixed, then we retain the concept of kth or $(n-k+1)$st order statistic.

There is a unified way to handle the distributions of $X_{k:n}$ and $X_{n-k+1:n}$. As a matter of fact, if we put

$$A_j(x) = \{X_j \geq x\} \quad \text{and} \quad B_j(x) = \{X_j < x\}, \tag{33}$$

then

$$\{X_{k:n} \geq x\} = \{\text{at most } k-1 \text{ of } B_j(x), 1 \leq j \leq n, \text{occur}\}$$

and

$$\{X_{n-k+1:n} < x\} = \{\text{at most } k-1 \text{ of } A_j(x), 1 \leq j \leq n, \text{occur}\},$$

where "at most zero events occurring" means that none of them occurs. Therefore, if $\nu_n(A, x)$ and $\nu_n(B, x)$ denote the number of $A_j(x)$ and $B_j(x)$, $1 \leq j \leq n$, respectively, which occur, then

$$P(X_{k:n} \geq x) = \sum_{t=0}^{k-1} P(\nu_n(B, x) = t) \tag{34}$$

1.4 BOUNDS ON THE DISTRIBUTION OF EXTREMES

and

$$P(X_{n-k+1:n} < x) = \sum_{t=0}^{k-1} P(\nu_n(A,x) = t). \tag{35}$$

Since our interest is centered about k fixed and n tending to $+\infty$, the number of terms in the sums above is fixed and thus we can concentrate on the individual terms. Taking this approach, we shall investigate the following general problem.

Let $C_1, C_2, \ldots, C_n$ be a sequence of events and put $\nu_n = \nu_n(C)$ for the number of $C_1, C_2, \ldots, C_n$ which occur. Let $S_{0,n} = 1$ and for $k \geq 1$,

$$S_{k,n} = \sum_{1 \leq i_1 < i_2 < \cdots < i_k \leq n} P(C_{i_1} C_{i_2} \cdots C_{i_k}). \tag{36}$$

For convenience, we permit $k > n$ when the sum above is empty and thus $S_{k,n} = 0$ for $k > n$. The problem is to set bounds on $P(\nu_n = t)$ in terms of $S_{k,n}, 0 \leq k \leq n$.

Notice that when C_j is one of the events defined in (33), then the terms of (36) become the functions (30) or (31). Furthermore, with the functions (31), (36) becomes (28). Hence, the theorems which follow generalize Theorem 1.3.3.

We first present three lemmas. The first one will justify our calling $S_{k,n}$ the kth binomial moment of ν_n.

Lemma 1.4.1. *For any $k \geq 0$,*

$$S_{k,n} = E\left\{ \binom{\nu_n}{k} \right\} = \sum_{r=k}^{n} \binom{r}{k} P(\nu_n = r).$$

Proof. Since both sides are equal to one for $k = 0$ and for $k > n$, only $1 \leq k \leq n$ needs proof. We turn to indicator variables. Let

$$I(C_{i_1} C_{i_2} \cdots C_{i_k}) = \begin{cases} 1 & \text{if } C_{i_1} C_{i_2} \cdots C_{i_k} \text{ occurs} \\ 0 & \text{otherwise.} \end{cases}$$

Thus, evidently,

$$S_{k,n} = E(J_{k,n}), \quad 1 \leq k \leq n,$$

where

$$J_{k,n} = \sum_{1 \leq i_1 < i_2 < \cdots < i_k \leq n} I(C_{i_1} C_{i_2} \cdots C_{i_k}).$$

Since each term contributes one or zero to $J_{k,n}$, one has to count the number of ones in the above sum. It is evident that those terms will be equal to one for which each C_{i_j} came from those v_n C_m's which occur. Hence

$$J_{k,n} = \binom{v_n}{k}, \quad 1 \leq k \leq n,$$

and thus taking expectations yields what was to be proved. ▲

Lemma 1.4.2. *Let $m \geq 1$ and $T \geq 0$ be integers. Then*

$$\sum_{k=0}^{T} (-1)^k \binom{m}{k} = (-1)^T \binom{m-1}{T}. \tag{37}$$

Proof. We prove by induction over T. If $T=0$, both sides are equal to one, hence (37) holds. Let us assume that (37) has been proved for a fixed T. Then, since

$$\sum_{k=0}^{T+1} (-1)^k \binom{m}{k} = \sum_{k=0}^{T} (-1)^k \binom{m}{k} + (-1)^{T+1} \binom{m}{T+1},$$

an application of the assumption of induction yields

$$\sum_{k=0}^{T+1} (-1)^k \binom{m}{k} = (-1)^T \binom{m-1}{T} + (-1)^{T+1} \binom{m}{T+1} = (-1)^{T+1} \binom{m-1}{T+1}.$$

Lemma 1.4.2 is thus established. ▲

Lemma 1.4.3. *For integers $n \geq 1$ and $0 \leq t \leq n$, $0 \leq a \leq n-t-1$,*

$$P(v_n = t) - \sum_{k=0}^{a} (-1)^k \binom{k+t}{t} S_{k+t,n}$$

$$= (-1)^{a+1} \sum_{r=t+a+1}^{n} \binom{r-t-1}{t} P(v_n = r).$$

Proof. Put

$$b_t(a) = \sum_{k=0}^{a} (-1)^k \binom{k+t}{t} S_{k+t,n}.$$

By Lemma 1.4.1

$$b_t(a) = \sum_{k=0}^{a} (-1)^k \binom{k+t}{t} \sum_{r=k+t}^{n} \binom{r}{k+t} P(\nu_n = r)$$

$$= \sum_{r=t}^{n} P(\nu_n = r) \sum_{k=0}^{T} (-1)^k \binom{k+t}{t} \binom{r}{k+t},$$

where $T = \min(a, r-t)$. Applying the identity

$$\binom{k+t}{t} \binom{r}{k+t} = \binom{r}{t} \binom{r-t}{k},$$

we get

$$b_t(a) = \sum_{r=t}^{n} P(\nu_n = r) \binom{r}{t} \sum_{k=0}^{T} (-1)^k \binom{r-t}{k}$$

$$= P(\nu_n = t) + \sum_{r=t+1}^{n} P(\nu_n = r) \binom{r}{t} \sum_{k=0}^{T} (-1)^k \binom{r-t}{k}.$$

An appeal to Lemma 1.4.2 thus yields

$$b_t(a) = P(\nu_n = t) + \sum_{r=t+1}^{n} (-1)^T \binom{r-t-1}{T} \binom{r}{t} P(\nu_n = r).$$

But, by definition of T, $\binom{r-t-1}{T} = 0$ for $r < a+t$, and for all other values of r, $T = a$. Hence

$$b_t(a) = P(\nu_n = t) + (-1)^a \sum_{r=t+a+1}^{n} \binom{r-t-1}{a} \binom{r}{t} P(\nu_n = r),$$

which is, in fact, the statement of Lemma 1.4.3. The proof is complete. ▲

We can now easily deduce from Lemma 1.4.3 the following theorem. It will play a basic role in the general theory of extremes.

Theorem 1.4.1. *Let $n \geq 1$ and $0 \leq t \leq n$ be integers. Then*

$$P(\nu_n = t) = \sum_{k=0}^{n-t} (-1)^k \binom{k+t}{t} S_{k+t,n}. \tag{38}$$

Furthermore, for any integer $s \geq 0$,

$$\sum_{k=0}^{2s+1}(-1)^k\binom{k+t}{t}S_{k+t,n}+\frac{2s+2}{n-t}\binom{2s+t+2}{t}S_{2s+t+2,n}$$

$$\leq P(\nu_n=t) \leq \sum_{k=0}^{2s}(-1)^k\binom{k+t}{t}S_{k+t,n}-\frac{2s+1}{n-t}\binom{2s+t+1}{t}S_{2s+t+1,n}.$$

(39)

Proof. The identity (38) is actually contained in Lemma 1.4.3. Indeed, if we apply Lemma 1.4.3 with $a=n-t-1$ and observe that $S_{n,n}=P(\nu_n=n)$, we get (38).

Turning to the inequalities (39), first recall that $S_{j,n}=0$ if $j>n$. Hence, the lower inequality for $2s+1\geq n-t$ becomes the identity (38). Consequently, we can assume that, for the lower inequality, $2s+1<n-t$, and $2s<n-t$ in the case of the upper inequality. With these values, however, we can apply Lemma 1.4.3.

Let $0\leq a=2s+1<n-t$. Then Lemma 1.4.3 yields

$$P(\nu_n=t)-\sum_{k=0}^{2s+1}(-1)^k\binom{k+t}{t}S_{k+t,n}=\sum_{r=t+2s+2}^{n}\binom{r-t-1}{2s+t}\binom{r}{t}P(\nu_n=r).$$

Therefore, for proving the lower inequality of (39), we have to show

$$\frac{2s+2}{n-t}\binom{2s+t+2}{t}S_{2s+t+2,n}\leq \sum_{r=t+2s+2}^{n}\binom{r-t-1}{2s+1}\binom{r}{t}P(\nu_n=r). \quad (40)$$

By applying Lemma 1.4.1, this inequality takes the form

$$\frac{2s+2}{n-t}\sum_{r=2s+t+2}^{n}\binom{2s+t+2}{t}\binom{r}{2s+t+2}P(\nu_n=r)$$

$$\leq \sum_{r=2s+t+2}^{n}\binom{r-t-1}{2s+1}\binom{r}{t}P(\nu_n=r).$$

Writing all binomial coefficients above in terms of factorials and simplifying by common factors, we get

$$\frac{1}{n-t}\sum_{r=2s+t+2}^{n}P(\nu_n=r)\leq \sum_{r=2s+t+2}^{n}\frac{1}{r-t}P(\nu_n=r).$$

This last inequality is evidently true, hence so is (40). The lower inequality in (39) is thus proved. The upper inequality can be proved in the same manner as the lower one, therefore we do not repeat the details. Theorem 1.4.1 is thus established. ▲

While Theorem 1.4.1 is a very useful tool in the theory of extremes, in numerical calculations the following disadvantage may arise. In a case when the terms $P(C_{i_1}C_{i_2}\cdots C_{i_k})$ of (36) are known only approximately, in $S_{k,n}$ alone there are $\binom{n}{k}$ error terms. While some of these errors are positive, others are negative, and thus several of them would cancel out; in actual calculations, however, signs of errors cannot be taken into account. Therefore, the errors in (39) may become so large that the result would be meaningless. The aim of the next theorem is to avoid this difficulty by limiting the number of terms in estimating $P(\nu_n = t)$.

We first introduce some notations. Let $H = \{1, 2, \ldots, n\}$. Let E_n be a subset of the set of ordered pairs (i,j) with $1 \leq i < j \leq n$. For a sequence $C_1, C_2, \ldots, C_n$ of events, let

$$S_{k,n}^* = \Sigma_k^* P(C_{i_1}C_{i_2}\cdots C_{i_k}), \quad 1 \leq k \leq n, \tag{41}$$

where Σ_k^* signifies summation over those subscripts $(i_1, i_2, \ldots, i_k)$ for which $1 \leq i_1 < i_2 < \cdots < i_k \leq n$ and no pairs of $(i_1, i_2, \ldots, i_k)$ belong to E_n. Furthermore, set

$$S_{k,n}^{**} = \Sigma_k^{**} P(C_{i_1}C_{i_2}\cdots C_{i_k}), \quad 1 \leq k \leq n, \tag{42}$$

where summation Σ_k^{**} is over those subscripts $(i_1, i_2, \ldots, i_k)$ for which $1 \leq i_1 < i_2 < \cdots < i_k \leq n$ and at most one pair (i_t, i_m) belongs to E_n.

Before going on, let us draw attention to the fact that if E_n is the empty set, then both (41) and (42) reduce to (36). In all other cases, $S_{k,n}^*$ and $S_{k,n}^{**}$ contain fewer terms than $S_{k,n}$. Their actual number of terms depends on E_n, which can be arbitrarily chosen.

We now prove the following result.

Theorem 1.4.2. *For any integers $n \geq 1$ and $m \geq 0$ with $2m + 1 \leq n$,*

$$1 - S_{1,n}^{**} + S_{2,n}^* - S_{3,n}^{**} + \cdots - S_{2m+1,n}^{**} \leq P(\nu_n = 0)$$
$$\leq 1 - S_{1,n}^* + S_{2,n}^{**} - S_{3,n}^* + \cdots + S_{2m,n}^{**}.$$

Remark 1.4.1. In order to avoid confusion, let us put down the rule for applying the stars in the superscripts. In both cases, the terms alternate in sign. In the lower estimate, the negative terms have double stars, the positive ones a single star. This rule is reversed in the upper estimate.

Remark 1.4.2. We have remarked that if E_n is chosen as the empty set, then $S_{k,n}^* = S_{k,n}^{**} = S_{k,n}$. In this case Theorem 1.4.2 is somewhat weaker than Theorem 1.4.1 with $t=0$, because the last term in both estimates is missing here. However, better estimates can be expected with a specific E_n than when E_n is arbitrary. We can, in fact, make other specifications of E_n in order to improve our estimates on $P(\nu_n = 0)$. For related results on $P(\nu_n = t)$, see Exercises 13 and 14.

Proof. We again turn to indicator variables. We shall prove that the corresponding inequalities hold for indicators, and thus integration leads to Theorem 1.4.2. Details are as follows. For $k \geq 1$, let

$$I(C_{i_1} C_{i_2} \cdots C_{i_k}) = \begin{cases} 1 & \text{if } C_{i_1} C_{i_2} \cdots C_{i_k} \text{ occurs} \\ 0 & \text{otherwise;} \end{cases}$$

$$J_{k,n}^* = \Sigma_k^* I(C_{i_1} C_{i_2} \cdots C_{i_k}) \qquad (43)$$

and

$$J_{k,n}^{**} = \Sigma_k^{**} I(C_{i_1} C_{i_2} \cdots C_{i_k}), \qquad (44)$$

where Σ_k^* and Σ_k^{**} are defined as in (41) and (42), respectively. Evidently,

$$S_{k,n}^* = E(J_{k,n}^*), \qquad S_{k,n}^{**} = E(J_{k,n}^{**}). \qquad (45)$$

Therefore, for Theorem 1.4.2 it suffices to show that, for $n \geq 1$ and $m \geq 0$ with $2m+1 \leq n$,

$$1 - J_{1,n}^{**} + J_{2,n}^* - \cdots - J_{2m+1,n}^{**} \leq I \leq 1 - J_{1,n}^* + J_{2,n}^{**} - \cdots + J_{2m,n}^{**}, \qquad (46)$$

when

$$I = \begin{cases} 1 & \text{if } \nu_n = 0, \\ 0 & \text{otherwise.} \end{cases} \qquad (47)$$

The inequalities of (46) can be proved by a simple combinatorial argument. Let us first observe that (46) holds if $\nu_n = 0$. Indeed, then $J_{k,n}^* = J_{k,n}^{**} = 0$ for all $k \geq 1$ and $I = 1$. Thus (46) becomes a trivial identity.

Now let $\nu_n = t \geq 1$. Let $j_1 < j_2 \cdots < j_t$ be those t subscripts which signify those C's which occur. Then the following terms contribute one each to $J_{k,n}^*$ and $J_{k,n}^{**}$, respectively. Let us first specify the nonzero terms of $J_{k,n}^*$. For the subscripts $(i_1, i_2, \ldots, i_k)$ of the general term of (43) the following should hold: $(i_1, i_2, \ldots, i_k)$ is a subset of $(j_1, j_2, \ldots, j_t)$ and no pairs of $(i_1, i_2, \ldots, i_k)$ belong to E_n. In case of $J_{k,n}^{**}$, $(i_1, i_2, \ldots, i_k)$ is again a subset of $(j_1, j_2, \ldots, j_t)$ and at most one pair from $(i_1, i_2, \ldots, i_k)$ may belong to E_n. Therefore, with the following notations, (46) takes a simple form. Let $h \subset H = \{1, 2, \ldots, n\}$. Let $N_n(h, a)$ be the number of subsets h_1 of h for which the following hold. The number $e(h_1)$ of elements of h_1 does not exceed a, and $e(h_1)$ is congruent to a mod 2. Furthermore, no two elements of h_1 belong to E_n. The definition of $M_n(h, a)$ is exactly the same as that of $N_n(h, a)$ except that in $M_n(h, a)$, each h_1 may contain at most one pair from E_n. With these notations, (46) becomes (we take the empty set as a set with even number of elements, and recall that $v_n = t \geq 1$)

$$N_n(h(t), 2m) - M_n(h(t), 2m+1) \leq 0, \qquad (48a)$$

$$M_n(h(t), 2m) - N_n(h(t), 2m-1) \geq 0, \qquad (48b)$$

where $h(t) = \{j_1, j_2, \ldots, j_t\}$. We shall prove (48a, b) by induction over t. For $t = 1$, E_n does not impose any condition, hence $N_n(h(1), 2m) = M_n(h(1), 2m) = 0$ and $N_n(h(1), 2m-1) = 0$, $M_n(h(1), 2m+1) = 1$ for any integer $m \geq 0$. Thus (48a) becomes an identity, while the left hand side of (48b) equals 1. Consequently, (48a, b) hold for $t = 1$. Assume now the validity of (48a, b) for all t_0 with $t \geq t_0 \geq 1$, and consider the left hand sides of (48a, b) for $t + 1$. When counting subsets of $h(t+1)$, we shall make distinctions whether a specific subset h_1 contains j_{t+1} or not, that is, whether $h_1 \subset h(t)$ or not. Hence

$$N_n(h(t+1), a) = N_n(h(t), a) + N_n^*(h(t), a-1) \qquad (49a)$$

and

$$M_n(h(t+1), a) = M_n(h(t), a) + M_n^*(h(t), a-1), \qquad (49b)$$

where the star signifies that we count those subsets of $h(t+1)$ which take $a-1$ elements from $h(t)$, and their ath element is j_{t+1}. Now, if $h^*(t)$ denotes that subset of $h(t)$ for which $(s, j_{t+1}) \notin E_n$ whenever $s \in h^*(t)$, then

$$N_n^*(h(t), a-1) = N_n(h^*(t), a-1) \qquad (50a)$$

and

$$M_n^*(h(t), a-1) = M_n(h^*(t), a-1) + R_n(t, a), \qquad (50b)$$

where $R_n(t, a)$ is the number of all remaining subsets, but we require only

that it satisfies
$$R_n(t,a) \geq 0. \tag{51}$$
Combining (49a, b), (50a, b), and (51), we get
$$N_n(h(t+1),2m) - M_n(h(t+1),2m+1)$$
$$\leq N_n(h(t),2m) - M_n(h(t),2m+1) - \{M_n(h^*(t),2m) - N_n(h^*(t),2m-1)\}.$$
But since $e(h(t)) = t$ and $e(h^*(t)) \leq t$, the assumption of induction yields
$$N_n(h(t+1),2m) - M_n(h(t+1),2m+1) \leq 0.$$
We similarly get
$$M_n(h(t+1),2m) - N_n(h(t+1),2m-1) \geq 0,$$
which completes the proof. ▲

Recall that our aim is to apply Theorems 1.4.1 and 1.4.2 to the specific events $A_j(x)$ and $B_j(x)$, defined in (33). Therefore, for these theorems one needs the structure of interdependence of the A's and B's or, equivalently, of the random variables X_j. If only a limited information is available on the X_j, then bounds other than given in Theorems 1.4.1 and 1.4.2 may provide better approximations. In the next theorem we give a lower bound on $P(\nu_n \geq 1)$, using only univariate and bivariate distributions. Again the theorem will be formulated for arbitrary events. We use our standard notation.

Theorem 1.4.3. *For integers $n \geq 1$ and $1 \leq k \leq n-1$,*
$$P(\nu_n \geq 1) \geq \frac{2}{k+1} S_{1,n} - \frac{2}{k(k+1)} S_{2,n}. \tag{52}$$

The maximum in k of the right hand side is attained with $k_0 = 1 + [2S_{2,n}/S_{1,n}]$, where $[y]$ signifies the largest integer not exceeding y.

Before giving the proof of Theorem 1.4.3, let us draw the attention of the reader to Exercise 17, where an optimal property of Theorem 1.4.3. is formulated.

Proof. By Lemma 1.4.1,
$$\frac{2}{k+1} S_{1,n} - \frac{2}{k(k+1)} S_{2,n} = E\left\{\frac{2\nu_n}{k+1} - \frac{\nu_n(\nu_n-1)}{k(k+1)}\right\}$$
$$= \frac{1}{k(k+1)} E\{\nu_n(2k+1-\nu_n)\}.$$

Since the parabola $x(2k+1-x)$ takes its maximum at $x=k+\frac{1}{2}$, $\nu_n(2k+1-\nu_n)$ takes its maximum at $\nu_n=k$ and $k+1$ in view of ν_n's being an integer. Hence

$$\frac{\nu_n(2k+1-\nu_n)}{k(k+1)} \leq 1,$$

which can also be written as

$$\frac{\nu_n(2k+1-\nu_n)}{k(k+1)} \leq I(\nu_n \geq 1), \tag{53}$$

where $I(\nu_n \geq 1)$ is the indicator variable of the event $\{\nu_n \geq 1\}$. By taking expectation in (53), we get (52).

The fact that the maximum in k of the right hand side of (52) is attained at the claimed value easily follows by checking that it increases up to k_0 and decreases thereafter. The proof is completed. ▲

Further inequalities are given among the Exercises for solution.
We now turn to applications of the inequalities of the present section.

1.5. ILLUSTRATIONS

We shall apply Theorems 1.4.1–1.4.3 to some specific classes of random variables $X_1, X_2, \ldots, X_n$. In all examples which follow, the events C_j are one of the two types defined in (33). Therefore, the $S_{k,n}$ of (36) are in terms of the multivariate distributions (30) and (31).

Let us start with a practical problem.

Example 1.5.1. Assume that a patient suffering from a specific disease will live Y time units. If a patient is treated by a specific drug, it adds U time units to his life. Hence, a treated patient will live $U+Y$ units, where Y varies with patients independently but U—which is also random— is the same random variable for each patient. That is, if we consider several patients with the same treatment, then the jth patient will live for $X_j = U + Y_j$ units, where now $U, Y_1, Y_2, \ldots$ are independent random variables and the Y's are identically distributed. Let us consider the distribution of the maximum of X_j for a given number of patients.

The same mathematical model can be used for an engineering problem. A manufacturer prepares his products for the roughest possible conditions. Assume that the jth product will last Y_j time units, where the $Y_1, Y_2, \ldots$ are i.i.d. random variables. A specific customer, however, uses the products under milder conditions, which adds U units to each Y_j. Thus the life length of product j is $X_j = U + Y_j$, where we considered those products which are used by the same customer. Here again, U is independent of the Y's. Our interest is $Z_n = \max(X_j, 1 \leq j \leq n)$.

The two above problems, and several others, can now be unified into the following mathematical model.

Let $Y_1, Y_2, \ldots, Y_{10}$, say, be independent and identically distributed random variables. Assume, however, that the Y's cannot be observed because, when in use, each Y_j suffers a random effect U. Hence, the observations come in the form of $X_j = U + Y_j$, $1 \leq j \leq 10$. Let U and the Y_j be independent and, for the example, we assume that both U and the Y_j have exponential distribution but with different parameters. Let $P(U < x) = 1 - e^{-2.5x}$ and $P(Y_j < x) = 1 - e^{-x}$, where $x > 0$. Our aim is to evaluate, or at least to estimate, the distribution $H_{10}(x)$ of Z_{10}, the maximum of the X_j.

Since we specified the structure of the X_j completely, we can determine the multivariate distribution of the X_j in any dimension. Hence, we can use Theorem 1.4.1. With the events $A_j = \{X_j \geq x\}$, $1 \leq j \leq 10$,

$$P(\nu_{10} = 0) = P(Z_{10} < x), \tag{54}$$

and thus (38) and (39) are applicable with $t = 0$. In numerical calculations (39) is always recommended, because it usually requires far less computation even for very high accuracy. In (39), we need the terms $S_{1,10}, S_{2,10}, \ldots$, which now take the form (28). For $S_{k,10}$ in (28), it is sufficient to determine

$$P(X_1 \geq x, X_2 \geq x, \ldots, X_k \geq x),$$

since the X_j are identically distributed.

By the elementary formula (see Appendix I, formula (A.1))

$$P(X_j \geq x, 1 \leq j \leq k) = \int_0^{+\infty} P(X_j \geq x, 1 \leq j \leq k | U = u) dP(U < u)$$

$$= \int_0^{+\infty} P(Y_j \geq x - u, 1 \leq j \leq k | U = u) 2.5 e^{-2.5u} du$$

$$= 2.5 \int_0^x e^{-k(x-u)} e^{-2.5u} du + 2.5 \int_x^{+\infty} e^{-2.5u} du$$

$$= \frac{2.5}{k - 2.5} e^{-kx} \left[e^{(k-2.5)x} - 1 \right] + e^{-2.5x}$$

$$= \frac{k}{k - 2.5} e^{-2.5x} - \frac{2.5}{k - 2.5} e^{-kx}.$$

1.5 ILLUSTRATIONS

Consequently,

$$S_{k,10} = \binom{10}{k}\frac{k}{k-2.5}e^{-2.5x} + \frac{2.5}{2.5-k}\binom{10}{k}e^{-kx}. \tag{55}$$

Let us choose $x = 5$.

We now compute the bounds in (39) sequentially. That is, when $S_{k,10}$ has been computed, we determine both bounds in (39) and then improve these bounds, if necessary, by moving to $S_{k+1,10}$. We terminate computation when the upper and lower bounds coincide up to two decimal digits. Since, by (54), $t = 0$, all binomial coefficients in (39) become one. Recall that $S_{0,n} = 1$. For easier reference, we list the bounds from (39), as we use them. The lower bounds are

$$1 - S_{1,10}; \quad 1 - S_{1,10} + \tfrac{2}{10}S_{2,10}; \quad 1 - S_{1,10} + S_{2,10} - S_{3,10},$$

while the upper bounds become

$$1 - \tfrac{1}{10}S_{1,10}; \quad 1 - S_{1,10} + S_{2,10}; \quad 1 - S_{1,10} + S_{2,10} - \tfrac{3}{10}S_{3,10}.$$

Now, by (55),

$$S_{1,10} = 0.1123.$$

Thus, the first terms in the above list yield

$$.8877 \leq P(Z_{10} < 5) \leq .9888.$$

Computing $S_{2,10}$ and applying the second terms in the list, we get

$$S_{2,10} = 0.0095$$

and

$$.8896 \leq P(Z_{10} < 5) \leq .8973.$$

Finally, since

$$S_{3,10} = 0.0025,$$

the last terms result in the bounds

$$.8948 \leq P(Z_{10} < 5) \leq .8965.$$

We add that if the effect of U had been ignored, that is, $U = 0$ had been taken and thus $X_j = Y_j$ for all j, we would have gotten

$$P(Z_{10} < 5) = (1 - e^{-5})^{10} = .934627.$$

This means that U has increased the chance of Z_{10}'s taking a value larger than five from .06537 to .104, which is a 59% increase. ▲

Example 1.5.2. Let X_j be the time to failure of the jth component of an equipment. Assume that each X_j is a unit exponential variate; that is, for each j,

$$P(X_j < x) = 1 - e^{-x}, \quad x > 0.$$

Consider a group of five components. We assume the structure is such that X_1, X_3 and X_5 are completely independent. In addition, X_2 is independent of both X_3 and X_4, and X_1 is independent of X_4. Any other combination has dependent components. We also specify the bivariate distributions of the X_j. For simplicity, let us use the same bivariate distribution for all dependent pairs. Let

$$P(X_1 < x, X_2 < y) = P(X_2 < x, X_5 < y) = P(X_3 < x, X_4 < y)$$

$$= P(X_4 < x, X_5 < y) = (1 - e^{-x})(1 - e^{-y})(1 - \tfrac{1}{2} e^{-x-y}). \quad (56)$$

No further assumption is made. Thus, we cannot compute the joint distribution of a group containing (X_1, X_2, X_5), or (X_3, X_4, X_5), or (X_2, X_4, X_5). Yet, we would like to estimate $P(W_5 \geq x)$.

We can appeal to Theorem 1.4.2 by specifying an E_5 by which (41) and (42) can be determined. Here, of course, the events $C_j = \{X_j < x\}$ and thus $\{\nu_5 = 0\} = \{W_5 \geq x\}$. Since $S_{k,n}^*$ and $S_{k,n}^{**}$ of (41) and (42) do not require terms when the subscripts contain more than one pair from E_5, and since our difficulty is exactly the computation of joint distributions which contain more than one pair from (56), we define

$$E_5 = \{(1,2), (2,5), (3,4), (4,5)\}.$$

We can now compute $S_{k,5}^*$ and $S_{k,5}^{**}$ for $1 \leq k \leq 5$. Evidently,

$$S_{1,5}^* = S_{1,5}^{**} = 5(1 - e^{-x}). \tag{57a}$$

For $k \geq 2$, we have to consider E_5 and (56). In $S_{k,5}^*$ each term is $(1 - e^{-x})^k$; hence we have to count their number of terms. We get

$$S_{2,5}^* = 6(1 - e^{-x})^2, \quad S_{3,5}^* = (1 - e^{-x})^3, \quad S_{4,5}^* = S_{5,5}^* = 0. \tag{57b}$$

In $S_{k,5}^{**}$ two types of terms are to be counted: those with no pairs from E_5 and those with exactly one term from E_5. Since the first kind of terms constituted $S_{k,5}^*$,

$$S_{k,5}^{**} = S_{k,5}^* + \text{contributions of the second kind of terms.}$$

1.5 ILLUSTRATIONS

The form of a second kind of term in $S_{k,5}^{**}$ is

$$(1-e^{-x})^k \left(1 - \tfrac{1}{2}e^{-2x}\right).$$

Therefore

$$S_{2,5}^{**} = 6(1-e^{-x})^2 + 4(1-e^{-x})^2\left(1 - \tfrac{1}{2}e^{-2x}\right), \qquad (58a)$$

$$S_{3,5}^{**} = (1-e^{-x})^3 + 6(1-e^{-x})^3\left(1 - \tfrac{1}{2}e^{-2x}\right), \qquad (58b)$$

and

$$S_{4,5}^{**} = S_{5,5}^{**} = 0. \qquad (58c)$$

For a numerical calculation, let us choose $x = 0.1$. We then estimate $P(W_5 \geq 0.1)$. By (57) and (58),

$$S_{1,5}^{*} = S_{1,5}^{**} = 0.475813, \qquad (59)$$

$$S_{2,5}^{*} = 0.054336, \qquad S_{2,5}^{**} = 0.07573, \qquad (60)$$

$$S_{3,5}^{*} = 0.000862, \qquad S_{3,5}^{**} = 0.00392,$$

and

$$S_{4,5}^{*} = S_{4,5}^{**} = S_{5,5}^{*} = S_{5,5}^{**} = 0.$$

Theorem 1.4.2 now gives

$$.57460 \leq P(W_5 \geq 0.1) \leq .59906. \qquad (61)$$

▲

Example 1.5.3. Let us apply Theorem 1.4.3 in the preceding example to estimate $P(W_5 \geq 0.1)$. As was pointed out, with the events $\{X_j < 0.1\}$,

$$P(W_5 \geq 0.1) = P(\nu_5 = 0) = 1 - P(\nu_5 \geq 1).$$

Hence, (52) will provide an upper bound on $P(W_5 \geq 0.1)$. Since $S_{1,5} = S_{1,5}^{**}$ and $S_{2,5} = S_{2,5}^{**}$, where $S_{k,5}^{**}$ denotes the numbers of Example 1.5.2, we can apply (59) and (60). We thus have

$$S_{1,5} = 0.4758129, \qquad S_{2,5} = 0.0757305.$$

Therefore, $k_0 = 1$. The estimate (52) becomes

$$P(W_5 \geq 0.1) \leq 1 - S_{1,5} + S_{2,5} = .599918. \qquad (62)$$

It is, of course, not surprising that (61) was a better estimate than (62). While (62) is the best possible that can be obtained in terms of $S_{1,5}$ and $S_{2,5}$ (see Exercise 17), in (61) we made use of further information on the X_j.

▲

1.6. SPECIAL PROPERTIES OF THE EXPONENTIAL DISTRIBUTION IN THE LIGHT OF EXTREMES

In a large proportion of applied investigations it is routinely assumed that waiting times to the first "occurrence" have exponential distribution. Here "occurrence" may mean the failure of an equipment, but also the death of a patient of a serious disease, the arrival of a bus at a station, the arrival of a customer at a service station, etc. Such a routine can develop either because of the mathematical simplicity of the distribution function $F(x) = 1 - e^{-x}$ or because there is some mathematical justification for it. In the present section we try to find some mathematical reasons for such a widespread usage.

We say that a random variable X is an exponential variate if, for some real numbers $a > 0$ and b,

$$P(X < x) = F(x) = \begin{cases} 1 - e^{-a(x-b)} & \text{if } x > b, \\ 0 & \text{if } x \leq b. \end{cases} \qquad (63)$$

One can reduce the exponential distribution to $b = 0$ by considering $X - b$. We also say that X is a unit exponential variate if $a = 1$.

The best known property of the exponential distribution is its lack of memory. Let us formulate this property first in the terminology of life length. We say that $X \geq 0$ lacks memory (or does not age) if, given that X has lived for x time units, the probability of its lasting for another y time units is the same value for all x. In mathematical terms,

$$P(X \geq y + x | X \geq x) = g(y),$$

a function of y only. Therefore, if we take $x = 0$, we get

$$P(X \geq y + x | X \geq x) = P(X \geq y). \qquad (64)$$

The fact that the exponential distribution (63) with $b = 0$ does have the lack-of-memory property (64) follows easily by substitution. Indeed, (64) is equivalent to

$$\frac{P(X \geq y + x)}{P(X \geq x)} = P(X \geq y)$$

1.6 SPECIAL PROPERTIES OF THE EXPONENTIAL DISTRIBUTION

for all x for which $P(X \geq x) > 0$, which in turn can be written as

$$G(x+y) = G(x)G(y), \qquad G(u) = 1 - F(u). \tag{64a}$$

The above equation (64a) evidently holds for (63) with $b=0$. The converse to this statement, although elementary, is very useful.

Theorem 1.6.1. *Let $X \geq 0$ and let X have a nondegenerate distribution function $F(x)$. If, for all $x \geq 0$ and $y \geq 0$, (64a) is satisfied, then $F(x)$ is exponential with $b=0$.*

Remark 1.6.1. The condition "all $x \geq 0$ and $y \geq 0$" can be relaxed considerably. The form above, however, suffices for our purposes.

Theorem 1.6.1 immediately follows from the next lemma.

Lemma 1.6.1. *Let $G(x)$ be monotonic and nonnegative for $x \geq 0$. Assume that, for all $x \geq 0$ and $y \geq 0$,*

$$G(x+y) = G(x)G(y). \tag{65}$$

Then, if $G(x)$ is not identically zero or one for $x \geq 0$, $G(x) = e^{ax}$ for all $x \geq 0$ with some real number a.

Evidently, $a > 0$ or $a < 0$ according as $G(x)$ is increasing or decreasing, respectively.

Proof of Lemma 1.6.1. We get by induction from (65) that, for any $x_i \geq 0$, $1 \leq i \leq n$, $n \geq 2$,

$$G(x_1 + x_2 + \cdots + x_n) = G(x_1)G(x_2)\cdots G(x_n). \tag{66}$$

First observe that $G(1) > 0$. Indeed, if $G(1) = 0$, then (66) would imply, by choosing $x_j = 1$ for each j and then $x_j = 1/n$ for each j, that for all $n \geq 1$

$$G(n) = G(1)^n = 0 \qquad \text{and} \qquad G(1) = G\left(\frac{1}{n}\right)^n = 0.$$

Since $G(x)$ is monotonic, $G(x) \equiv 0$, which was excluded from our investigation.

Let $x_1 = x_2 = \cdots = x_n = x > 0$ in (66). We then get

$$G(nx) = G^n(x). \tag{67}$$

If $x = 1/m$ with a positive integer m, then (67) yields

$$G\left(\frac{n}{m}\right) = G^n\left(\frac{1}{m}\right). \tag{68}$$

Hence, in particular, if $n = m$,

$$G(1) = G^n\left(\frac{1}{n}\right) \quad \text{or} \quad G\left(\frac{1}{n}\right) = G^{1/n}(1). \tag{69}$$

Combining (68) and (69), we have

$$G\left(\frac{n}{m}\right) = G^{n/m}(1), \qquad n, m \geqslant 1 \text{ integer,}$$

which can be restated as

$$G(x) = e^{ax}, x \text{ rational,} \tag{70}$$

where $a = \log G(1)$. Let now y be irrational. Let x_1 and x_2 be rational numbers with $x_1 < y < x_2$. Since $G(x)$ is monotonic,

$$e^{ax_1} = G(x_1) \leqslant G(y) \leqslant G(x_2) = e^{ax_2} \tag{71a}$$

if $G(x)$ is increasing, while

$$e^{ax_2} \leqslant G(y) \leqslant e^{ax_1} \tag{71b}$$

if $G(x)$ is decreasing. Letting $x_1 \to y$ and $x_2 \to y$, the outer terms of (71a, b) tend to e^{ay}, and thus the middle term $G(y) = e^{ay}$. The proof is completed. ▲

Proof of Theorem 1.6.1. By Lemma 1.6.1, $G(x) = 1 - F(x)$ is either identical to zero or to one or $G(x) = e^{ax}$ for all $x \geqslant 0$. The first two cases are not possible, however. With $G(x) \equiv 0$, $F(x)$ is degenerate at 0, which was excluded by the assumptions. On the other hand, $G(x) \equiv 1$ for $x \geqslant 0$ would make $F(x) \equiv 0$, which is not a proper distribution function. Hence, $G(x) = e^{ax}$, $x \geqslant 0$, and, since it is decreasing, $a < 0$. The theorem is established. ▲

Theorem 1.6.1 is very practical. It says that if time to the first "occurrence" is not affected by the passing of time, then this random waiting period is exponentially distributed. For example, it is reasonable to assume that the probability distribution of the time to the first road accident by a specific car (the insurance company's interest) in good condition remains the same whether a policy is purchased today or some time later. Hence, the insurance company can use the exponential distribution to compute the premium. Similarly, warranty periods can be determined by using the exponential distribution. The manufacturer assumes that all parts of the sold equipment function properly for a certain time period. Hence, if within this period the equipment malfunctions, then its aging was not the cause, but rather the assumption of having provided good parts was wrong.

1.6 SPECIAL PROPERTIES OF THE EXPONENTIAL DISTRIBUTION

But if aging was not the reason, then time to first failure is exponential in view of Theorem 1.6.1. Therefore, the expected cost of replacement can be computed in advance.

There is another direct consequence of (64). In the course of the proof of Lemma 1.6.1 we obtained that, for $x>0$ and $n \geq 1$,

$$G(nx) = G^n(x). \qquad (67)$$

This equation can be interpreted by (10) as follows. Let $X_1, X_2, \ldots, X_n$ be i.i.d. random variables with common distribution function $F(x) = 1 - G(x)$. Let W_n be the minimum of the X_j. Then, by (10) and (67),

$$P(W_n \geq x) = P(X_1 \geq nx) \qquad (67a)$$

or, by putting $y = nx$,

$$P(nW_n \geq y) = P(X_1 \geq y). \qquad (67b)$$

This equation says that, if the lack of memory holds, then nW_n has the same distribution as the X_j had. Although the lack of memory implies that the distribution in question is exponential, there may be other distributions satisfying (67b). As it turns out, there are no additional distributions with this property. In this regard, we prove the following result.

Theorem 1.6.2. *Let $X_1, X_2, \ldots, X_n$ be i.i.d. random variables with common distribution function $F(x)$. Let $E(X_1)$ be finite. Assume that, for all $n \geq 1$,*

$$nE(W_n) = E(X_1). \qquad (72)$$

Then $X_1 \geq 0$ almost surely, and, if not degenerate at zero, then $F(x)$ is exponential. In particular, if, for all $n \geq 1$ and all $y \geq 0$, (67b) is satisfied, then $E(X_1)$ is finite and (72) holds for all $n \geq 1$; hence, either $F(x)$ is degenerate at zero or $F(x)$ is exponential.

Proof. We first prove that $P(X_1 \geq 0) = 1$. We shall arrive at this conclusion by showing that the assumption $P(X_1 < 0) > 0$ would lead to $E(W_n) \leq \eta < 0$ for all sufficiently large n, where η does not depend on n. Then (72) would imply $E(X_1) = -\infty$, which contradicts the assumption that it is finite.

Let us now give the details. We start with the formula (see (A11) of Appendix I)

$$E(W_n) = \int_{-\infty}^{+\infty} x\, dP(W_n < x) = \int_{-\infty}^{0} x\, dP(W_n < x) + \int_{0}^{+\infty} P(W_n \geq x)\, dx.$$

Since

$$P(W_n < x) = 1 - [1 - F(x)]^n = 1 - G^n(x),$$

the formula above becomes

$$E(W_n) = \int_{-\infty}^{0} x\,dP(W_n < x) + \int_{0}^{+\infty} G^n(x)\,dx. \tag{72a}$$

Let us estimate the two terms on the right hand side of (72a). For arbitrary $T > 0$,

$$\int_{0}^{+\infty} G^n(x)\,dx = \int_{0}^{T} G^n(x)\,dx + \int_{T}^{+\infty} G^n(x)\,dx$$

$$\leqslant TG^n(0) + G^{n-1}(0) \int_{T}^{+\infty} G(x)\,dx. \tag{72b}$$

Since $E(X_1)$ is finite, the integral

$$\int_{0}^{+\infty} x\,dF(x) = \int_{0}^{+\infty} G(x)\,dx < +\infty.$$

We can therefore choose T such that the last integral in (72b) remains smaller than one. We thus get from (72b),

$$\int_{0}^{+\infty} G^n(x)\,dx \leqslant G^{n-1}(0)[TG(0) + 1], \tag{72c}$$

where T is fixed and does not depend on n. On the other hand, if $P(X_1 < 0) > 0$, then there is an integer $m \geqslant 1$ such that $P(X_1 < -1/m) = p_m > 0$. Hence,

$$\int_{-\infty}^{0} x\,dP(W_n < x) \leqslant -\frac{1}{m} \int_{-\infty}^{-1/m} dP(W_n < x)$$

$$= -\frac{1}{m} P\left(W_n < -\frac{1}{m}\right)$$

$$= -\frac{1}{m}\left[1 - G^n\left(-\frac{1}{m}\right)\right] = -\frac{1}{m}[1 - (1 - p_m)^n],$$

which, since $p_m > 0$, becomes less than $-1/2m$ for all sufficiently large n. This fact, combined with the estimate (72c) and with the equation (72a), yields that, for n sufficiently large, $E(W_n) \leqslant -1/3m$, say. As was pointed out, this contradicts (72), which establishes $P(X_1 \geqslant 0) = 1$.

Let us now turn to (72). Assume that $P(X_1 = 0) \neq 1$. Let us write $E(W_n)$

1.6 SPECIAL PROPERTIES OF THE EXPONENTIAL DISTRIBUTION

in the form

$$E(W_n) = \int_0^{+\infty} x\, d\{1 - [1 - F(x)]^n\} = n\int_0^1 F^{-1}(y)(1-y)^{n-1}\, dy, \quad (72d)$$

where we made the substitution $y = F(x)$, by which $d\{1 - (1-y)^n\} = n(1-y)^{n-1}\, dy$ and $x = F^{-1}(y)$. We want to compare our distribution $F(x)$ with the exponential distribution $F_1(x) = 1 - e^{-ax}$, $x > 0$, when $a > 0$ is defined as $1/a = E(X_1)$. (Such an a can be found, since $P(X_1 \geq 0) = 1$ and $P(X_1 = 0) \neq 1$, hence $E(X_1) > 0$.) Notice that, for the exponential distribution, (72) holds (see the remark following (67b) or (26a) in Example 1.3.1), and

$$F_1^{-1}(y) = -\frac{1}{a}\log(1-y).$$

Hence, applying (72) and (72d) with our distribution $F(x)$ as well as with $F_1(x)$, we get

$$E(X_1) = n^2 \int_0^1 F^{-1}(y)(1-y)^{n-1}\, dy = -\frac{n^2}{a}\int_0^1 (1-y)^{n-1}\log(1-y)\, dy.$$

Let us rewrite the last equation as

$$\int_0^1 \left[F^{-1}(y) + \frac{1}{a}\log(1-y)\right](1-y)^{n-1}\, dy = 0, \quad n \geq 1. \quad (73)$$

It is a well known result (see Appendix II) that the validity of (73) for all $n \geq 1$ implies

$$F^{-1}(y) + \frac{1}{a}\log(1-y) \equiv 0,$$

from which

$$F(x) = 1 - e^{-ax}, \quad x \geq 0.$$

This establishes the main part of the statement.

For turning to the particular case, we first remark that (67b) evidently implies (72). What remains, therefore, is to show that (67b) implies that $E(X_1)$ is finite. We have seen that, by (72), and thus by (67b) as well, $P(X_1 \geq 0) = 1$. Thus

$$0 \leq E(X_1) = \int_0^{+\infty} x\, dF(x) = \sum_{k=1}^{+\infty} \int_{k-1}^k x\, dF(x) \leq \sum_{k=1}^{+\infty} k[F(k) - F(k-1)],$$

which, in view of

$$F(k) - F(k-1) = [1 - F(k-1)] - [1 - F(k)],$$

can be rearranged as

$$0 \leq E(X_1) \leq \sum_{k=1}^{+\infty} [1 - F(k-1)].$$

But, by (72b) and (67a),

$$1 - F(k-1) = [1 - F(1)]^{k-1} = G(1)^{k-1},$$

and thus

$$0 \leq E(X_1) \leq \sum_{k=1}^{+\infty} G(1)^{k-1},$$

which is finite if $G(1) = 1 - F(1) < 1$. However, if $G(1) = 1$, then, by (67b), $G(n) = 1$ for all $n \geq 1$. $G(x)$ being nonincreasing, this would imply that $G(x) \equiv 1$ for all $x \geq 0$, or, $F(x) \equiv 0$ for all $x \geq 0$, which is not a proper distribution function. This completes the proof. ▲

Theorem 1.6.2 is of theoretical value only, since it requires to check (72) or (67b) for infinitely many values of n. It is, however, significant because it shows that a very simple property of the minimum determines the distribution of the population. It will also serve as a major tool in proving another practical theorem in Section 3.12.1, which also leads to the exponential distribution.

The condition expressed in (72) is a very sensitive tool for determining the population distribution. Indeed, if we take the condition that, for all $n \geq 1$,

$$(n+1) E(W_n) = E(X_1), \tag{74}$$

where $E(X_1)$ is assumed finite, we can then conclude that $F(x) = x/a$ for $0 \leq x \leq a$, for some $a > 0$; that is, the population is uniform over the interval $(0, a)$. This claim can be proved in the same manner as we proved the first part of Theorem 1.6.2, or it can be deduced as a corollary to a general statement: The sequence $\{E(W_n): n \geq 1\}$ uniquely determines the distribution of the population. A similar statement is true with $E(Z_n)$ or with more general sequences (see Exercises 23 and 24).

The simple property of lack of memory turned out to be equivalent to a distributional property (67b) of the minimum, which is unique for the exponential distribution. We shall now turn to the maximum and establish another type of unique property of the exponential variates.

Let $X_1, X_2, \ldots, X_n$ be i.i.d. random variables with common distribution function $F(x)$. Assume that $F(x)$ is differentiable and let $f(x) = F'(x)$. Let

1.6 SPECIAL PROPERTIES OF THE EXPONENTIAL DISTRIBUTION

$X_{1:n} \leq X_{2:n} \leq \cdots \leq X_{n:n}$ be the order statistics of the X_j. We can determine the joint density function of all order statistics $X_{r:n}$, $1 \leq r \leq n$, by the following simple argument. The joint density of the X_j, $1 \leq j \leq n$, is, by assumption,

$$f(x_1)f(x_2)\cdots f(x_n).$$

Since the n values X_j, $1 \leq j \leq n$, can be rearranged in increasing order in $n!$ ways (ties can be neglected in view of continuity), every specific $X_{r:n}$, $1 \leq r \leq n$, could have come from $n!$ different samples X_j, $1 \leq j \leq n$. Hence, the density of the vector $X_{r:n}$, $1 \leq r \leq n$, is

$$n!f(x_1)f(x_2)\cdots f(x_n), \qquad 0 \leq x_1 < x_2 < \cdots < x_n.$$

For the exponential distribution $F(x) = 1 - e^{-ax}$, $x > 0$, for which $f(x) = ae^{-ax}$, $x \geq 0$, this becomes

$$n!a^n \exp\bigl[-a(x_1+x_2+\cdots+x_n)\bigr], \qquad 0 \leq x_1 < x_2 < \cdots < x_n. \tag{75}$$

Putting $U_1 = X_{1:n}$, $U_j = X_{j:n} - X_{j-1:n}$ for $j \geq 2$, we get that, for each j, $U_j \geq 0$. Their joint density is obtained from (75) by the substitution $u_1 = x_1$, $u_j = x_j - x_{j-1}$ for $j \geq 2$. Hence, it is

$$n!a^n \exp\left[-a \sum_{j=1}^{n} (n-j+1)u_j\right]$$

$$= \prod_{j=1}^{n} (n-j+1)a \exp\bigl[-a(n-j+1)u_j\bigr], \qquad u_j \geq 0,\ 1 \leq j \leq n. \tag{76}$$

From (76) we see that the random variables

$$U_1 = X_{1:n}, \qquad U_j = X_{j:n} - X_{j-1:n} \qquad \text{for } j \geq 2 \tag{77}$$

are independent exponential variates. The parameter of U_j is $a(n-j+1)$. Hence, the variables $(n-j+1)U_j = V_j$ are i.i.d. In particular,

$$Z_n = X_{n:n} = \sum_{j=1}^{n} U_j = \sum_{j=1}^{n} \frac{V_j}{n-j+1}. \tag{78}$$

This suggests that perhaps extreme value theory can be reduced to weighted sums of i.i.d. random variables. The following theorem tells us that it cannot: a representation like (78) is unique for the exponential distribution under the sole assumption that the distribution is continuous.

Theorem 1.6.3. *Let $X_1, X_2, \ldots, X_n$ be i.i.d. with common continuous distribution function $F(x)$. Assume that the random variables U_j, $j \geq 1$, of (77) are independent. Then $F(x) = 1 - \exp[-a(x-b)]$ with some constants $a > 0$ and b, where $x \geq b$.*

Remark 1.6.2. Since adding a constant b to each X_j does not affect the independence of the terms of (77), the result in Theorem 1.6.3 actually is a characterization theorem (an "if and only if" statement) in view of the fact that we have checked the independence of the U_j for the exponential distribution.

Proof. We first prove that the independence of the U_j implies that, for all x, $F(x) < 1$. Namely, assume that, for a finite a, $P(X_j \leq a) = 1$, $1 \leq j \leq n$. Let ω be the smallest of all a's with such property; that is, for any $\varepsilon > 0$,

$$P(X_j < \omega) = 1, \qquad P(X_j \geq \omega - \varepsilon) > 0.$$

Then, for all $n \geq 1$ and for any $\varepsilon > 0$,

$$P(U_1 \geq \omega - \varepsilon) = [1 - F(\omega - \varepsilon)]^n > 0. \tag{79}$$

On the other hand, by the independence of U_1 and U_2 and by the choice of ω, for any $\varepsilon > 0$,

$$P(U_1 \geq \omega - \varepsilon) P(U_2 > \varepsilon) = P(U_1 \geq \omega - \varepsilon, U_2 > \varepsilon)$$

$$\leq P(X_{1:n} \geq \omega - \varepsilon, X_{2:n} > \omega)$$

$$\leq P(X_j > \omega) \quad \text{for some} \quad j$$

$$\leq \sum_{j=1}^{n} P(X_j > \omega) = 0.$$

In view of (79), we thus have that, for any $\varepsilon > 0$,

$$P(U_2 > \varepsilon) = P(X_{2:n} - X_{1:n} > \varepsilon) = 0.$$

Such relation is possible only for degenerate distributions, which are excluded by the assumption that $F(x)$ is continuous. Hence, our claim of $F(x) < 1$ for all x follows.

Next we observe that if $U_1, U_2, \ldots, U_j$ are independent, then, evidently, so are U_1 and $T_j = U_2 + U_3 + \cdots + U_j = X_{j:n} - X_{1:n}$, $j \geq 2$. We can now easily arrive at the conclusion of the present theorem by looking at an arbitrary integral in terms of T_j, $j \geq 2$. Let C be an $(n-1)$-dimensional set and consider the probability that

$$A = \{T_2, T_3, \ldots, T_n) \in C\}.$$

Since the T_j, $j \geq 2$, are independent of U_1,

$$P(A) = P(A|U_1=t) = \int_C dP(T_j < t_j, 2 \leq j \leq n | U_1 = t),$$

where t is a possible value of $U_1 = X_{1:n}$, and thus of X_j, $j \geq 1$. Let $B = \{(t_2, t_3, \ldots, t_n): 0 \leq t_2 \leq t_3 \leq \cdots \leq t_n\}$. Since the vector $(T_2, T_3, \ldots, T_n) \in B$,

$$P(A|U_1 = t) = \int_{CB} dP(T_j < t_j, 2 \leq j \leq n | U_1 = t).$$

But on B,

$$dP(T_j < t_j, 2 \leq j \leq n | U_1 = t) = dP(X_{j:n} < t_j + t, 2 \leq j \leq n | X_{1:n} = t)$$

$$= (n-1)! \, dP(X_j < t_j + t, 2 \leq j \leq n | X_{1:n} = t)$$

$$= (n-1)! \prod_{j=2}^{n} \frac{dF(t_j + t)}{1 - F(t)},$$

where t is a possible value of $X_{1:n}$. Since we have seen that $1 - F(t) > 0$ for all t, the restriction on t is that

$$0 < P(X_{1:n} < t) = 1 - [1 - F(t)]^n,$$

that is, that $F(t) > 0$. Therefore, from now on, we assume that

$$t > \alpha(F) = \min\{x : F(x) > 0\}. \tag{80}$$

With such a t, we can combine our previous formulas for $P(A)$. These yield

$$P(A) = \int_{CB} \prod_{j=2}^{n} \frac{dF(t_j + t)}{1 - F(t)}.$$

Since $P(A)$ does not depend on t, neither can the integral above. But the set CB does not contain t either, and thus

$$\prod_{j=2}^{n} \frac{dF(t_j + t)}{1 - F(t)} \tag{81}$$

must not depend on t. Since here, for each j, the same function F occurs at varying points, (81) does not depend on t only if an individual factor $dF(t_j+t)/[1-F(t)]$ is not a function of t. In other words, for $s\geqslant 0$,

$$\int_0^s \frac{dF(t_j+t)}{1-F(t)} = \frac{F(s+t)-F(t)}{1-F(t)} \tag{82}$$

is the same value for all t satisfying (80). We thus immediately obtain that $\alpha(F)$ is finite. As a matter of fact, if $\alpha(F)=-\infty$, then in (82) we could let $t\to-\infty$, which would imply for all t,

$$\frac{F(s+t)-F(t)}{1-F(t)} = \lim_{t=-\infty} \frac{F(s+t)-F(t)}{1-F(t)} = 0,$$

which is not possible for a distribution function. Hence $\alpha(F)=b$ is finite. By continuity of $F(x)$, $F(b)=0$. Let $t\to b$ in (82). We get

$$\frac{F(s+t)-F(t)}{1-F(t)} = \lim_{t=b} \frac{F(s+t)-F(t)}{1-F(t)} = F(x+b),$$

where $s\geqslant 0$ and $t\geqslant b$. If we put $G(x)=1-F(x)$, the equation above becomes

$$G(s+t)=G(t)G(s+b), \quad s\geqslant 0, \quad t\geqslant b. \tag{83}$$

One more transformation $G^*(x)=G(x+b)$ in (83) results in the following equation, where we replaced $t-b=u$:

$$G^*(s+u)=G^*(u)G^*(s), \quad s\geqslant 0, \quad u\geqslant 0.$$

An appeal to Corollary 1.6.1 thus yields that, for $x>0$, $G^*(x)=e^{-ax}$ with some $a>0$. Hence

$$F(x)=1-G(x)=1-G^*(x-b)=1-e^{-a(x-b)}, \quad x\geqslant b.$$

Since $F(b)=0$, $F(x)=0$ for $x<b$. The theorem is established. ▲

We pointed out a theoretical consequence of Theorem 1.6.3 just before it was formulated. It should be emphasized, however, that its content is also a very valuable tool for the applied scientist. Let us first look at a specific problem and then extend it to a general model.

Consider n policyholders of an insurance company who belong to the same risk group. Then the time to the first accident for these individuals can be considered as i.i.d. random variables $X_j, 1\leqslant j\leqslant n$, where the common distribution $F(x)$ is continuous. The actual accident reports (or claims) arrive at the insurance company in increasing order; that is, the company actually observes $X_{j:n}$, $1 \leq j \leq n$. The company can conclude by Theorem 1.6.3 that $F(x) = 1 - e^{-ax}$, $a > 0$, $x \geq 0$, if it is reasonable from their experience that time intervals between claims are independent (in this special case $b = 0$ because $X_j \geq 0$ follows from the nature of the problem).

Notice that the following abstract model is typified by the above problem. We have n individuals (not necessarily human) who act independently of each other. Their action is observed by an outsider and is terminated by the occurrence of a specific event. Let the time to this occurrence be distributed identically for each individual with a continuous distribution function $F(x)$. The observer records the times when the events in question occur. If the time intervals between these occurrences are independent, then $F(x)$ is exponential.

We conclude this section with a very important remark. We have seen three characterizations of the exponential distribution. Each, however, can be restated in a number of equivalent forms which characterize other distributions. Namely, if $h(x)$ is a strictly monotonic function and $Y_j = h(X_j)$, then $X_j, 1 \leq j \leq n$, are i.i.d. if, and only if, $Y_j, 1 \leq j \leq n$, are i.i.d. Therefore, if a property characterizes the common distribution function $F(x)$ of the X_j, $1 \leq j \leq n$, then this property can be restated for Y_j, $1 \leq j \leq n$, and their distribution $D(x)$ will be characterized. $D(x)$ is, of course, different from $F(x)$. One example for this possibility is formulated below as a corollary; others are found in Exercises 25 and 26.

Corollary 1.6.2. *Let $Y_1, Y_2, \ldots, Y_n$ be i.i.d. nonnegative random variables with continuous distribution function $D(x)$. Let $Y_{r:n}$ be the rth order statistic of the Y_j, $1 \leq j \leq n$. Then the random variables*

$$R_{r:n} = \frac{Y_{r:n}}{Y_{r+1:n}}, \quad 1 \leq r \leq n, \quad Y_{n+1:n} = 1, \tag{84}$$

are independent if, and only if, $D(x) = Cx^a$ for $x \in (0, A)$, where $c > 0$, $a > 0$ are arbitrary constants.

Proof. Since $Y_j \geq 0$, we can take logarithms. Define $X_j = -\log Y_j$, $1 \leq j \leq n$. Then, for $1 \leq r \leq n$,

$$X_{r:n} = -\log Y_{n-r+1:n}, \quad U_{r:n} = X_{r:n} - X_{r-1:n} = \log R_{n-r:n},$$

where $X_{0:n} = 0$. Thus the independence of $R_{r:n}$ in (84) is equivalent to the independence of the differences $U_{r:n}$, $1 \leq r \leq n$. We can therefore apply Theorem 1.6.3, which yields

$$P(-\log Y_j < x) = 1 - \exp[-a(x - b)], \quad a > 0, \quad b \text{ finite}.$$

Hence, for $x>0$,

$$D(x)=P(Y_j<x)=P(-\log Y_j>-\log x)=e^{ab}x^a,$$

where $a>0$, b finite, and $\log x<-b$. That is, with $C=e^{ab}$ and $A=e^{-b}$, $D(x)=Cx^a$ on $(0,A)$. On the other hand, Remark 1.6.2 implies that $R_{r:n}$, $1\leq r\leq n$, are indeed independent for this specific distribution $D(x)$. The corollary is established. ▲

1.7. SURVEY OF THE LITERATURE

The applied models mentioned in the introductory sections will be analyzed in Chapter 3. Hence, the survey here is restricted to Sections 1.3–1.6.

Section 1.3 is an elementary introduction to the subject matter, and its examples are selected to show what can be expected. Although this section is restricted to i.i.d variables, the simple bounds obtained are of general nature. Indeed, a recent discovery by Galambos (1975a) implies that if only binomial moments $S_{k,n}$ are used in estimating the distribution of extremes, then the same accuracy can be achieved whether the variables are i.i.d. or not. For the accurate statement, see Exercise 8. Therefore, in order to obtain bounds on the distribution of Z_n or W_n or the other extremes, which can show the real difference between independent and dependent cases, terms other than the $S_{k,n}$ should be used. One useful set of inequalities is given in Theorem 1.4.2, which is due to A. Rényi (1961). Its extension, partly formulated in Exercise 13, was obtained by Galambos (1966). These are very powerful tools to obtain limit theorems (see Chapter 3). For estimations, the inequality of Exercise 18, due to E. G. Kounias (1968), is valuable. Evidently, one can relabel the terms of the sequence, and thus the first term does not have any special role in it. This inequality was recently extended by D. Hunter (1976). For other inequalities see Kounias (1968) and S. Kounias and J. Marin (1976). One of the results of Kounias (1968), who generalizes work of S. Gallot (1966), is that if **P** denotes the vector $P(C_j)$ and Q is any of the generalized inverses of the matrix whose entries are $P(C_iC_j)$, then

$$P(\nu_n\geq 1)\geq \mathbf{P}Q\mathbf{P},$$

where **P** is written once as a column, once as a row.

If individual terms of probabilities of intersections are not available but the moments $S_{k,n}$ are known, then the inequality of Exercise 18 becomes a special case of (39). These inequalities are new here. The case $t=0$, first obtained by M. Sobel and V. R. R. Uppuluri (1972) for exchangeable events, was extended to arbitrary events by Galambos (1975a). It is shown in Galambos and R. Mucci (1978) that arbitrary t can be reduced to $t=0$.

Sometimes very simple forms of the inequalities of the present chapter provide useful tools in statistics. For the implications of the first two inequalities of Exercise 6, see the paper by R. L. Dykstra et al. (1973).

If one uses only $S_{1,n}$ and $S_{2,n}$, the best lower bound for $P(\nu_n \geq 1)$ has been found by S. M. Kwerel (1975a). The actual bound was known earlier, but its extremal property is new in Kwerel's work. It is the content of Theorem 1.4.3, proved here by a new method and originally due to Dawson and Sankoff (1967). Its extremal property as stated in Exercise 17 is established by Galambos (1977b). The simple method suggested by Exercises 16 and 17 should prove powerful for other results as well. As pointed out in Exercise 15, the bound of Theorem 1.4.3 covers a previously known inequality which was discovered by Chung and Erdös (1952) and reobtained by P. Whittle (1959). This form is extended by Galambos (1969) in several directions, one of which is contained in Exercise 16. The inequalities of Exercise 19 and 20, proved by Éva Galambos (1965) and R. M. Meyer (1969), turned out to be very useful in the asymptotic theory of multivariate extremes (see Chapter 5).

There are several methods of proof for inequalities discussed here. The method of indicators, due to M. Loève (1942), was reobtained in another form by Rényi (1958). This formulation made it possible to obtain methods of proof of quadratic Boolean inequalities $\sum d_{ij} P(B_i) P(B_j) \geq 0$ (see Exercise 1 for definition), first by Galambos and Rényi (1968) and, in a refined form, in Galambos (1969). Other general methods, partly mentioned earlier, are given by Galambos (1975a) and (1977b), Galambos and Mucci (1978), and Kwerel (1975a,b,c). The method of the book differs from that of all quoted papers.

Pioneers in this subject area were M. Fréchet (1940) and K. Jordan (1927). L. Takács (1958) gives a good account of the early history and shows several interesting applications. Later, Takács (1967b) extends the formula (38) to the distribution of the occurrences in an infinite sequence of events.

For general formulas of moments and distributions of order statistics, we refer to a book by H. A. David (1970), which, however, is not a prerequisite for the present book.

The short section on characterizations is an introduction only. It serves to show that there are cases when the population distribution is well determined, so that the statistician does not face the difficult problem of choosing a distribution which may result in a huge error in decisions. This statement is clarified and justified by results of the forthcoming chapters. Because of its limited coverage here, the theory of characterizations is not covered by the bibliography. For characterizations by order statistics, the reader is referred to Galambos (1975b,c) and to a more detailed account by Galambos and S. Kotz (1978). The theorems contained in Section 1.6 are due to P. V. Sukhatme (1937), J. S. Huang (1974), Z. Govindarajulu (1966), and H. J. Rossberg (1960).

We conclude by stressing the value of formula (87) of Exercise 21. Although it is very simple to prove, its consequence is remarkable: $E(Z_n)$ becomes almost as large for i.i.d. variables for most distributions as it can be for arbitrary systems with a given marginal distribution. This observation, and the evaluation of its consequences, is due to T. L. Lai and H. Robbins (1976).

1.8. EXERCISES

1. (The method of indicators) Let $C_1, C_2, \ldots, C_n$ be events and B_j, $1 \leq j \leq m$, be so-called Boolean functions of the C's. This means that B_j can be expressed by a finite number of the operations union, intersection, and taking complements. Let $I(B_j)$ be the indicator of B_j; thus it equals one or zero according as B_j occurs or fails to occur. Show that an inequality

$$\sum_{j=1}^{m} d_j P(B_j) \geq 0, \qquad d_j \text{ real}, \tag{85}$$

holds for arbitrary choice of $C_1, C_2, \ldots, C_n$ if, and only if,

$$\sum_{j=1}^{m} d_j I(B_j) \geq 0$$

with probability one.

2. Use the above method to prove

$$\frac{S_{k,n}}{\binom{n}{k}} \geq \frac{S_{k+1,n}}{\binom{n}{k+1}}, \qquad k \geq 1,$$

where $S_{k,n}$ is defined by (36).

3. Prove the following identities for binomial coefficients.

(i) $$k\binom{n}{k} = n\binom{n-1}{k-1}, \qquad n \geq k \geq 1.$$

(ii) $$\binom{N}{T}\binom{N-T}{n} = \binom{N}{N-T}\binom{N-T}{n} = \binom{N}{n}\binom{N-n}{T}.$$

4. Prove the binomial theorem:

$$(a+b)^n = \sum_{k=0}^{n} \binom{n}{k} a^k b^{n-k}.$$

5. Using the method of indicators, give a new proof for Theorem 1.4.1.

6. Write out in details the inequalities (39) for $s=0,1,2$, and 3. As a particular case, conclude that

$$S_{1,n} - S_{2,n} \leqslant P(\nu_n \geqslant 1) \leqslant S_{1,n} - \frac{2}{n} S_{2,n} \leqslant S_{1,n}.$$

Finally deduce that, for an infinite sequence $A_1, A_2, \ldots$ of events,

$$P\left(\bigcup_{j=1}^{+\infty} A_j\right) \leqslant \sum_{j=1}^{+\infty} P(A_j).$$

7. Let $X_1, X_2, \ldots, X_n$ be random variables with $F_j(x) = P(X_j < x)$. Put $Z_n = \max(X_1, X_2, \ldots, X_n)$. Deduce from Exercise 6 that if

$$S_{1,n} = \sum_{j=1}^{n} P(X_j \geqslant x) \leqslant 1$$

and $P(Z_n \geqslant x) = S_{1,n}$, then the events $\{X_j \geqslant x\}$, $1 \leqslant j \leqslant n$, are mutually exclusive.

8. For a sequence $C_1, C_2, \ldots, C_n$ of events, let B_j be the event that exactly j of the events C_t, $1 \leqslant t \leqslant n$, occur. Assume that the inequalities,

$$\sum_{k=0}^{n} a_k S_{k,n} \leqslant P(B_j) \leqslant \sum_{k=0}^{n} b_k S_{k,n}, \tag{86}$$

where a_k and b_k, $1 \leqslant k \leqslant n$, are constants, not depending on n, have been established in the following special case: $C_1, C_2, \ldots, C_n$ are independent and $P(C_j) = p$ for all j. Prove that (86) then holds for arbitrary events C_t, $1 \leqslant t \leqslant n$.

[J. Galambos (1975a)]

9. Extend the criterion of the preceding exercise by letting one coefficient vary with n which is monotonic in n.

10. Give a new proof of Theorem 1.4.1 by the method of Exercise 9.

11. Give new proofs for Theorem 1.4.3 by the methods of Exercises 1 and 8.

12. Prove the following extension of the criterion of Exercise 8. Let $a_k = a_k(n)$ and $b_k = b_k(n)$ depend on n in arbitrary manner. Assume that (86) has been established in the special case of Exercise 8 for all choices $a_k(N)$ and $b_k(N), N \geqslant n$, of the coefficients. Then (86) holds for arbitrary events.

[J. Galambos and R. Mucci (1978)]

13. Let $S_{k,n}^*$ and $S_{k,n}^{**}$ be defined by (41) and (42). Prove that, for any $n \geqslant 1$ and $m \geqslant 0$ with $2m+1 \leqslant n$,

$$S_{1,n}^* - 2S_{2,n}^{**} + 3S_{3,n}^* - \cdots - 2mS_{2m,n}^{**} \leqslant P(\nu_n = 1)$$

$$\leqslant S_{1,n}^{**} - 2S_{2,n}^* + 3S_{3,n}^{**} - \cdots + (2m+1)S_{2m+1,n}^{**}.$$

Extend the formula by appropriately extending the definition of $S_{k,n}^{**}$ to obtain bounds on $P(\nu_n = t)$, $t > 1$.

[J. Galambos (1966)]

14. Use the inequalities of Exercise 13 to set bounds on $P(X_{2:5} \geqslant 0.1)$ in Example 1.5.2, where $X_{2:5}$ is the time to the second failure of components if the first failure did not require replacement.

15. Show that Theorem 1.4.3 implies the following inequality:

$$P(\nu_n \geqslant 1) \geqslant \frac{S_{1,n}^2}{2S_{2,n} + S_{1,n}}.$$

[K. L. Chung and P. Erdös (1952)]

16. For real numbers $y_1 \geqslant y_2 \geqslant \cdots \geqslant y_n \geqslant 0$, define

$$S_k = \sum_{j=k}^{n} \binom{j-1}{k-1} y_j, \quad k \geqslant 1.$$

Show that

$$y_k \geqslant \frac{(S_1 - k + 1)S_k}{(k+1)S_{k+1} + kS_k}.$$

Show that if $y_k = P(\nu_n \geqslant k)$, then S_k is the binomial moment $S_{k,n}$ of ν_n.

17. Apply the method of the preceding exercise to prove the following result. If, for arbitrary sequence $C_1, C_2, \ldots, C_n$ of events,

$$P(\nu_n \geq 1) \geq aS_{1,n} + bS_{2,n}$$

and a and b are not of the form $a = 2/(k+1), b = 2/k(k+1)$ with some $1 \leq k \leq n-1$, then there is an integer $1 \leq k \leq n-1$ such that

$$\frac{2}{k+1} S_{1,n} - \frac{2}{k(k+1)} S_{2,n} > aS_{1,n} + bS_{2,n}.$$

Consequently, Theorem 1.4.3 is the best linear lower bound in terms of $S_{1,n}$ and $S_{2,n}$.

[J. Galambos (1977b)]

18. By the method of indicators, or otherwise, prove that, for arbitrary events $C_1, C_2, \ldots, C_n$,

$$P(\nu_n \geq 1) \leq S_{1,n} - \sum_{j=2}^n P(C_1 C_j).$$

[E. G. Kounias (1968)]

19. Let $A_{jt}, 1 \leq j \leq m, 1 \leq t \leq n$, be a double indexed sequence of events. Let $\nu_{j,n}$ denote the number of events $A_{jt}, 1 \leq t \leq n$, that occur. Let

$$S(u_1, u_2, \ldots, u_m) = \Sigma P \left[\bigcap_{j=1}^m \bigcap_{s=1}^{u_j} A_{j, t(j,s)} \right],$$

where the summation Σ is for $1 \leq j \leq m$ and for all $1 \leq t(j,1) < t(j,2) < \cdots < t(j, u_j) \leq n$. Finally, let

$$f(\mathbf{u}; \mathbf{k}; r) = (-1)^{r-K} S(\mathbf{u}) \prod_{j=1}^m \binom{u_j}{k_j},$$

where $\mathbf{u} = (u_1, u_2, \ldots, u_m)$, $\mathbf{k} = (k_1, k_2, \ldots, k_m)$, and $K = k_1 + k_2 + \cdots + k_m$. By the method of indicators, or otherwise, prove that, for arbitrary integers $k_j \geq 0$ and $M \geq 0$,

$$P(\nu_{j,n} = k_j, 1 \leq j \leq m) \leq \sum_{r=K}^{K+2M} \sum_{\mathbf{u}} f(\mathbf{u}; \mathbf{k}; r),$$

where $\Sigma_{\mathbf{u}}$ signifies summation over all vectors $\mathbf{u} = (u_1, u_2, \ldots, u_m)$, $u_j \geq 1$, such that $u_1 + \cdots + u_m = r$.

[Éva Galambos (1965)]

20. With the notations of the preceding exercise prove that

$$P(v_{j,n}=k_j, 1\leqslant j\leqslant m) \geqslant \sum_{r=K}^{K+2M+1} \sum_{\mathbf{u}} f(\mathbf{u};\mathbf{k};r).$$

[R. M. Meyer (1969)]

21. Let $X_1, X_2, \ldots, X_n$ be arbitrary random variables. Let $J(x)=x$ if $x>0$ and zero if $x\leqslant 0$. Show that, for any real number a,

$$Z_n = \max(X_1,\ldots,X_n) \leqslant a + \sum_{j=1}^n J(X_j - a).$$

Hence, if the X's are identically distributed,

$$E(Z_n) \leqslant a + n\int_a^{+\infty}[1-F(x)]dx. \qquad (87)$$

Show that if the integral on the right hand side is finite, then the right hand side is minimized at

$$a = a_n = \inf\{x : F(x) \geqslant 1 - 1/n\}.$$

[T. L. Lai and H. Robbins (1976)]

22. Use the inequality (87) to show that if $F(x)$ is the standard normal distribution, then, whatever be the interdependence of the X's,

$$E(Z_n) \leqslant (2\log n - \log\log n)^{1/2}, \qquad n\geqslant 3.$$

[T. L. Lai and H. Robbins (1976)]

23. Let $X_1, X_2, \ldots, X_n$ be i.i.d. random variables. Show that the sequence $E(X_{r:n})$, $1\leqslant r\leqslant n$, $n\geqslant 1$, uniquely determines the population distribution $F(x)$.

24. With the notations of the preceding exercise, prove that, for any $F(x)$ and for $1\leqslant r<n, n\geqslant 1$,

$$(n-r)E(X_{r:n}) + rE(X_{r+1:n}) = nE(X_{r:n-1}).$$

Hence, conclude that if $r(n)$ is an arbitrary function of n with $1\leqslant r(n)\leqslant n$, then $E(X_{r(n):n})$, $n\geqslant 1$, uniquely determines $F(x)$.

25. Prove that if $X_1, X_2, \ldots, X_n$ are i.i.d. and, for all $n\geqslant 1$, $P(W_n + \log n \geqslant y) = P(X_1 \geqslant y)$, then $P(X_1 < x) = 1 - \exp(-e^x)$. [Hint: Apply Theorem 1.6.2 and the monotonic transformation $\exp(X_j)$.]

26. Restate all theorems of Section 1.6 for all the continuous distributions you know by first transforming a random variable X with distribution function $F(x)$ to the exponential variate $Y = -\log F(X)$.

CHAPTER 2

Weak Convergence for Independent and Identically Distributed Variables

Throughout this chapter $X_1, X_2, \ldots, X_n$ denote independent and identically distributed random variables. We put

$$F(x) = P(X_j < x). \tag{1}$$

Furthermore, as before,

$$Z_n = \max(X_1, X_2, \ldots, X_n) \tag{2}$$

and

$$W_n = \min(X_1, X_2, \ldots, X_n). \tag{3}$$

By assumption

$$H_n(x) = P(Z_n < x) = F^n(x) \tag{4}$$

and

$$L_n(x) = P(W_n < x) = 1 - (1 - F(x))^n. \tag{5}$$

We seek conditions on $F(x)$ to guarantee the existence of sequences a_n, $b_n > 0$, and/or c_n, $d_n > 0$ of constants such that, as $n \to +\infty$,

$$\lim H_n(a_n + b_n x) = H(x) \tag{6}$$

and

$$\lim L_n(c_n + d_n x) = L(x) \tag{7}$$

exist for all continuity points of $H(x)$ and $L(x)$, respectively, where $H(x)$ and $L(x)$ are nondegenerate distribution functions.

Such convergence will be called weak convergence of distribution functions or of random variables. That is, a sequence U_n of random variables, or their distribution functions $R_n(x)$, are said to converge weakly if, as $n \to +\infty$, $\lim R_n(x) = R(x)$ exists for all continuity points x of the limit $R(x)$. With this term, our aim in this chapter is therefore to find conditions on $F(x)$ under which Z_n (or W_n) can be normalized by constants a_n and $b_n > 0$ so that $(Z_n - a_n)/b_n$ (or $(W_n - c_n)/d_n$) converges weakly to a nondegenerate distribution.

In view of (4) and (5), the expressions (6) and (7) are equivalent to

$$\lim F^n(a_n + b_n x) = H(x) \tag{6a}$$

and

$$\lim [1 - F(c_n + d_n x)]^n = 1 - L(x), \tag{7a}$$

where the limits are for $n \to +\infty$ and in the sense of weak convergence.

Corollaries 1.3.1 and 1.3.2 contain an answer to our problem. In a set of theorems, we now make these corollaries more specific. Namely, we shall give rules for the construction of the sequences a_n, b_n, c_n, and d_n as well as find criteria for $F(x)$ under which (6a) and (7a) hold. In addition, we shall make $H(x)$ and $L(x)$ explicit.

Before starting with this program, let us introduce two notations which will be used throughout this book. We say that $\alpha(F)$, defined as

$$\alpha(F) = \inf\{x : F(x) > 0\}, \tag{8}$$

is the lower endpoint of the distribution function $F(x)$. Similarly, the upper endpoint $\omega(F)$ of $F(x)$ is defined by

$$\omega(F) = \sup\{x : F(x) < 1\}. \tag{9}$$

Evidently, $\alpha(F)$ is either $-\infty$ or finite and $\omega(F)$ is either $+\infty$ or finite. For example, for the exponential distribution $F(x) = 1 - e^{-x}$, $x > 0$, $\alpha(F) = 0$ and $\omega(F) = +\infty$. If $F(x)$ is the distribution function of a random variable, taking the values 0 and 1 only, then $\alpha(F) = 0$ and $\omega(F) = 1$. As a final example, we take $F(x) = 1/(1 + e^{-x})$, for which $\alpha(F) = -\infty$ and $\omega(F) = +\infty$.

The notations (1)–(9) will be used repeatedly without any further reference to their location. The reader is advised to become familiar with them.

2.1. LIMIT DISTRIBUTIONS FOR MAXIMA AND MINIMA: SUFFICIENT CONDITIONS

As in the previous chapter, we state separately theorems for maxima and minima. However, recall that theorems on maxima and minima are equivalent, as emphasized at the end of Section 1.2. Indeed, we shall obtain results on the minimum from those on the maximum by transforming the random variables X_j into $(-X_j)$, by which maximum becomes minimum and vice versa.

We first state three theorems on the maximum and then give proofs. Their counterparts on the minimum follow afterward.

Theorem 2.1.1. *Let $\omega(F) = +\infty$. Assume that there is a constant $\gamma > 0$ such that, for all $x > 0$, as $t \to +\infty$,*

$$\lim \frac{1 - F(tx)}{1 - F(t)} = x^{-\gamma}. \tag{10}$$

Then there is a sequence $b_n > 0$ such that, as $n \to +\infty$,

$$\lim P(Z_n < b_n x) = H_{1,\gamma}(x),$$

where

$$H_{1,\gamma}(x) = \begin{cases} \exp(-x^{-\gamma}) & \text{if } x > 0, \\ 0 & \text{otherwise.} \end{cases} \tag{11}$$

The normalizing constant b_n can be chosen as

$$b_n = \inf\left\{ x : 1 - F(x) \leq \frac{1}{n} \right\}. \tag{12}$$

Theorem 2.1.2. *Let $\omega(F)$ be finite. Assume that the distribution function $F^*(x) = F(\omega(F) - 1/x)$, $x > 0$, satisfies condition (10) of the preceding theorem. Then there are sequences a_n and $b_n > 0$ such that, as $n \to +\infty$,*

$$\lim P(Z_n < a_n + b_n x) = H_{2,\gamma}(x),$$

where

$$H_{2,\gamma}(x) = \begin{cases} 1 & \text{if } x \geq 0, \\ \exp(-(-x)^\gamma) & \text{if } x < 0. \end{cases} \tag{13}$$

The normalizing constants a_n and b_n can be chosen as $a_n = \omega(F)$ and

$$b_n = \omega(F) - \inf\left\{x : 1 - F(x) \leq \frac{1}{n}\right\}. \tag{14}$$

Theorem 2.1.3. *Assume that, for some finite a,*

$$\int_a^{\omega(F)} (1 - F(y)) \, dy < +\infty. \tag{15}$$

For $\alpha(F) < t < \omega(F)$, define

$$R(t) = (1 - F(t))^{-1} \int_t^{\omega(F)} (1 - F(y)) \, dy. \tag{16}$$

Assume that, for all real x, as $t \to \omega(F)$,

$$\lim \frac{1 - F(t + xR(t))}{1 - F(t)} = e^{-x}. \tag{17}$$

Then there are sequences a_n and $b_n > 0$ such that, as $n \to +\infty$,

$$\lim P(Z_n < a_n + b_n x) = H_{3,0}(x),$$

where

$$H_{3,0}(x) = \exp(-e^{-x}), \quad -\infty < x < +\infty. \tag{18}$$

The normalizing constants a_n and b_n can be chosen as

$$a_n = \inf\left\{x : 1 - F(x) \leq \frac{1}{n}\right\} \tag{19}$$

and

$$b_n = R(a_n). \tag{20}$$

Our notation $H_{1,\gamma}(x)$, $H_{2,\gamma}(x)$, and $H_{3,0}(x)$ of the limit distributions conforms to the following parametric form of von Mises. Let $H_\tau(x)$ be a distribution function and, for those x's for which $0 < H_\tau(x) < 1$, let

$$H_\tau(x) = \exp\{-(1 + \tau x)^{-1/\tau}\},$$

where τ is a given real number. For $\tau = 0$, $H_\tau(x)$ is defined as $\lim H_\tau(x)$, where $\tau \to 0$. Thus, apart from a change of the origin and a change in the unit on the x-axis, $H_\tau(x)$ yields $H_{1,\gamma}(x)$, $H_{2,\gamma}(x)$ and $H_{3,0}(x)$ according as $\tau > 0$, $\tau < 0$, or $\tau = 0$, respectively, where, for $\tau \neq 0$, $\gamma = |1/\tau|$.

Although not contained in the statements, we add here that Theorems 2.1.1–2.1.3 exhaust all possibilities for the existence of the asymptotic distribution of maxima of i.i.d. variables. In other words, if $F(x)$ does not fall into any of the three categories of the above theorems, then there are no normalizing constants a_n and b_n for which (6) would hold. This fact will be shown in Section 2.4. It should also be emphasized that the constants a_n and b_n are not unique. One possible choice is given in each theorem for a_n and b_n. The relation of other possible choices to our specific ones will be discussed in Section 2.2.

Proof of Theorem 2.1.1. We shall reduce our theorem to Corollary 1.3.1. For that purpose we choose $a_n = 0$ and b_n as defined in (12). With these choices, we shall show that, as $n \to +\infty$,

$$\lim n(1 - F(b_n x)) = \begin{cases} x^{-\gamma} & \text{if } x > 0, \\ +\infty & \text{for } x < 0. \end{cases} \tag{21}$$

When (21) is proved, Corollary 1.3.1 implies the conclusion of the present theorem.

Let us first remark that the assumption $\omega(F) = +\infty$ implies that b_n of (12) tends to $+\infty$. Hence, for $x < 0$, $b_n x \to -\infty$ and thus $1 - F(b_n x) \to 1$. Consequently, (21) is proved for $x < 0$.

Now let $x > 0$. Applying (10) with $t = b_n$, which tends to $+\infty$ with n, we get that, as $n \to +\infty$,

$$\lim n(1 - F(b_n x)) = \lim n(1 - F(b_n)) \frac{1 - F(b_n x)}{1 - F(b_n)}$$

$$= x^{-\gamma} \lim n(1 - F(b_n)).$$

Therefore, (21) will be proved, and thus the proof completed, if we show that, for $n \to +\infty$,

$$\lim n(1 - F(b_n)) = 1. \tag{22}$$

For proving (22), we first observe that, in view of (12),

$$1 - F(b_n + 0) \leq \frac{1}{n} \leq 1 - F(b_n)$$

or equivalently,

$$1 \leq n(1 - F(b_n)) \leq \frac{1 - F(b_n)}{1 - F(b_n + 0)}. \tag{23}$$

But, since $1-F$ decreases,
$$1-F(b_n+0) \geqslant 1-F(b_n x), \quad x>1.$$

Applying this last inequality in (23) and appealing to (10) again, we get
$$1 \leqslant n(1-F(b_n)) \leqslant \frac{1-F(b_n)}{1-F(b_n x)} \leqslant (1-\varepsilon)x^\gamma, \tag{24}$$

where $\varepsilon \to 0$ as $n \to +\infty$. Since $x>1$ is arbitrary, (24) implies (22). The theorem is established. ▲

Proof of Theorem 2.1.2. We first apply Theorem 2.1.1 to random variables with common distribution $F^*(x)$, which is defined for all $x>0$. Since $\omega(F^*)=+\infty$ and (10) applies to F^*, the conclusion of Theorem 2.1.1, stated in the form of (6a), says that, for any $x>0$, as $n \to +\infty$,
$$\lim F^{*n}(b_n^* x) = \lim F^n\left(\omega(F) - \frac{1}{b_n^* x}\right) = H_{1,\gamma}(x),$$

where
$$b_n^* = \inf\left\{x : 1-F^*(x) \leqslant \frac{1}{n}\right\}$$
$$= \inf\left\{x : 1-F\left[\omega(F) - \frac{1}{x}\right] \leqslant \frac{1}{n}\right\}$$
$$= \left(\omega(F) - \inf\left\{x : 1-F(x) \leqslant \frac{1}{n}\right\}\right)^{-1}.$$

Hence, with $a_n = \omega(F)$ and $b_n = 1/b_n^*$, as $n \to +\infty$,
$$\lim F^n\left(a_n - \frac{b_n}{x}\right) = H_{1,\gamma}(x), \quad x>0,$$

or
$$\lim F^n(a_n + b_n x) = H_{1,\gamma}\left(-\frac{1}{x}\right), \quad x<0.$$

But, for $x<0$, $H_{1,\gamma}(-1/x) = H_{2,\gamma}(x)$. The proof is now completed by adding that, since $a_n = \omega(F)$ and $b_n > 0$, $F(a_n+b_n x) \equiv 1$ for $x>0$. Hence, as

$n \to +\infty$,

$$\lim F^n(a_n + b_n x) = 1, \ x > 0.$$

▲

Proof of Theorem 2.1.3. For the proof we choose a_n and b_n by (19) and (20), respectively. We shall follow the method of proof of Theorem 2.1.1, which in turn was reduced to Corollary 1.3.1. Observing that a_n of (19) tends to $\omega(F)$ as $n \to +\infty$, (17) implies that, as $n \to +\infty$,

$$\lim \frac{1 - F(a_n + b_n x)}{1 - F(a_n)} = e^{-x} \tag{25}$$

for all x. Therefore, for arbitrary x, as $n \to +\infty$,

$$\lim n(1 - F(a_n + b_n x)) = \lim n(1 - F(a_n)) \frac{1 - F(a_n + b_n x)}{1 - F(a_n)}$$

$$= e^{-x} \lim n[1 - F(a_n)].$$

Consequently, if we show that, as $n \to +\infty$,

$$\lim n(1 - F(a_n)) = 1, \tag{26}$$

Corollary 1.3.1 immediately yields Theorem 2.1.3. For proving (26), we write the definition of a_n in (19) in detail, and, by a trick, we appeal to (25). More precisely, by (19)

$$1 - F(a_n + 0) \leqslant \frac{1}{n} \leqslant 1 - F(a_n).$$

On the other hand, for any $\varepsilon > 0$,

$$1 - F(a_n + \varepsilon b_n) \leqslant 1 - F(a_n + 0).$$

These two sets of inequalities can be combined to read

$$1 \leqslant n(1 - F(a_n)) \leqslant \frac{1 - F(a_n)}{1 - F(a_n + \varepsilon b_n)} \to e^{\varepsilon},$$

as $n \to +\infty$, as was obtained in (25). Since $\varepsilon > 0$ is arbitrary, (26) now follows. The proof is completed. ▲

As stated earlier, Theorems 2.1.1–2.1.3 can be restated for minima by considering $(-X_j)$ instead of X_j. We give below all these theorems in detail, since their content is very basic to the asymptotic theory of extremes. They evidently do not require proof.

Theorem 2.1.4. *Let $\alpha(F) = -\infty$. Assume that there is a constant $\gamma > 0$ such that, for all $x > 0$, as $t \to -\infty$,*

$$\lim \frac{F(tx)}{F(t)} = x^{-\gamma}. \tag{27}$$

Then there is a sequence $d_n > 0$ such that, as $n \to +\infty$,

$$\lim P(W_n < d_n x) = L_{1,\gamma}(x),$$

where

$$L_{1,\gamma}(x) = \begin{cases} 1 - \exp(-(-x)^{-\gamma}) & \text{if } x < 0 \\ 1 & \text{if } x > 0. \end{cases} \tag{28}$$

The normalizing constant d_n can be chosen as

$$d_n = \sup\left\{ x : F(x) \leq \frac{1}{n} \right\}. \tag{29}$$

Theorem 2.1.5. *Let $\alpha(F)$ be finite. Assume that the distribution function $F^*(x) = F(\alpha(F) - 1/x)$, $x < 0$, satisfies condition (27). Then there are sequences c_n and $d_n > 0$ such that, as $n \to +\infty$,*

$$\lim P(W_n < c_n + d_n x) = L_{2,\gamma}(x),$$

where

$$L_{2,\gamma}(x) = \begin{cases} 1 - \exp(-x^{\gamma}) & \text{if } x > 0, \\ 0 & \text{if } x < 0. \end{cases} \tag{30}$$

The normalizing constants c_n and d_n can be chosen as $c_n = \alpha(F)$ and

$$d_n = \sup\left\{ x : F(x) \leq \frac{1}{n} \right\} - \alpha(F). \tag{31}$$

Theorem 2.1.6. *Assume that, for some finite a,*

$$\int_{\alpha(F)}^{a} F(y) \, dy < +\infty \tag{32}$$

For $t > \alpha(F)$, define

$$r(t) = \frac{1}{F(t)} \int_{\alpha(F)}^{t} F(y)\,dy. \tag{33}$$

Assume that, for all real x, as $t \to \alpha(F)$,

$$\lim \frac{F(t + xr(t))}{F(t)} = e^x. \tag{34}$$

Then there are sequences c_n and $d_n > 0$ such that, as $n \to +\infty$,

$$\lim P(W_n < c_n + d_n x) = L_{3,0}(x),$$

where

$$L_{3,0}(x) = 1 - \exp(-e^x), \quad -\infty < x < +\infty. \tag{35}$$

The normalizing constants c_n and d_n can be chosen as

$$c_n = \sup\left\{x : F(x) \leq \frac{1}{n}\right\} \tag{36}$$

and

$$d_n = r(c_n). \tag{37}$$

We turn now to the discussion of other possible choices for the normalizing constants a_n, b_n, c_n, and d_n. For examples to Theorems 2.1.1–2.1.6 with specific distributions, see Section 2.3.

2.2. OTHER POSSIBILITIES FOR THE NORMALIZING CONSTANTS IN THEOREMS 2.1.1–2.1.6

In each theorem of the preceding section we gave specific choices for the normalizing constants a_n, b_n, c_n, and d_n. There are, however, other possibilities for these constants. We do not even claim that the choices given so far are the simplest ones. Their advantage is that the rules given for their calculation are generally easy to apply to all distributions for which a limit law is presented.

That normalizing constants, in general, cannot be unique can be seen from the following simple discussion. Let Y_n be a sequence of random variables and assume that, with some constants C_n and $D_n > 0$, as $n \to +\infty$,

$$\lim P(Y_n < C_n + D_n x) = G(x) \tag{38}$$

for all continuity points of $G(x)$, where $G(x)$ is a nondegenerate distribution function. Writing

$$P(Y_n < C_n + D_n x) = P\left(\frac{Y_n}{D_n} - \frac{C_n}{D_n} < x\right),$$

one immediately sees that, if $D_n \to +\infty$, say, then modifying C_n by a quantity C_n^* which satisfies $C_n^*/D_n \to 0$, as $n \to +\infty$, will have no effect in the limit (38). Similarly, one expects that D_n can be replaced by D_n^* satisfying $D_n^*/D_n \to 1$ as $n \to +\infty$. These observations are made precise in the following lemma. We formulate the lemma in terms a bit more general than we need here; this will, however, not complicate the proof. Later we shall use it in this general form.

Lemma 2.2.1. *Let U_n and δ_n be two sequences of random variables. Assume that there is a distribution function $G(x)$ such that, for all of its continuity points x, as $n \to +\infty$,*

$$\lim P(U_n < x) = G(x). \tag{39}$$

Furthermore, assume that, for every $\varepsilon > 0$, as $n \to +\infty$,

$$\lim P(|\delta_n| > \varepsilon) = 0. \tag{40}$$

Then, as $n \to +\infty$,

$$\lim P(U_n + \delta_n < x) = G(x) \tag{41}$$

for all continuity points x of $G(x)$.

Proof. Let $\varepsilon > 0$. Write

$$P(U_n + \delta_n < x) = P(U_n + \delta_n < x, |\delta_n| < \varepsilon) + P(U_n + \delta_n < x, |\delta_n| \geq \varepsilon). \tag{42}$$

Since

$$P(U_n + \delta_n < x, |\delta_n| \geq \varepsilon) \leq P(|\delta_n| \geq \varepsilon),$$

this term tends to 0 as $n \to +\infty$, in view of (40). Turning to the first term on the right hand side of (42), we make the following estimates:

$$P(U_n + \delta_n < x, |\delta_n| < \varepsilon) \leq P(U_n < x + \varepsilon, |\delta_n| < \varepsilon)$$
$$\leq P(U_n < x + \varepsilon) \tag{43}$$

2.2 THE NORMALIZING CONSTANTS IN THEOREMS 2.1.1–2.1.6

and

$$P(U_n + \delta_n < x, |\delta_n| < \varepsilon) \geq P(U_n < x - \varepsilon, |\delta_n| < \varepsilon)$$
$$= P(U_n < x - \varepsilon) - P(U_n < x - \varepsilon, |\delta_n| \geq \varepsilon). \quad (44)$$

The last term in (44) can again be estimated by $P(|\delta_n| \geq \varepsilon)$, which tends to zero by another appeal to (40). Therefore, if ε is such that $x + \varepsilon$ and $x - \varepsilon$ are continuity points of $G(x)$, (39) and (42)–(44) imply that, as $n \to +\infty$,

$$G(x - \varepsilon) \leq \liminf P(U_n + \delta_n < x) \leq \limsup P(U_n + \delta_n < x) \leq G(x + \varepsilon).$$

Letting $\varepsilon \to 0$, (41) follows for continuity points x of $G(x)$. The lemma is proved. ▲

We can now easily prove our claims about the constants in (38).

Lemma 2.2.2. *Let Y_n be a sequence of random variables. Let C_n and $D_n > 0$ be sequences of real numbers for which (38) holds for all continuity points x of $G(x)$. Let C_n^* and $D_n^* > 0$ be two additional sequences of real numbers which satisfy, as $n \to +\infty$,*

$$\lim \frac{(C_n - C_n^*)}{D_n} = 0 \qquad (45)$$

and

$$\lim \frac{D_n}{D_n^*} = 1. \qquad (46)$$

Then (38) holds for all continuity points x of $G(x)$, when C_n and/or D_n are replaced by C_n^* and/or D_n^*, respectively.

Proof. To show that we can replace C_n by C_n^*, we apply Lemma 2.2.1 with $U_n = (Y_n - C_n)/D_n$ and $\delta_n = (C_n - C_n^*)/D_n$. Assumption (45) implies (40), and thus (41) holds. This is exactly our claim, since

$$U_n + \delta_n = \frac{Y_n - C_n^*}{D_n}.$$

Turning to replacing D_n by D_n^*, we write

$$\frac{Y_n - C_n}{D_n^*} = \frac{Y_n - C_n}{D_n} + \frac{Y_n - C_n}{D_n}\left(\frac{D_n}{D_n^*} - 1\right).$$

Putting $U_n = (Y_n - C_n)/D_n$ and

$$\delta_n = \frac{Y_n - C_n}{D_n}\left(\frac{D_n}{D_n^*} - 1\right),$$

we have to show that (40) holds. Let $\varepsilon > 0$. Let $M > 1$ be such that $\pm \varepsilon M$ are continuity points of $G(x)$. Let n_0 be such that, for all $n \geq n_0$,

$$\left|\frac{D_n}{D_n^*} - 1\right| \leq \frac{1}{M}.$$

Such $n_0 = n_0(M)$ can be found for arbitrary $M > 0$ by (46). Then, for all $n \geq n_0$,

$$P(|\delta_n| \geq \varepsilon) \leq P\left(\left|\frac{Y_n - C_n}{D_n}\right| \geq \varepsilon M\right).$$

Letting $n \to +\infty$ and then $M \to +\infty$, we get (40). Lemma 2.2.1 now implies that the limiting distribution of $(Y_n - C_n)/D_n^*$ is also $G(x)$. Since C_n and C_n^* can be interchanged in this latter argument, the proof is completed. ▲

So far we have discussed the question that, if we have found normalizing constants in (38) for a general limit distribution problem, then to what extent can these constants be changed without affecting the limit? As we shall see in the next section, the simple conclusion of Lemma 2.2.2 is very useful in making the constants of Theorems 2.1.1–2.1.6 simpler and neater. We now focus our attention on the structure of the sequences a_n, b_n, c_n, and d_n occurring in (6) and (7).

Theorem 2.2.1. *Let a_n and $b_n > 0$ be sequences of real numbers for which (6) holds. Then, for arbitrary integer $m \geq 1$, the limits, as $n \to +\infty$,*

$$\lim \frac{a_{nm} - a_n}{b_n} = A_m \tag{47}$$

and

$$\lim \frac{b_{nm}}{b_n} = B_m > 0, \tag{48}$$

exist and are finite. Furthermore,

$$H^m(A_m + B_m x) = H(x). \tag{49}$$

2.2 THE NORMALIZING CONSTANTS IN THEOREMS 2.1.1–2.1.6

Before proving the theorem, let us remark that (49) uniquely determines the limits A_m and B_m for each specific $H(x)$ (Section 2.4).

For the proof of Theorem 2.2.1, we need the following general result.

Lemma 2.2.3. *Let $F_n(y)$ be a sequence of distribution functions. Let C_n, $D_n > 0$, ρ_n and $\tau_n > 0$ be sequences of real numbers such that, as $n \to +\infty$,*

$$\lim F_n(C_n + D_n x) = G(x), \qquad \lim F_n(\rho_n + \tau_n x) = T(x) \qquad (50)$$

for all continuity points x of the limits, where $G(x)$ and $T(x)$ are nondegenerate distribution functions. Then, as $n \to +\infty$, the limits

$$\lim \frac{\tau_n}{D_n} = B \neq 0, \qquad \lim \frac{\rho_n - C_n}{D_n} = A \qquad (51)$$

are finite, and

$$T(x) = G(A + Bx). \qquad (52)$$

Proof. Select two points x_1, x_2 and y_1, y_2 of continuity for each of $G(x)$ and $T(x)$ such that

$$0 < G(x_1) \leqslant G(x_2) < 1 \quad \text{and} \quad T(y_1) < G(x_1), \quad T(y_2) > G(x_2).$$

Then, by (50), for n sufficiently large,

$$\rho_n + \tau_n y_1 \leqslant C_n + D_n x_1 \leqslant C_n + D_n x_2 \leqslant \rho_n + \tau_n y_2. \qquad (53)$$

Taking differences of the middle terms and of the outermost terms, we get

$$D_n(x_2 - x_1) \leqslant \tau_n(y_2 - y_1)$$

or

$$\frac{D_n}{\tau_n} \leqslant \frac{y_2 - y_1}{x_2 - x_1}. \qquad (54)$$

On the other hand, the first two terms in (53) yield

$$\frac{\rho_n - C_n}{D_n} \leqslant x_1 - \left(\frac{\tau_n}{D_n}\right) y_1. \qquad (55)$$

Since x_1, x_2, y_1, y_2 are fixed, (54) implies that D_n/τ_n remains bounded as $n \to +\infty$. Interchanging the roles of $G(x)$ and $T(x)$ in the argument, leading to (53), we similarly get that τ_n/D_n remains bounded. An appeal to (55) now shows that $(\rho_n - C_n)/D_n$ is bounded, and, by the symmetry of the roles of $T(x)$ and $G(x)$, we can conclude that $(C_n - \rho_n)/D_n$ is bounded as well. Let us now take a subsequence n_t of n for which (51) holds. The limit B is indeed different from zero, since its reciprocal is finite in view of the preceding argument. Let $\varepsilon > 0$ be arbitrary and n_t sufficiently large. Then, by the choice of n_t,

$$(B-\varepsilon)D_{n_t} \leqslant \tau_{n_t} \geqslant D_{n_t}(B+\varepsilon)$$

and

$$C_{n_t} + D_{n_t}(A-\varepsilon) \leqslant \rho_{n_t} \leqslant C_{n_t} + D_{n_t}(A+\varepsilon).$$

Hence, for $x > 0$,

$$F_{n_t}\left(C_{n_t} + AD_{n_t} + BD_{n_t}x - \varepsilon(x+1)D_{n_t}\right) \leqslant F_{n_t}(\rho_{n_t} + \tau_{n_t}x)$$
$$\leqslant F_{n_t}\left(C_{n_t} + AD_{n_t} + BD_{n_t}x + \varepsilon(x+1)D_{n_t}\right).$$

Therefore, if x and ε are such that $A + Bx - \varepsilon(x+1)$ as well as $A + Bx + \varepsilon(x+1)$ are continuity points of $G(x)$, (50) implies that, as $n_t \to +\infty$,

$$G(A + Bx - \varepsilon(x+1)) \leqslant \liminf F_{n_t}(\rho_{n_t} + \tau_{n_t}x)$$
$$\leqslant \limsup F_{n_t}(\rho_{n_t} + \tau_{n_t}x) \leqslant G(A + Bx + \varepsilon(x+1)).$$

Finally, if x is a continuity point of $T(x)$ and $A + Bx$ is a continuity point of $G(x)$, then letting $\varepsilon \to 0$ yields (52). Although we assumed $x > 0$, the effect of $x < 0$ in the above argument is only that εx will be replaced by $-\varepsilon x$, and thus (52) follows again. However, (52) uniquely determines A and B. Consequently, for every subsequence n_t for which (51) holds, the limits A and B are the same values; thus (51) holds. This completes the proof. ▲

Proof of Theorem 2.2.1. Let $F(x)$ be a distribution function satisfying (6a). Let $m > 1$ be a fixed integer. Then, as $n \to +\infty$,

$$\lim F^{nm}(a_{nm} + b_{nm}x) = H(x)$$

or

$$\lim F^n(a_{nm} + b_{nm}x) = H^{1/m}(x).$$

Comparing this with (6a), we have the situation of Lemma 2.2.3–namely, $F_n = F^n$, $C_n = a_n$, $D_n = b_n$, $\rho_n = a_{nm}$, $\tau_n = b_{nm}$, $G(x) = H(x)$, and $T(x) = H^{1/m}(x)$. The conclusion of Lemma 2.2.3 with this special case is exactly what was to be proved. ▲

We also state the counterpart of Theorem 2.2.1 for minima. This can be deduced from Theorem 2.2.1 by our usual transformation, or one can easily prove it by a direct appeal to Lemma 2.2.3. We omit the details of proof.

Theorem 2.2.2. *Let c_n and $d_n > 0$ be sequences of real numbers for which (7) holds. Then, for arbitrary integer $m \geq 1$, the limits, as $n \to +\infty$,*

$$\lim \frac{c_{nm} - c_n}{d_n} = A_m^* \tag{56}$$

and

$$\lim \frac{d_{nm}}{d_n} = B_m^* \neq 0, \tag{57}$$

exist and are finite. Furthermore,

$$1 - L(x) = \{1 - L(A_m^* + B_m^* x)\}^m. \tag{58}$$

As a side result in Theorems 2.2.1–2.2.2 we obtained that the class of limit distributions $H(x)$ and $L(x)$ for maxima and minima, respectively, is the set of solutions of the functional equations (49) and (58). This result will help us in settling our claim that Theorems 2.1.1–2.1.6 exhaust all possibilities for our problem stated at (6) and (7). We shall discuss this in more detail in Section 2.4. Let us now turn to some examples through special distributions.

2.3. THE ASYMPTOTIC DISTRIBUTION OF THE MAXIMUM AND MINIMUM FOR SOME SPECIAL DISTRIBUTIONS

We shall apply our results of the previous two sections for some specific population distributions. With the following examples we have the double aim of presenting results for the most important distributions as well as working out details as exercises for our theory obtained so far. Therefore, in each case, we shall appeal to Theorems 2.1.1–2.1.6 and 2.2.1–2.2.2, even though, for some distributions, a direct solution of (6) and (7) would be simple. Since we do not want to repeat the examples of Chapter 1, the reader is referred to Examples 1.3.1 and 1.3.2, where the exponential distribution is introduced. We leave as Exercise 10 the reobtaining of the normalizing constants through Theorems 2.1.1–2.1.6.

In order to reduce the number of parameters in the distributions to be considered, let us remark that, if a and $b > 0$ are constants, then

$$\max(a+bX_1, a+bX_2, \ldots, a+bX_n) = a + b\max(X_1, X_2, \ldots, X_n)$$

and, for $b < 0$, the left hand side equals

$$a + b\min(X_1, X_2, \ldots, X_n).$$

A similar relation also holds for the minimum of the linear transformations $a + bX_j$, $1 \leq j \leq n$. One can therefore easily modify the normalizing constants, to be obtained below, when a change in location (a) or in scale (b) is needed.

2.3.1. The Uniform Distribution on $(0, 1)$

Let the common distribution function $F(x)$ be defined

$$F(x) = \begin{cases} 0 & \text{for } x < 0, \\ x & \text{for } 0 \leq x \leq 1, \\ 1 & \text{for } x > 1. \end{cases}$$

Thus, $\alpha(F) = 0$ and $\omega(F) = 1$. Hence, for Z_n, either Theorem 2.1.2 or 2.1.3 is to be applied. Let us check (17). Since

$$\frac{1 - (t + xR(t))}{1 - t} = 1 - x\frac{R(t)}{1 - t}$$

cannot tend to e^{-x} as $t \to 1$, (17) fails and thus Theorem 2.1.2 is the only possibility. We introduce, for $x > 0$,

$$F^*(x) = F\left(1 - \frac{1}{x}\right) = 1 - \frac{1}{x} \qquad \text{if } x > 1.$$

Turning to (10), we find that, for $x > 0$, as $t \to +\infty$,

$$\lim \frac{1 - F^*(tx)}{1 - F^*(t)} = \lim \frac{t}{tx} = x^{-1}.$$

Hence, Theorem 2.1.2 applies with $\gamma = 1$. It follows that, since

$$\inf\left\{x : 1 - F(x) \leq \frac{1}{n}\right\} = 1 - \frac{1}{n},$$

$$\lim P\left(Z_n < 1 + \frac{1}{n}x\right) = H_{2,1}(x) \qquad (n \to +\infty),$$

2.3 THE MAXIMUM AND MINIMUM FOR SOME SPECIAL DISTRIBUTIONS

which, in this special case, becomes $H_{2,1}(x) = e^x$ for $x < 0$ and $H_{2,1}(x) = 1$ for $x \geq 0$.

Turning to W_n, again either Theorem 2.1.5 or Theorem 2.1.6 applies. Starting with Theorem 2.1.5, we have to check (27) with $F^*(x) = F(-1/x)$, $x < 0$, which becomes $F^*(x) = -1/x$ for $x < -1$. Thus, for all $x > 0$, as $t \to -\infty$

$$\lim \frac{F^*(tx)}{F^*(t)} = x^{-1}.$$

Consequently, Theorem 2.1.5 applies with $\gamma = 1$. By (31), $d_n = 1/n$, $c_n = \alpha(F) = 0$. Hence, as $n \to +\infty$,

$$\lim P(nW_n \leq x) = L_{2,1}(x),$$

where $L_{2,1}(x) = 1 - e^{-x}$ for $x > 0$ and 0 for $x \leq 0$.

2.3.2. The Standard Normal Distribution

We turn now to the distribution

$$F(x) = (2\pi)^{-1/2} \int_{-\infty}^{x} e^{-y^2/2} dy,$$

where x is arbitrary. Thus $\alpha(F) = -\infty$ and $\omega(F) = +\infty$. Now either Theorem 2.1.1 or Theorem 2.1.3 applies to Z_n. We shall in fact show that Theorem 2.1.3 applies. More specifically, we shall conclude that, with

$$a_n = (2 \log n)^{1/2} - \frac{\frac{1}{2}(\log \log n + \log 4\pi)}{(2 \log n)^{1/2}} \tag{59}$$

and

$$b_n = (2 \log n)^{-1/2}, \tag{60}$$

$(Z_n - a_n)/b_n$ converges weakly to $H_{3,0}(x)$. For this, we first find an asymptotic expression for $1 - F(x)$ as $x \to +\infty$. Let $x > 0$. Integrating by parts, we get

$$(2\pi)^{1/2}(1 - F(x)) = \int_{x}^{+\infty} e^{-u^2/2} du = \int_{x}^{+\infty} (ue^{-u^2/2}) u^{-1} du$$

$$= -u^{-1} e^{-u^2/2} \Big]_{x}^{+\infty} - \int_{x}^{+\infty} e^{-u^2/2} u^{-2} du,$$

that is,

$$(2\pi)^{1/2}(1-F(x)) - x^{-1}e^{-x^2/2} = -\int_x^{+\infty} e^{-u^2/2} u^{-2} du, \quad x > 0.$$

We can continue this method for obtaining as many terms as we wish to approximate $1 - F(x)$. For example, one more step results in the inequalities

$$\frac{1}{(2\pi)^{1/2}}\left(\frac{1}{x} - \frac{1}{x^3}\right)e^{-x^2/2} < 1 - F(x)$$

$$< \frac{1}{(2\pi)^{1/2} x} e^{-x^2/2}, \quad x > 0. \tag{61}$$

We thus have, as $x \to +\infty$,

$$\lim x(1 - F(x))e^{x^2/2} = (2\pi)^{-1/2}. \tag{62}$$

We may pause here to observe that Theorem 2.1.1 indeed fails for the normal distribution. That is, by (62), the limit in (10) is not finite. Hence, we should turn to Theorem 2.1.3. By the upper estimate in (61), (15) evidently holds with, say, $a = 1$. The next step is to check the validity of (17). For this we approximate $R(t)$ of (16). By integrating (61), we get, for $t > 0$,

$$e^{-t^2/2}t^{-2} - 3\int_t^{+\infty} e^{-y^2/2} y^{-3} dy \leq \int_t^{+\infty} (1 - F(y)) dy$$

$$\leq e^{-t^2/2}t^{-2} - 2\int_t^{+\infty} e^{-y^2/2} y^{-3} dy. \tag{63}$$

Since

$$\int_t^{+\infty} e^{-y^2/2} y^{-3} dy \leq t^{-3} \int_t^{+\infty} e^{-y^2/2} dy \leq t^{-4} e^{-t^2/2},$$

where the last step was obtained by the upper inequality of (61), (62) and (63) yield

$$R(t) = \frac{1}{t} + 0(t^{-3}), \quad t \to +\infty. \tag{64}$$

Hence, by (62), as $t \to +\infty$,

$$\lim \frac{1 - F(t + xR(t))}{1 - F(t)} = \lim \frac{te^{t^2/2}}{(t + xR(t))\exp\left\{\frac{1}{2}(t + xR(t))^2\right\}}$$

$$= \lim \frac{t}{t + xR(t)} \exp\left\{-xR(t)\left(t + \frac{1}{2}xR(t)\right)\right\} = e^{-x}.$$

2.3 THE MAXIMUM AND MINIMUM FOR SOME SPECIAL DISTRIBUTIONS

Consequently, Theorem 2.1.3 does apply, and thus the limit distribution of $(Z_n - a_n)/b_n$ is indeed $H_{3,0}(x)$. It remains to show that the expressions (59) and (60) can be used for a_n and b_n, respectively. For showing this, we shall apply the formulas (19) and (20) as well as Lemma 2.2.2. The latter tells us how accurately we should solve the relation in (19). Since $F(x)$ is continuous, (19) reduces to $1 - F(a_n) = 1/n$. Hence, the logarithm of (61) with $x = a_n$ yields

$$\log n + \log(1 - a_n^{-2}) < \tfrac{1}{2}a_n^2 + \log a_n + \tfrac{1}{2}\log(2\pi) < \log n. \qquad (65)$$

We shall determine a_n as a sum of terms which are smaller and smaller in magnitude. Evidently, the first term in a_n is $(2\log n)^{1/2}$. This fact already implies that we can choose b_n by the formula (60). As a matter of fact, by (20) and (64), any further terms in a_n, which are of smaller magnitude than $(\log n)^{1/2}$, will modify b_n to b_n^* to such an extent only that $b_n/b_n^* \to 1$ as $n \to +\infty$. Therefore, by Lemma 2.2.2, we are free to choose either b_n or b_n^*. Let us therefore establish that our choice for b_n is formula (60). Another appeal to Lemma 2.2.2 now tells us that, in view of (45), if we neglect a term in a_n which, when divided by b_n, tends to zero, the limit distribution is not affected. Hence, we have to expand a_n into terms which are not of smaller magnitude than $(\log n)^{-1/2}$. It is easily achieved by substituting $a_n = (2\log n)^{1/2} - k/(2\log n)^{1/2}$ in (65). We immediately get (59).

Since the standard normal distribution $F(x)$ is symmetric about zero, $c_n = -a_n$ and $d_n = b_n$ can be used as normalizing constants for W_n. The limiting distribution is $L_{3,0}(x)$ of (35).

2.3.3. The Lognormal Distribution

If $X > 0$ and $\log X$ is a normal variate, then X is said to have a lognormal distribution. Thus, in standard form,

$$F(x) = (2\pi)^{-1/2} \int_{-\infty}^{\log x} e^{-u^2/2} du, \qquad x > 0. \qquad (66)$$

We can therefore easily find normalizing constants a_n and $b_n > 0$ such that $(Z_n - a_n)/b_n$ converges weakly. Let first a_n^* and b_n^* be the normalizing constants (59) and (60) for the standard normal distribution. Then, by definition, as $n \to +\infty$,

$$P(\log Z_n < a_n^* + b_n^* x) \to \exp(-e^{-x}). \qquad (67)$$

This, of course, does not provide directly our desired form of $P(Z_n < a_n + b_n x)$. However, since b_n^*, given as b_n in (60), tends to zero as $n \to +\infty$, by the Taylor expansion, for some $|v| \leq 1$,

$$\exp(a_n^* + b_n^* x) = \exp(a_n^*)\left[1 + b_n^* x + v(b_n^* x)^2\right]$$
$$= \exp(a_n^*) + b_n^* \exp(a_n^*)(1 + v b_n^* x) x.$$

As remarked, $1 + v b_n^* x \to 1$ as $n \to +\infty$. Hence, by Lemma 2.2.2, as $n \to +\infty$,

$$\lim P(Z_n < \exp(a_n^* + b_n^* x)) = \lim P(Z_n < a_n + b_n x)$$

with $a_n = \exp(a_n^*)$ and $b_n = b_n^* \exp(a_n^*)$. The limit distribution, as obtained in (67), is $\exp(-e^{-x})$.

The minimum W_n can similarly be reduced to the case of the standard normal distribution. We immediately get that, as $n \to +\infty$,

$$\lim P(W_n < c_n + d_n x) = 1 - \exp(-e^x),$$

where $c_n = \exp(-a_n^*) = 1/a_n$ and $d_n = b_n^* \exp(-a_n^*)$.

2.3.4. The Cauchy Distribution

If the common distribution $F(x)$ is given by

$$F(x) = \frac{1}{2} + \frac{1}{\pi} \arctan x,$$

then $\alpha(F) = -\infty$ and $\omega(F) = +\infty$. Hence, for Z_n, either Theorem 2.1.1 or Theorem 2.1.3 can be applied. But since integration by parts gives

$$\int_0^{+\infty} \left[1 - F(x)\right] dx = x(1 - F(x))\Big|_0^{+\infty} + \int_0^{+\infty} x \frac{1}{\pi(1+x^2)} dx = +\infty,$$

(15) fails. We have therefore to consider Theorem 2.1.1. Applying L'Hospital's rule, as $t \to +\infty$,

$$\lim \frac{1 - F(tx)}{1 - F(t)} = \lim \frac{(1+t^2)x}{1+(tx)^2} = x^{-1}.$$

We can thus apply Theorem 2.1.1 with $\gamma = 1$. We obtain that, as $n \to +\infty$,

$$\lim P(Z_n < b_n x) = \exp\left(-\frac{1}{x}\right), \qquad x > 0,$$

where b_n can be chosen as

$$\frac{1}{2} - \frac{1}{\pi}\arctan b_n = \frac{1}{n},$$

or

$$b_n = \tan\left(\frac{\pi}{2} - \frac{\pi}{n}\right). \tag{68}$$

The symmetry of $F(x)$ about zero again makes it unnecessary to investigate W_n in detail. Since

$$P(W_n < -x) = P(Z_n > x), \qquad x > 0,$$

we get that, as $n \to +\infty$,

$$\lim P(W_n < d_n x) = 1 - \exp\left(\frac{1}{x}\right), \qquad x < 0,$$

where $d_n = b_n$ of (68).

2.4. NECESSARY CONDITIONS FOR WEAK CONVERGENCE

In this section we settle two mathematical problems. They can be summarized by saying that Theorems 2.1.1–2.1.6 exhaust all the possibilities for the existence of asymptotic distribution for maxima and minima, after suitable normalization, in the i.i.d. case. Before giving the exact statements, however, let us look at the definition (6) of the existence of a limit law $H(x)$ for Z_n. The argument will, of course, apply to W_n as well.

In (6), we are seeking sequences a_n and $b_n > 0$ of numbers such that $(Z_n - a_n)/b_n$ has a limit law $H(x)$. However, the sequences a_n and b_n are not unique, as we saw in Section 2.2. Nevertheless, $H(x)$ cannot vary much with varying a_n and b_n. We obtained in Lemma 2.2.3 that if, for some sequences a_n and $b_n > 0$, as $n \to +\infty$,

$$\lim P(Z_n < a_n + b_n x) = H(x)$$

and if, for other sequences A_n and $B_n > 0$,

$$\lim P(Z_n < A_n + B_n x) = H^*(x),$$

then $H^*(x) = H(A + Bx)$ with some constants A and $B > 0$. Such a situation does occur, as the following example shows. Let $X_1, X_2, \ldots, X_n$ be unit exponential variates; that is, their common distribution function

$F(x) = 1 - e^{-x}$, $x \geq 0$. Then (see Example 1.3.1), as $n \to +\infty$,

$$\lim P(Z_n < \log n + x) = \exp(-e^{-x}).$$

Hence, $a_n = \log n$, $b_n = 1$, and $H(x) = \exp(-e^{-x})$. Choose now $A_n = 3 + \log n$ and $B_n = 2$, say. We get, for $n \to +\infty$,

$$H^*(x) = \lim P(Z_n < A_n + B_n x)$$
$$= \lim P[Z_n < a_n + b_n(2x+3)]$$
$$= H(2x+3).$$

Therefore, we cannot speak of a unique limiting distribution but of a whole family. This family will be referred to as distributions of the same type, which is the content of the following definition.

Definition 2.4.1. We say that the distribution functions $H(x)$ and $H^*(x)$ are of the same type if there are real numbers A and $B > 0$ such that

$$H^*(x) = H(A + Bx).$$

For making the formulation of the results of the present section simpler, we introduce another concept.

Definition 2.4.2. Let $H(x)$ be a nondegenerate distribution function which is a possible limit in (6). Then we say that $F(x)$ is in the domain of attraction of $H(x)$ if there are sequences a_n and $b_n > 0$ such that, as $n \to +\infty$,

$$\lim F^n(a_n + b_n x) = H(x).$$

Furthermore, if $L(x)$ is a nondegenerate distribution function occurring in (7), then $F(x)$ is said to be in the domain of attraction of $L(x)$ if, for some sequences c_n and $d_n > 0$, as $n \to +\infty$,

$$\lim [1 - F(c_n + d_n x)]^n = 1 - L(x).$$

If we replace x by $A + Bx$, $B > 0$, in either of the limits above, we get that if $F(x)$ is in the domain of attraction of $H(x)$ (or $L(x)$), $F(x)$ is in the domain of attraction of any function which is of the same type as $H(x)$ (or $L(x)$). In other words, the domains of attraction of two functions are identical if the two functions in question are of the same type.

We can now formulate the results of this section.

2.4 NECESSARY CONDITIONS FOR WEAK CONVERGENCE

Theorem 2.4.1. *There are only three types of nondegenerate distributions $H(x)$ satisfying (6). These are $H_{1,\gamma}(x)$ of (11), $H_{2,\gamma}(x)$ of (13), and $H_{3,0}(x)$ of (18).*

Theorem 2.4.2. *There are only three types of nondegenerate distributions $L(x)$ satisfying (7). These are $L_{1,\gamma}(x)$ of (28), $L_{2,\gamma}(x)$ of (30), and $L_{3,0}(x)$ of (35).*

Theorem 2.4.3. *The distribution function $F(x)$ is in the domain of attraction of*

(i) $H_{1,\gamma}(x)$ *if, and only if, $\omega(F) = +\infty$ and (10) holds;*
(ii) $H_{2,\gamma}(x)$ *if, and only if, $\omega(F) < +\infty$ and*

$$F^*(x) = F\left[\omega(F) - \frac{1}{x}\right], \qquad x > 0$$

satisfies (10); and
(iii) $H_{3,0}(x)$ *if, and only if, (15) is finite and (17) holds.*

The final result in this sequence is as follows.

Theorem 2.4.4. *$F(x)$ is in the domain of attraction of*

(i) $L_{1,\gamma}(x)$ *if, and only if, $\alpha(F) = -\infty$ and (27) holds;*
(ii) $L_{2,\gamma}(x)$ *if, and only if, $\alpha(F) > -\infty$ and the function*

$$F^*(x) = F\left[\alpha(F) - \frac{1}{x}\right], \qquad x < 0$$

satisfies (27); and
(iii) $L_{3,0}(x)$ *if, and only if, (32) is finite and (34) holds.*

As pointed out several times, the theory for minima is equivalent to that for maxima. Hence, we shall prove only Theorems 2.4.1 and 2.4.3 in detail. Theorems 2.4.2 and 2.4.4 will then follow by our turning to the sequence $\{-X_j\}$ of random variables instead of $\{X_j\}$.

Proof of Theorem 2.4.1. We have established in Theorem 2.2.1 that if $H(x)$ is a nondegenerate limit in (6), then it satisfies the equation

$$H^m(A_m + B_m x) = H(x), \qquad m \geq 1, \tag{69}$$

where A_m and $B_m > 0$ are suitable constants. Let us determine the solutions of (69) in several steps.

(i) We first find those solutions $H(x)$ of (69) for which $B_m = 1$ for all $m \geq 1$; that is, with some constants A_m,

$$H^m(A_m + x) = H(x), \qquad m \geq 1. \tag{69a}$$

Since $H(x)$ is nondecreasing, $0 = A_1 \leq A_2 \leq \cdots$. There is at least one m for which $A_m > 0$; otherwise (69a) would become $H^m(x) = H(x)$ for all x, and thus $H(x) = 0$ or 1 for all x, contrary to our assumption that $H(x)$ is nondegenerate. It is also evident that, as $m \to +\infty$,

$$\lim A_m = \omega(H). \tag{70}$$

As a matter of fact if $A_m \leq q < \omega(H)$ for all $m \geq 1$, then, by (69a), for all $x < \omega(H) - q$, $H(x) = 0$. But if $H(x) = 0$ for $x = x^*$, then $H(x) = 0$ for $x_1 = x^* \pm A_m$, $m \geq 1$. Letting x_1 take the role of x^*, we would get $H(x) = 0$ for $x_2 = x^* \pm 2A_m$, $m \geq 1$, and then, by induction, $H(x) = 0$ for $x_k = x^* \pm kA_m$, $k \geq 1$, would follow. Since there is one m with $A_m > 0$, $H(x_k) = 0$, $k \geq 1$, and the monotonicity of $H(x)$ would imply $H(x) \equiv 0$ for all x. This is, however, not possible for a proper distribution function. We have thus proved (70) as well as $H(x) > 0$ for all x. A similar argument shows that $H(x) < 1$ for all x and thus $\omega(H) = +\infty$. We can therefore take logarithms in (69a). We get

$$m \log H(A_m + x) = \log H(x), \qquad m \geq 1, \quad -\infty < x < +\infty, \tag{71}$$

where

$$0 = A_1 \leq A_2 \leq \cdots; \qquad A_m \to +\infty \text{ with } m. \tag{72}$$

From (71) and (72) we shall deduce that the function

$$G(x) = \log H(x), \qquad G^*(x) = \frac{G(x)}{G(0)} \tag{73}$$

satisfies the equation

$$G^*(x+y) = G^*(x)G^*(y), \qquad \text{for all } x, y. \tag{74}$$

For this purpose, consider an arbitrary number $z > 0$. By (72), we can define a unique m with $A_m \leq z < A_{m+1}$. Since $G(x)$ is nondecreasing,

$$G(A_m + x) \leq G(z + x) \leq G(A_{m+1} + x). \tag{75}$$

Applying (75) with an arbitrary x and with $x = 0$, we get, in view of $G(x) < 0$,

$$\frac{G(A_m + x)}{G(A_{m+1})} \geq \frac{G(z + x)}{G(z)} \geq \frac{G(A_{m+1} + x)}{G(A_m)},$$

2.4 NECESSARY CONDITIONS FOR WEAK CONVERGENCE

which, by (71), becomes

$$\frac{(m+1)G(x)}{mG(0)} \geq \frac{G(z+x)}{G(z)} \geq \frac{mG(x)}{(m+1)G(0)}.$$

If $z \to +\infty$, then $m \to +\infty$. Hence

$$\lim_{z=+\infty} \frac{G(z+x)}{G(z)} = \frac{G(x)}{G(0)} = G^*(x).$$

Let now x and y be arbitrary numbers. Then

$$G^*(x+y) = \lim_{z=+\infty} \frac{G(z+x+y)}{G(z)}$$

$$= \lim_{z=+\infty} \frac{G(z+x+y)}{G(z+x)} \cdot \frac{G(z+x)}{G(z)}$$

$$= G^*(y)G^*(x),$$

which is exactly (74). Notice that $G(x) < 0$ and thus $G^*(x) > 0$. Furthermore, $G^*(x)$ is nondecreasing. Hence, by (74), Lemma 1.6.1 implies that, for $x \geq 0$, $G^*(x) = e^{-ax}$ for some $a > 0$. Once again applying Lemma 1.6.1 to $G^*(-x)$, $x \geq 0$, we get $G^*(x) = e^{-bx}$, $x \leq 0$, for some $b > 0$. Since $G^*(0) = 1$, (74) yields, with $x = -y$, that $a = b$. Hence

$$\log H(x) = G(x) = G(0)G^*(x) = G(0)e^{-ax}$$

with some $a > 0$, where x is arbitrary. If we write $G(0) = -e^{-c}$,

$$H(x) = \exp(-e^{-ax-c}), \qquad a > 0,$$

which is of the type of $H_{3,0}(x)$.

(ii) We now turn to the case of (69), when there is one $M > 1$ such that $B_M < 1$. We first show that $\omega(H)$ is finite. In fact we show that

$$\omega(H) = \frac{A_M}{1 - B_M}. \tag{76}$$

Indeed, if $x \geq A_M/(1 - B_M)$, then $x \geq A_M + B_M x$. Since $H(u)$ is nondecreasing,

$$H(x) \geq H(A_M + B_M x).$$

But then by (69)

$$H^M(x) \leq H(x) = H^M(A_M + B_M x) \leq H^M(x),$$

and thus $H(x) = 1$ for $x \geq A_M/(1 - B_M)$. In order to complete the proof of (76), we have to show that, for $x < A_M/(1 - B_M)$, $H(x) < 1$. However, if this were not true, then $\omega(H) < A_M/(1 - B_M)$ or, equivalently, $\omega(H) < A_M + B_M \omega(H)$. We could then choose two numbers y and x with

$$\omega(H) < y < A_M + B_M x < A_M + B_M \omega(H). \tag{77}$$

Hence, by (69),

$$H(x) = H^M(A_M + B_M x) \geq H^M(y) = 1, \tag{78}$$

where the last equation follows from the definition of $\omega(H)$. The inequality in (78) would thus give $H(x) = 1$. But, by the choice of x in (77), $x < \omega(H)$, for which $H(x) = 1$ is not possible. Consequently, (76) is valid.

We also deduce that, under the condition of $B_M < 1$, $\alpha(H) = -\infty$. Since $H(x)$ is nondegenerate,

$$\alpha(H) < \omega(H) = \frac{A_M}{1 - B_M}.$$

Hence, if $\alpha(H) > -\infty$, just as in (77), we could find an x with

$$x < \alpha(H) < A_M + B_M x, \tag{79}$$

by simply taking any x with $\alpha(H) < A_M + B_M x < A_M + B_M \alpha(H)$. For such an x, however,

$$0 = H[\alpha(H)] \geq H(x) = H^M(A_M + B_M x),$$

and thus $H(A_M + B_M x) = 0$. This contradicts the definition of $\alpha(H)$ in view of the upper inequality of (79).

The inequalities in the preceding argument can be reversed, and we can thus reach the following conclusions. If $B_s > 1$ for some $s > 1$, then $\alpha(H)$ is finite and $\omega(H) = +\infty$. Hence, if $B_M < 1$ for some $M > 1$, then $B_s \leq 1$ for all $s > 1$. We add that $B_s = 1$ for $s > 1$ is also impossible if $B_M < 1$ for some $M > 1$. Indeed, we would have, for any $x < \omega(H)$,

$$H(x) = H^s(x + A_s), \tag{80}$$

from which it is evident that $A_s > 0$. The application of (80) with $x < \omega(H) < x + A_s$ would yield $H(x) = 1$, contradicting the definition of $\omega(H)$. We have thus shown that, if $B_M < 1$ for some $M > 1$, then $B_m < 1$ for all $m > 1$.

2.4 NECESSARY CONDITIONS FOR WEAK CONVERGENCE

But then (76) can be applied with arbitrary $m > 1$ and we get

$$\frac{A_m}{1 - B_m} = \omega(H) \quad \text{for } m > 1. \tag{81}$$

We can now reduce (69) to (69a) by introducing the following function. Let

$$G(z) = H[\omega(H) - e^{-z}], \quad z \text{ real.}$$

Then, with $C_m = -\log B_m$, (69) and (81) yield, for $m \geq 1$,

$$G^m(C_m + z) = H^m[\omega(H) - B_m e^{-z}],$$
$$= H^m\{A_m + B_m[\omega(H) - e^{-z}]\}$$
$$= H[\omega(H) - e^{-z}] = G(z),$$

which is (69a). The solution in part (i) gives

$$G(z) = \exp(-e^{-az-c}), \quad a > 0,$$

from which

$$H(z) = G\left[\log \frac{1}{\omega(H) - z}\right] = \exp\{-e^{-c}[\omega(H) - z]^a\},$$

where $z < \omega(H)$ and $a > 0$. This is evidently of the same type as $H_{2,\gamma}(x)$ of (13).

(iii) There remains the case of (69), when there is an $m > 1$ with $B_m > 1$. As remarked in the course of (ii), (see the discussion between (79) and (80)), by reversing the inequalities, leading to (76), one can get that $\alpha(H)$ is finite and $\omega(H) = +\infty$. We then get, as in (ii), that necessarily $B_m > 1$ for all $m > 1$. The transformation $G(z) = H[\alpha(H) + e^z]$ again leads to (69a), from which only one type of solution is obtained for (69), namely, that of $H_{1,\gamma}(x)$.

In (i)–(iii) we have seen that (69) has only three types of solutions in $H(x)$. As remarked at the beginning of the proof, the limit $H(x)$ of (6) is necessarily one of the solutions of (69). Theorems 2.1.1–2.1.3 show that all of these three types are indeed possible, hence the proof is completed. ▲

Proof of Theorem 2.4.3. The part "if" in each of (i)–(iii) has been proved, namely in Theorems 2.1.1–2.1.3. Hence, in each case, the part "only if" needs proof.

Proof of part (i). We assume that $F(x)$ is such that, with suitable sequences a_n and $b_n > 0$, as $n \to +\infty$,

$$\lim F^n(a_n + b_n x) = H_{1,\gamma}(x). \tag{82}$$

From (82), we shall conclude that $\omega(F) = +\infty$ and that (10) holds. We shall do this by turning to a subsequence $\{n(k)\}$ of n in (82), on which we can show that we may take $a_{n(k)} = 0$. Then, by an elementary argument, by which we showed (74), we shall deduce (10). The details are as follows.

Let us apply Theorem 2.2.1 with $m = 2$, say. Since the limits A_2 and B_2 satisfy

$$H_{1,\gamma}^2(A_2 + B_2 x) = H_{1,\gamma}(x),$$

we have $A_2 = 0$ and $B_2 = 2^{1/\gamma}$. Hence, by (82), (47), and (48), as $n \to +\infty$,

$$\lim \frac{a_{2n} - a_n}{b_n} = 0$$

and

$$\lim \frac{b_{2n}}{b_n} = 2^{1/\gamma}.$$

Putting $n(k) = 2^k$, the above relations can be rewritten as

$$\lim_{k = +\infty} \frac{a_{n(k+1)} - a_{n(k)}}{b_{n(k)}} = 0, \quad \lim_{k = +\infty} \frac{b_{n(k+1)}}{b_{n(k)}} = 2^{1/\gamma}. \tag{83}$$

Writing $a_{n(0)} = 0$ and

$$\frac{a_{n(k)}}{b_{n(k)}} = \sum_{j=0}^{k-1} \frac{a_{n(j+1)} - a_{n(j)}}{b_{n(k)}} = \sum_{j=0}^{k-1} \frac{a_{n(j+1)} - a_{n(j)}}{b_{n(j)}} \frac{b_{n(j)}}{b_{n(k)}}$$

$$= \sum_{j=0}^{M} \cdots + \sum_{j=M+1}^{k-1} \cdots = \Sigma_1 + \Sigma_2, \tag{84}$$

we choose M so that, for any $\varepsilon > 0$ and for all $j > M$,

$$\left| \frac{a_{n(j+1)} - a_{n(j)}}{b_{n(j)}} \right| < \varepsilon,$$

which can be done in view of (83). On the other hand, since $\gamma > 0$, $2^{-1/\gamma} <$

2.4 NECESSARY CONDITIONS FOR WEAK CONVERGENCE

1, the second limit in (83) implies that, for any q with $2^{-1/\gamma} < q < 1$,

$$\frac{b_{n(j)}}{b_{n(k)}} < Cq^{n(k)-n(j)}, \qquad k \geq j,$$

where $C > 0$ is a suitable constant. Hence,

$$|\Sigma_1| \leq C_M q^{n(k)-n(M)} \to 0, \qquad \text{as } k \to +\infty,$$

for any fixed M. Furthermore,

$$|\Sigma_2| < \varepsilon \sum_{j=M+1}^{k-1} Cq^{n(k)-n(j)} < C\varepsilon \sum_{t=0}^{+\infty} q^t = \frac{C\varepsilon}{1-q}.$$

These estimates yield, by (84), that, as $k \to +\infty$,

$$\lim \frac{a_{n(k)}}{b_{n(k)}} = 0.$$

Consequently, we can apply Lemma 2.2.2 to (82) for the subsequence $n = n(k)$ with $C_n = a_n, C_n^* = 0$, and $D_n = b_n$, which yield, for $k \to +\infty$,

$$\lim F^{n(k)}(b_{n(k)}x) = H_{1,\gamma}(x). \tag{85}$$

Since $b_{n(k)} > 0$ and $0 < H_{1,\gamma}(x) < 1$ for all $x > 0$, $\omega(F) = +\infty$. Otherwise we could choose an $x > 0$ such that $b_{n(k)}x > \omega(F)$ for all k, contradicting (85). In addition, the second limit in (83) implies that, for k sufficiently large, $b_{n(k+1)} > b_{n(k)}$ and that, as $k \to +\infty, b_{n(k)} \to +\infty$. Hence, for t sufficiently large, we can determine k with $n(k) \leq t < n(k+1)$, and then, for $x > 0$,

$$F(b_{n(k)}x) \leq F(tx) \leq F(b_{n(k+1)}x). \tag{86}$$

These inequalities are preserved if we take logarithms. Noting that, for y sufficiently large, $\log F(y)$ is defined and it is negative, the application of (86) with $x > 0$ and with $x = 1$ yields, for sufficiently large t

$$\frac{\log F(b_{n(k)}x)}{\log F(b_{n(k)})} \geq \frac{\log F(tx)}{\log F(t)} \geq \frac{\log F(b_{n(k+1)}x)}{\log F(b_{n(k+1)})}. \tag{87}$$

Taking logarithms in (85) and letting t, or equivalently, k, tend to $+\infty$, (87) results in

$$\lim_{t = +\infty} \frac{\log F(tx)}{\log F(t)} = x^{-\gamma}.$$

This is equivalent to (10), because for any y for which $0 < F(y)$,

$$\log F(y) = \log\{1 - [1 - F(y)]\} = -[1 - F(y)] + \nu[1 - F(y)]^2,$$

where $|\nu| \leqslant 1$ if $F(y) > \tfrac{1}{2}$. Hence, if $F(y) < 1$ for all y, then as $y \to +\infty$,

$$\frac{-\log F(y)}{1 - F(y)} \to 1.$$

This completes the proof of part (i). ▲

Proof of Theorem 2.4.3(ii). We now assume that $F(x)$ is such that there are sequences a_n and $b_n > 0$ of real numbers such that, as $n \to +\infty$,

$$\lim F^n(a_n + b_n x) = H_{2,\gamma}(x). \tag{88}$$

We first deduce that $\omega(F)$ is finite. We again appeal to Theorem 2.2.1, and it will suffice to take $m = 2$, say. Since

$$H_{2,\gamma}^2(A_2 + B_2 x) = H_{2,\gamma}(x)$$

yields $A_2 = 0$ and $B_2 = 2^{1/\gamma}$, the limits (47) and (48) now become

$$\lim_{n = +\infty} \frac{a_{2n} - a_n}{b_n} = 0, \qquad \lim_{n = +\infty} \frac{b_{2n}}{b_n} = 2^{-1/\gamma}, \qquad \gamma > 0.$$

For the subsequence $n(k) = 2^k$, we thus have

$$\lim_{k = +\infty} \frac{a_{n(k+1)} - a_{n(k)}}{b_{n(k)}} = 0, \qquad \lim_{k = +\infty} \frac{b_{n(k+1)}}{b_{n(k)}} = 2^{-1/\gamma}, \qquad \gamma > 0. \tag{89}$$

The essential difference between (83) and (89) is that the second limit in (83) was larger than one, while in (89) it is smaller. Hence, it is immediate here that $b_{n(k)} \to 0$ as $k \to +\infty$. Also, an easy estimate yields

$$a_{n(k+m)} - a_{n(k)} \to 0, \qquad k \to +\infty, \quad m \to +\infty,$$

which implies that $a_{n(k)} \to a$ as $k \to +\infty$. Finally, it also follows that, as $k \to +\infty$,

$$\lim \frac{a - a_{n(k)}}{b_{n(k)}} = 0.$$

Therefore, Lemma 2.2.2 is applicable, which yields that, in (88), we can change the constants when n is restricted to the subsequence $n(k)$. We get

$$\lim_{k=+\infty} F^{n(k)}(a+b_{n(k)}x) = H_{2,\gamma}(x). \tag{90}$$

Since $H_{2,\gamma}(0)=1$, we have from (90) that $F(a)=1$. Hence $\omega(F) \leq a$. It is immediate that, in fact, $\omega(F)=a$: since $b_{n(k)} \to 0$ as $k \to +\infty$, and since $H_{2,\gamma}(x)<1$ for $x<0$, $F(x)<1$ for $x<a$. Equation (90) can now be reduced to (85) by the transformation $F^*(x) = F(a-1/x), x>0$. As we have seen, (85) implies that $F^*(x)$ satisfies (10), which was to be proved. ▲

2.5. THE PROOF OF PART (iii) OF THEOREM 2.4.3

Since the proof is somewhat lengthy, we devote a whole section to it. We split up the proof into several steps, some of which are of independent interest. As before, the part "only if" remains to be proved. That is, we start with the assumption that there are sequences A_n and $B_n > 0$ such that, as $n \to +\infty$,

$$\lim F^n(A_n + B_n x) = H_{3,0}(x) = \exp(-e^{-x}). \tag{91}$$

From this we deduce that (15) and (17) hold.

Throughout the proof, we use the function

$$G^*(x) = \sup\{y : F(y) \leq 1-x\}, \, 0 < x < 1. \tag{92}$$

Step 1. In (91), we can take $a_n = A_n$ and $b_n = B_n$, where

$$a_n = G^*\left(\frac{1}{n}\right), \, b_n = G^*\left(\frac{1}{ne}\right) - a_n. \tag{93}$$

A rough argument shows that the claim (93) is quite apparent. If we apply (91) with $x=0$, we get

$$F^n(A_n) \sim \frac{1}{e} \sim \left(1-\frac{1}{n}\right)^n,$$

from which it is clear that $F(A_n)$ is "close" to $1-1/n$. Similarly, putting $x=1$ in (91), we obtain

$$F^n(A_n + B_n) \sim \exp\left(-\frac{1}{e}\right) \sim \left(1-\frac{1}{ne}\right)^n,$$

which "suggests" the second formula of (93) (recall Lemma 2.2.2, by which certain asymptotic formulas for normalizing constants can be replaced by equalities). However, for making this argument precise, we need careful estimates. If we want to apply Lemma 2.2.2, we have to show that if (91) holds, then, as $n \to +\infty$,

$$\lim \frac{a_n - A_n}{B_n} = 0, \qquad \lim \frac{B_n}{b_n} = 1. \tag{94}$$

For proving (94), observe that, by (92) and (93),

$$F(a_n) \leqslant 1 - \frac{1}{n} \leqslant F(a_n + 0). \tag{95}$$

On the other hand, we get from (91) with $x=0$ that, for arbitrary $\varepsilon > 0$ and $\eta > 0$, we can choose n_0 so that, for all $n \geqslant n_0$,

$$F^n(A_n - B_n \varepsilon) < H_{3,0}(-\varepsilon) + \eta, \qquad F^n(A_n + B_n \varepsilon) > H_{3,0}(\varepsilon) - \eta. \tag{96}$$

In order to make (95) and (96) comparable, we deduce from (96), by suitable choice of ε and η,

$$F^n(A_n - B_n \varepsilon) < \left(1 - \frac{1}{n}\right)^n < F^n(A_n + B_n \varepsilon). \tag{97}$$

Let z be defined by

$$H_{3,0}(z) = \left(1 - \frac{1}{n}\right)^n.$$

That is,

$$z = -\log\left\{n \log\left[\left(1 - \frac{1}{n}\right)^{-1}\right]\right\} = -\log n \left[\log\left(1 + \frac{1}{n-1}\right)\right].$$

Hence, applying the Taylor formula

$$\log(1+y) = y + vy^2, \qquad |v| \leqslant 1, \quad |y| \leqslant \tfrac{1}{2}, \tag{98}$$

we get

$$z = -\log\left[n\left(\frac{1}{n-1} + \frac{v}{(n-1)^2}\right)\right] = -\log\left(1 + \frac{v^*}{n}\right) = \frac{v^{**}}{n}$$

where $|v^{**}| \leqslant 3$ for $n \geqslant 2$. Therefore, for any fixed $\varepsilon > 0$, $-\varepsilon < z < \varepsilon$ for large

2.5   THE PROOF OF PART (iii) OF THEOREM 2.4.3

n. Thus, if we choose $\eta > 0$ in (96) such that both inequalities

$$H_{3,0}(\varepsilon) - H_{3,0}(z) > \eta, \qquad H_{3,0}(z) - H_{3,0}(-\varepsilon) > \eta$$

hold (this can be done because $H_{3,0}(x)$ is strictly increasing), then (97) holds for large values of n.

Combining (95) and (97), we get

$$A_n - B_n \varepsilon \leqslant a_n \leqslant A_n + B_n \varepsilon.$$

This is equivalent to the first limit in (94) on account of $\varepsilon > 0$ being arbitrary. The second limit in (94) follows similarly if we start with $x = 1$ in (91). This proves Step 1.   ▲

Step 2. Let $F(x)$ be in the domain of attraction of $H_{3,0}(x)$. Then, for any real x, as $t \to \omega(F)$,

$$\lim \frac{1 - F[t + xh(t)]}{1 - F(t)} = e^{-x}, \tag{99}$$

where

$$h(t) = G^* \left[\frac{1 - F(t)}{e} \right] - t. \tag{100}$$

Equation (99) is similar to (17); the only difference is that $h(t) \neq R(t)$. We shall, however, show later that $h(t)/R(t) \to 1$ as $t \to \omega(F)$ and that $R(t)$ can indeed take the place of $h(t)$ in (99).

For relating $h(t)$ to b_n of (93), notice that both are related to the function $G^*(x/e) - G^*(x)$: $x = 1/n$ yields b_n, while $x = 1 - F(t)$ leads to $h(t)$.

For proving (99), we have to prove it for an arbitrary sequence $t_n \to \omega(F)$ as $n \to +\infty$. We first consider the special sequence $t_n = a_n$ of (93), which evidently tends to $\omega(F)$ as $n \to +\infty$. Since the right hand side of (91) is positive for all x, taking logarithms is permitted for large n. We thus get from (91), for $n \to +\infty$,

$$\lim n \log F(a_n + b_n x) = -e^{-x}. \tag{101}$$

In particular, as $n \to +\infty$,

$$\lim F(a_n + b_n x) = 1, \tag{102}$$

and thus, by (98),

$$-\log F(a_n + b_n x) = -\log\{1 - [1 - F(a_n + b_n x)]\} \sim 1 - F(a_n + b_n x).$$

We can thus rewrite (101) as

$$\lim_{n=+\infty} n[1-F(a_n+b_n x)] = e^{-x}. \tag{101a}$$

The special case $x=0$ yields

$$n[1-F(a_n)] \to 1, \qquad (n \to +\infty), \tag{103}$$

and thus a further equivalent form of (101) is

$$\lim_{n=+\infty} \frac{1-F(a_n+b_n x)}{1-F(a_n)} = e^{-x}. \tag{104}$$

This result will suffice for proving (99) for an arbitrary sequence $t_n \to \omega(F)$ as $n \to +\infty$. Let now t_n be an arbitrary sequence which tends to $\omega(F)$ as $n \to +\infty$. Let us define the unique integer $m = m(t_n)$ by the inequalities

$$m[1-F(t_n)] \le 1 < (m+1)[1-F(t_n)].$$

Then, on one hand,

$$\lim_{n=+\infty} m[1-F(t_n)] = 1 \tag{105}$$

and, on the other hand, by (92) and (93),

$$a_m \le t_n \le a_{m+1} \tag{106}$$

and

$$b_m + a_m - a_{m+1} \le h(t_n) \le b_{m+1} + a_{m+1} - a_m. \tag{107}$$

If we rewrite (104) with $m = m(t_n)$ for n, then (103) and (105) imply that, in the denominator, we can replace a_m by t_n. We thus have

$$\lim_{n=+\infty} \frac{1-F(a_m+b_m x)}{1-F(t_n)} = e^{-x}, \qquad m = m(t_n). \tag{108}$$

Notice that (108) is a limit theorem on distributions. That is, if we define

$$F_n(z) = \begin{cases} 1 - \dfrac{1-F(z)}{1-F(t_n)} & \text{if } z > t_n, \\ 0 & \text{otherwise,} \end{cases} \tag{109}$$

then $F_n(z)$ is a distribution function. Hence, (108) is a limiting form for

2.5 THE PROOF OF PART (iii) OF THEOREM 2.4.3

$F_n(a_m + b_m x)$, where $m = m(t_n)$. Thus, by Lemma 2.2.2, (108) implies (99) if we show that, for $m = m(t_n)$, as $n \to +\infty$,

$$\lim \frac{a_m - t_n}{b_m} = 0, \quad \lim \frac{h(t_n)}{b_m} = 1.$$

For these limits, on account of (106) and (107), it suffices to show that, as $m \to +\infty$,

$$\lim \frac{a_{m+1} - a_m}{b_m} = 0, \quad \lim \frac{b_{m+1}}{b_m} = 1. \tag{110}$$

The limits of (110) are immediate from (91) and (102), when (91) is applied with $a_m = A_m$ and $b_m = B_m$ of (93). Indeed, combining (91) and (102), we get, as $m \to +\infty$,

$$\lim F^m(a_m + b_m x) = \lim F^m(a_{m+1} + b_{m+1} x) = H_{3,0}(x).$$

An appeal to Lemma 2.2.3 yields (110), which completes the proof of Step 2. ▲

Step 3. Let $F(x)$ be in the domain of attraction of $H_{3,0}(x)$. Let $z_n > 0$ be a sequence which tends to zero as $n \to +\infty$. Then, for $G^*(x)$ of (92), and for $u > 0$, as $n \to +\infty$,

$$\lim \frac{G^*(z_n u) - G^*(z_n)}{G^*(z_n/e) - G^*(z_n)} = -\log u. \tag{111}$$

We first establish (111) for $z_n = 1/n, n = 1, 2, \ldots$.

Let $s > 1$. Then (91), with $a_n = A_n$, $b_n = B_n$ and $x = y \log s$, yields that, as $n \to +\infty$,

$$\lim F^n(a_n + y b_n \log s) = \exp[-\exp(-y \log s)] = \exp(-s^{-y}). \tag{112}$$

On the other hand, if, for some sequences a_n^* and $b_n^* > 0$, as $n \to +\infty$,

$$\lim F^n(a_n^* + b_n^* y) = \exp(-s^{-y}), \tag{113}$$

then, similarly as in Step 1, it can be shown that we can take

$$a_n^* = G^*\left(\frac{1}{n}\right) = a_n, \quad b_n^* = G^*\left(\frac{1}{ns}\right) - a_n.$$

Hence, in view of (112) and (113), Lemma 2.2.3 implies

$$\lim_{n=+\infty} \frac{b_n^*}{b_n \log s} = 1,$$

which is (111) with $u = 1/s < 1$ and $z_n = 1/n$. But if (111) holds for $z_n = 1/n$, then it holds for any subsequence $z_n = 1/m_n$, where m_n is an arbitrary sequence of positive integers. Consequently, if $z_n \to 0$ is an arbitrary sequence, then we have established (111) for $0 < u < 1$ and for $1/m_n$ taking the place of z_n, where m_n is the integer part of $1/z_n$. The fact that we can now take z_n itself follows easily from inequalities similar to (107), where $m = m_n$ and $h(t_n)$ is replaced by $G^*(z_n u) - G^*(z_n)$.

In order to get rid of the restriction that $u < 1$, first observe that (111) is trivial for $u = 1$. If $u > 1$, then, putting $t_n' = t_n u$,

$$\frac{G^*(t_n u) - G^*(t_n)}{G^*(t_n/e) - G^*(t_n)} = \frac{G^*(t_n') - G^*(t_n'/u)}{G^*(t_n'/ue) - G^*(t_n'/e)}$$

$$= \frac{G^*(t_n') - G^*(t_n'/u)}{G^*(t_n') - G^*(t_n'/e)} : \frac{G^*(t_n'/ue) - G^*(t_n'/e)}{G^*(t_n') - G^*(t_n'/e)}.$$

Here we can appeal to (111) with $1/u < 1$ after plugging $G^*(t_n') - G^*(t_n')$ into the numerator of the second term. We thus get (111) for $u > 1$, which terminates the proof of Step 3. ▲

Step 4. Let $F(x)$ be continuous and strictly increasing for all $x_0 \leq x \leq \omega(F)$ with some $-\infty < x_0 < \omega(F)$. Furthermore, let $F(x)$ be in the domain of attraction of $H_{3,0}(x)$. Then (15) and (17) hold.

Set

$$g(z) = G^*\left(\frac{z}{e}\right) - G^*(z).$$

We shall show that, for $0 < x < 1$,

$$\int_0^x g(z) \, dz < +\infty. \tag{114}$$

From (114) it is immediate that $G^*(z)$ itself is integrable on the interval $(0, x)$. Hence the function

$$k(x) = \frac{1}{x} \int_0^x G^*(z) \, dz - G^*(x)$$

2.5 THE PROOF OF PART (iii) OF THEOREM 2.4.3

is well defined on $(0, 1)$. Observing that

$$k(x) = R(t) \quad \text{with} \quad x = 1 - F(t), \tag{115}$$

we immediately get that the integral in (15) is finite because so is $k(x)$.

Let us establish (114). With the substitution $t = 1/v$

$$\int_0^x g(t)\,dt = \int_{1/x}^{+\infty} v^{-2} g\left(\frac{1}{v}\right) dv.$$

We put $g_1(v) = v^{-2} g(1/v)$. Since, by (111), for $u > 0$,

$$\lim_{z=0} \frac{g(zu)}{g(z)} = \lim_{z=0} \frac{G^*(zu/e) - G^*(z) + G^*(z) - G^*(zu)}{G^*(z/e) - G^*(z)}$$

$$= -\log \frac{u}{e} + \log u = 1, \tag{116}$$

we get for $g_1(v)$

$$\lim_{v=+\infty} \frac{g_1(4v)}{g_1(v)} = \lim_{v=+\infty} \frac{(4v)^{-2} g(1/4v)}{v^{-2} g(1/v)} = \frac{1}{16}.$$

Hence, for all $v \geqslant v_0$,

$$g_1(4v) \leqslant \tfrac{1}{8} g_1(v), \quad \text{say.} \tag{117}$$

Since $g_1(v)$ is continuous and finite on any interval $0 < a < b < +\infty$, for proving (114) it suffices to prove that, for a fixed $m \geqslant 1$,

$$\int_{4^m}^{+\infty} g_1(v)\,dv = \sum_{k=m}^{+\infty} \int_{4^k}^{4^{k+1}} g_1(v)\,dv < +\infty. \tag{118}$$

But, by (117),

$$\int_{4^k}^{4^{k+1}} g_1(v)\,dv = 4 \int_{4^{k-1}}^{4^k} g_1(4z)\,dz \leqslant \frac{1}{2} \int_{4^{k-1}}^{4^k} g_1(z)\,dz,$$

from which induction yields

$$\int_{4^k}^{4^{k+1}} g_1(v)\,dv \leqslant \left(\frac{1}{2}\right)^{k-m} \int_{4^m}^{4^{m+1}} g_1(v)\,dv,$$

where m is a fixed integer such that $4^m \geqslant v_0$, introduced at (117). We thus get a convergent series in (118), which proves (114). As was remarked earlier, (115) now follows.

For proving (17), we shall prove that, as $t \to \omega(F)$, $R(t)/h(t) \to 1$, where $h(t)$ is the function defined in (100). Then the application of Lemma 2.2.2 to the function $F_n(z)$ of (109) reduces (99) to (17).

Notice that $g(z) = h(t)$ if $z = 1 - F(t)$. Hence, what remains to be proved is the relation

$$\lim_{t=\omega(F)} \frac{R(t)}{h(t)} = \lim_{z=0} \frac{k(z)}{g(z)} = 1.$$

The first equation is evident by the relation of $R(t)$ and $h(t)$ to $k(z)$ and $g(z)$, respectively. Consequently, the last equation needs proof, which easily follows by the following observation. If we substitute $zs = y$ in the integral below, we get

$$k(z) = \frac{1}{z} \int_0^z G^*(y) dy - G^*(z) = \int_0^1 G^*(zs) ds - G^*(z),$$

and thus

$$\frac{k(z)}{g(z)} = \int_0^1 \frac{G^*(zs) - G^*(z)}{G^*(z/e) - G^*(z)} ds.$$

Since the integrand above as a function of s is strictly monotonic, Lebesgue's dominated convergence theorem (Theorem A.I.5 of Appendix I), combined with (111), yields

$$\lim_{z=0} \frac{k(z)}{g(z)} = \int_0^1 (-\log s) ds = 1.$$

This completes the proof of Step 4. ▲

Step 5. Let $F(x)$ be in the domain of attraction of $H_{3,0}(x)$. Then (15) and (17) hold.

This is the last step in the proof of Theorem 2.4.3 (iii). This will be achieved by showing that there is a strictly increasing continuous function $F^*(x)$ which belongs to the domain of attraction of $H_{3,0}(x)$ and for which, as $x \to \omega(F)$,

$$\lim \frac{1-F(x)}{1-F^*(x)} = 1. \tag{119}$$

If such an $F^*(x)$ does exist, then, by Step 4, we know that (15) and (17)

2.5   THE PROOF OF PART (iii) OF THEOREM 2.4.3

hold for it. In addition, by (119), for any $\varepsilon > 0$ and for all $x \geq x_0 = x_0(\varepsilon)$,

$$(1-\varepsilon)[1 - F^*(x)] < 1 - F(x) < (1+\varepsilon)[1 - F^*(x)]. \tag{120}$$

Integration of the upper inequality of (120) yields that the validity of (15) for F^* implies that for F. Furthermore, if we integrate both inequalities in (120), we get

$$\lim_{t = \omega(F)} \frac{\int_t^{\omega(F)} [1 - F(x)] dx}{\int_t^{\omega(F)} [1 - F^*(x)] dx} = 1. \tag{121}$$

Since, from (119), $\omega(F) = \omega(F^*)$, this last limit relation can be rewritten as, for $t \to \omega(F)$,

$$\lim \frac{R(t, F)}{R(t, F^*)} = 1$$

where $R(t, F)$ and $R(t, F^*)$ signify the function in (16) when it is calculated for F and F^*, respectively. Hence, one more appeal to Lemma 2.2.2 with the function $F_n(z)$ of (109) yields that (17) can be proved for $F(x)$ either in terms of $R(t, F)$ or in terms of $R(t, F^*)$. But, with $R(t, F^*)$, (17) follows for $F(x)$ from the conclusion of Step 4 and from (119).

It remains therefore to show the existence of $F^*(x)$ with the stated properties.

Let $t_1, t_2, \ldots$ be the points of discontinuity of $F(t)$. For each t_j, let us construct the closed intervals $[t_j, s_j]$, where s_j is defined in terms of the function $h(t)$ of (100) as follows. First choose a sequence $x_j = x(t_j) > 0$ which tends to zero as $t_j \to \omega(F)$ (e.g., if $\omega(F) = +\infty$, we can choose $x_j = 1/t_j$, while, for finite $\omega(F)$, x_j can be $\omega(F) - t_j$). Then let $s_1 = t_1 + x_1 h(t_1)$. We then define the sequence s_j sequentially. If $t_1 < t_2 < s_1$, then $s_2 = t_2$. Otherwise $s_2 = t_2 + x_2 h(t_2)$. Similarly, if $s_1, s_2, \ldots, s_{j-1}$ have been defined, then $s_j = t_j$, if t_j belongs to one of the intervals (t_k, s_k), $1 \leq k \leq j-1$. Otherwise $s_j = t_j + x_j h(t_j)$. Let us keep those intervals, for which $t_j < s_j$. These can now be arranged in such a way that the endpoints t_j are increasing. Take any j_0 in this sequence for which $F(t_{j_0}) > 0$ and again we drop terms, namely, those for which $t_j < t_{j_0}$. Hence, we have the sequence $[t_{j_m}, s_{j_m}], m = 0, 1, \ldots$, of intervals such that $t_{j_m} < t_{j_{m+1}}$, and if a point of discontinuity of $F(t)$ is larger than t_{j_0}, then it belongs to one of the intervals $[t_{j_m}, s_{j_m}]$.

Let us now define $F_1^*(t)$ as a continuous distribution function with the following properties. Let $F_1^*(t)$ be arbitrary but continuous strictly increasing for $t < t_{j_0}$. Furthermore, let $F_1^*(t) = F(t)$ if $t \notin (t_{j_m}, s_{j_m})$ for any $m \geq 0$. Finally, on the intervals $[t_{j_m}, s_{j_m}]$, $F^*(t)$ is defined as a linear function. Let T be the union of all intervals $[t_{j_m}, s_{j_m}]$. Then $F_1^*(t)$ is continuous and $1 - F(t) = 1 - F_1^*(t)$ for $t \notin T$. Furthermore,

$$F(t_{j_m}) \leq F_1^*(t) \leq F[t_{j_m} + x_{j_m} h(t_{j_m})], \qquad t_{j_m} \leq t \leq s_{j_m}$$

—that is, for $t_{j_m} \leq t \leq s_{j_m}$,

$$1 - F(t_{j_m}) \geq 1 - F_1^*(t) \geq 1 - F[t_{j_m} + x_{j_m} h(t_{j_m})].$$

On the other hand, on this same interval, since F is nondecreasing,

$$1 - F(t_{j_m}) \geq 1 - F(t) \geq 1 - F[t_{j_m} + x_{j_m} h(t_{j_m})].$$

Taking ratios, and letting $t \to \omega(F)$, (99) leads to (119) with $F_1^*(t)$ (notice that (99) is applicable with $x_j \to 0$, since both sides of (99) are decreasing in x). The function $F_1^*(t)$ is continuous but not necessarily strictly increasing. But we can repeat the above construction to get rid of parts of $F_1^*(t)$ which are parallel to the t-axis. We thus get an $F^*(t)$ which is both continuous and strictly increasing. Furthermore, (119) applies.

To show that $F^*(t)$ also belongs to the domain of attraction of $H_{3,0}(x)$, we apply (119) and (99). They imply that, if $t_n \to \omega(F)$, then

$$\lim \frac{1 - F^*(t_n + xh_n)}{1 - F^*(t_n)} = e^{-x}, \qquad (122)$$

where $h_n = h(t_n)$ with function $h(t)$ defined in (100). Hence, if t_n is such that $[1 - F^*(t_n)]^{-1} = m$ is an integer, then, writing $t_n = a_m$ and $h_n = b_m$, (122) becomes

$$\lim_{m = +\infty} m[1 - F^*(a_m + b_m x)] = e^{-x}.$$

We thus get by Corollary 1.3.1 that $F^*(x)$ is indeed in the domain of attraction of $H_{3,0}(x)$. This completes the proof of Step 5 as well as of Theorem 2.4.3(iii). ▲

2.6. ILLUSTRATIONS

We shall work out a number of examples to illustrate the results of the previous sections.

Example 2.6.1. Let

$$F(x) = 1 - \frac{1}{\log x}, \qquad x \geq e.$$

2.6 ILLUSTRATIONS

In Example 1.3.3 we referred to Chapter 2 for showing that, with $F(x)$, there are no sequences a_n and $b_n > 0$ such that $(Z_n - a_n)/b_n$ would have a limit law. This can now be decided by applying Theorems 2.4.1 and 2.4.3. Theorem 2.4.1 says that, in order to have a limit law for $(Z_n - a_n)/b_n$, one of the criterions of Theorem 2.4.3 should apply to $F(x)$. Since $\omega(F) = +\infty$, either (i) or (iii) should apply. However, part (iii) is not applicable, since (15) fails, which would require that $1/\log x$ be integrable on $(e, +\infty)$. On the other hand, part (i) fails as well because, for $x > 0$,

$$\lim_{t=+\infty} \frac{1-F(tx)}{1-F(t)} = \lim_{t=+\infty} \frac{\log t}{\log tx} = 1,$$

which contradicts (10).

Example 2.6.2. Let $X_1, X_2, \ldots, X_n$ be i.i.d. random variables with common distribution function $F(x)$ of the previous example. Then, for $Y_j = \log X_j$, $1 \leq j \leq n$, the maximum Z_n^* of $Y_1, Y_2, \ldots, Y_n$ has an asymptotic distribution in the sense of (6).

Indeed,

$$P(Y_j < x) = P(X_j < e^x) = 1 - \frac{1}{x}, \qquad x \geq 1.$$

Hence, (10) applies with $\gamma = 1$. We thus have from Theorem 2.2.1, as $n \to +\infty$,

$$P(Z_n^* < nx) \to \exp\left(-\frac{1}{x}\right), \qquad x > 0. \tag{123}$$

This fact is, however, not of much practical help concerning Z_n, the maximum of X_j, $1 \leq j \leq n$. Since, evidently, $Z_n^* = \log Z_n$, (123) yields, for $0 < y < x$,

$$P(e^{ny} < Z_n < e^{nx}) \sim \exp\left(-\frac{1}{x}\right) - \exp\left(-\frac{1}{y}\right).$$

For example, with $y = \frac{1}{2}$ and $x = 15$ we get

$$P(e^{n/2} < Z_n < e^{15n}) \sim .80,$$

where the bounds on Z_n are too distant for any practical use. As an example, we take $n = 10$, when the above approximation becomes $P(148 < Z_{10} < 1.394 \times 10^{65}) \sim .80$. These bounds on Z_{10}, of course, have no practical value. This example clearly indicates that if the logarithms of observations have a nice tendency, it may have no relevance to the original observations. ▲

Example 2.6.3. A statistical sample was collected on a random quantity X. The experimenter assumes that X is a standard normal variate and performs a goodness of fit test. Let us assume that the test supports his assumption. However, the true distribution of X is that of $\sigma^{-1}(Y^\sigma - 1)$, where Y is a lognormal variate (we use our definition of Section 2.3) and σ is so small that the statistical test could not detect the difference (see Exercise 9). Let us compare the experimenter's conclusion on Z_{50} with the actual situation, if $\sigma = 0.1$.

The experimenter argues with the normal distribution. Hence, he computes a_{50} and b_{50} from (59) and (60), respectively. He gets $a_{50} = 2.1009$ and $b_{50} = 0.3575$ and thus concludes that

$$P(Z_{50} < 2.1009 + 0.3575x) \sim \exp(-e^{-x}).$$

The correct formula, however, is obtained if we start with the lognormal distribution. Let Z_{50}^* be the maximum of 50 i.i.d. lognormal variates. Then, from Section 2.3.3,

$$P(Z_{50}^* < 8.1734 + 2.9220x) \sim \exp(-e^{-x}).$$

Since, by assumption,

$$Z_{50}^* = (0.1 Z_{50} + 1)^{10},$$

the formula above gives

$$P\left[Z_{50} < 10(8.1734 + 2.9220x)^{0.1} - 10\right] \sim \exp(-e^{-x}).$$

Thus, if we calculate $P(Z_{50} < 2.6)$ from this last expression, we get $x = 0.6544$ and thus

$$P(Z_{50} < 2.6) \sim .5947.$$

On the other hand, the experimenter's answer is

$$P(Z_{50} < 2.6) \sim .7807.$$

It is important to emphasize that the huge error on the part of the experimenter is not negligence: it is due to the fact that goodness of fit tests can hardly distinguish two distributions which are uniformly close to each other. On the other hand, maxima increase distinctly for different distributions. We shall return to this difficulty once again in Chapter 3. ▲

Example 2.6.4. Let X be the random "life length" of a product. Evidently, $X \geq 0$ and thus $P(X < x) = F(x)$ satisfies $F(0) = 0$. Let

$X_1, X_2, \ldots, X_n$ be independent observations on X. The first complaint about the product is at time W_n, the minimum of the X_j. Let us analyze the distribution of W_n.

If $(W_n - c_n)/d_n$, with some constants c_n and $d_n > 0$, has a limiting distribution, then Theorems 2.4.2 and 2.4.4 imply that the limiting distribution is either $L_{2,\gamma}(x)$ or $L_{3,0}(x)$. Which of these distributions applies depends on the way $F(x)$ tends to zero as $x \to 0^+$ (notice that, by $\alpha(F) = 0$, (32) is automatically satisfied). The condition for $L_{2,\gamma}(x)$ now is the validity of the relation

$$\lim_{t=0^+} \frac{F(tx)}{F(t)} = x^\gamma, \qquad x > 0, \quad \gamma > 0, \tag{124}$$

while for $L_{3,0}(x)$ the limit (34) should apply. Since applied scientists are more inclined to accept (124) than (34) (and, indeed, several, although not all, frequently used continuous distributions $F(x)$ with $F(0) = 0$ do satisfy (124)),

$$L_{2,\gamma}(x) = 1 - \exp(-x^\gamma), \qquad x > 0, \quad \gamma > 0, \tag{125}$$

received general acceptance as the limiting distribution of the minimum of life lengths. (For further development of this argument to justify that the actual distribution of the time to failure of an equipment with a large number of components is $L_{2,\gamma}(x)$, see Section 3.12.) The family $L_{2,\gamma}[(x-\alpha)/\beta]$, when α and $\beta > 0$ are arbitrary parameters, is called Weibull distributions. In recent years this family has become one of the basic distributions in engineering applications, partly but not entirely for the reasons discussed above. ▲

Notice that $L_{2,1}(x)$ is the unit exponential distribution. We have seen that if $F(x)$ is exponential, then so is $P(W_n < x)$ for any fixed n (consequently, its limit as well, when suitably normalized). There are, however, several other distributions $F(x)$ with $F(0) = 0$, for which the limit of $(W_n - c_n)/d_n$ is $L_{2,1}(x)$. As one example, take $F(x) = 2\Phi(x) - 1$, where $\Phi(x)$ is the standard normal distribution. By L'Hospital's rule, (124) is immediate with $\gamma = 1$. Hence, by Theorem 2.1.5, we can take $c_n = \alpha(F) = 0$ and $d_n = \Phi^{-1}(\frac{1}{2} + 1/2n)$. With this d_n,

$$P(W_n < d_n x) \sim 1 - e^{-x}, \qquad x > 0.$$

Example 2.6.5. For a fast diverging subsequence n_k of the positive integers, we can construct a continuous distribution function $F(x)$ such that, with suitable constants a_k and $b_k > 0$, $(Z_{n_k} - a_k)/b_k$ is asympototically uniformly distributed on the interval $(0, 1)$. In other words, if $X_1, X_2, \ldots, X_n$

are i.i.d. with common distribution $F(x)$, then as $k \to +\infty$,

$$\lim P(Z_{n_k} < a_k + b_k x) = \begin{cases} 0 & \text{if } x < 0, \\ x & \text{if } 0 \leq x < 1, \\ 1 & \text{if } x \geq 1. \end{cases} \tag{126}$$

Before carrying out the construction of $F(x)$, let us relate the meaning of this example to the theorems on the asymptotic distribution of Z_n. This example shows that if we do not require the existence of a limiting distribution of Z_n in the sense of (6) but rather a limit (126) on a specific subsequence n_k, then distributions other than $H_{1,\gamma}(x)$, $H_{2,\gamma}(x)$, and $H_{3,0}(x)$ can serve as "limiting distribution" for Z_n. This is mathematically an interesting fact, but it has no negative effect on the way of applying the asymptotic extreme value theory in practice. Namely, in all applications, we let the sample size n increase in an arbitrary manner rather than jumping from a value to a considerably larger one.

We carry out the construction of $F(x)$ for a specific sequence n_k. Let

$$n_k = 10^{m(k)}, \qquad m(k) = 2^k + 2^{k-1}, \qquad k \geq 1.$$

We shall construct an $F(x)$ which is concentrated on the interval $(0, 1)$. That is, $F(0) = 0$ and $F(1) = 1$. For $0 \leq x < 1$ we define $F(x)$ as follows. Let us divide the interval $(0, 1)$ into the subintervals $I_k = [1 - 2^{-k}, 1 - 2^{-k-1})$, $k = 0, 1, 2, \ldots$. We define $F(x)$ on each I_k as the $(1/n_k)$th power of a linear function:

$$F(x) = (\alpha_k + \beta_k x)^{1/n_k}, \qquad x \in I_k, \quad \beta_k > 0, \quad k = 0, 1, 2, \ldots.$$

In the choices of α_k and β_k we want to guarantee that $F(x)$ be continuous and that (126) should hold. These can be achieved by the following sequences:

$$\alpha_k = u_k - 2^{k+1}(v_k - u_k)(1 - 2^{-k}), \qquad \beta_k = 2^{k+1}(v_k - u_k), \quad k \geq 1,$$

where

$$u_k = (1 - 10^{-2^k})^{n_k}, \qquad v_k = (1 - 10^{-2^{k+1}})^{n_k}.$$

In addition, let $\alpha_0 = 0$ and β_0 be such that $\frac{1}{2}\beta_0 = (\alpha_1 + \frac{1}{2}\beta_1)^{1/n_1}$. By substitution, we can easily check that $F(x)$ is indeed continuous. We also claim that (126) holds with

$$a_k = -\frac{\alpha_k}{\beta_k}, \qquad b_k = \frac{1}{\beta_k}, \qquad k \geq 1.$$

In other words, with the above a_k and b_k we have to show that

$$F^{n_k}(a_k + b_k x) \to x, \quad 0 < x < 1. \tag{126a}$$

(by monotonicity of F, the limits for $x < 0$ and for $x \geq 1$ then become 0 and 1, respectively). Notice that, by definition, if

$$a_k + b_k x = \frac{x - \alpha_k}{\beta_k} \in I_k,$$

then $F^{n_k}(a_k + b_k x) = x$. Hence, for proving (126a), we have to show that, for any $0 < x < 1$ and for large k, $(x - \alpha_k)/\beta_k \in I_k$. However,

$$\frac{x - \alpha_k}{\beta_k} \in I_k \quad \text{if, and only if,} \quad u_k < x < v_k,$$

which is valid for any $0 < x < 1$ and for large k, because of the limits below. By choice, as $k \to +\infty$,

$$v_k = \exp\left[n_k \log(1 - 10^{-2^{k+1}})\right] \sim \exp(-10^{-2^{k-1}}) \sim 1$$

and

$$u_k = \exp\left[n_k \log(1 - 10^{-2^k})\right] \sim \exp(-10^{2^{k-1}}) \to 0,$$

where we need Taylor's formula for $\log(1 - z)$ as well as the specific form of n_k. We have thus shown the validity of (126). ▲

2.7. FURTHER RESULTS ON DOMAINS OF ATTRACTION

Mainly for historical reasons, we prove two theorems which give sufficient conditions for an absolutely continuous distribution function $F(x)$ to belong to the domain of attraction of a possible limiting distribution of extremes.

Theorem 2.7.1. *Let $F(x)$ be a distribution function with $\omega(F) = +\infty$ and for which there is a real number x_1 such that, for all $x \geq x_1, f(x) = F'(x)$ exists. If, as $x \to +\infty$,*

$$\lim \frac{xf(x)}{1 - F(x)} = \gamma \tag{127}$$

exists, where $0 < \gamma < +\infty$, then $F(x)$ belongs to the domain of attraction of $H_{1,\gamma}(x)$.

Proof. We shall deduce our theorem from Theorem 2.1.1. Let us put

$$g(x) = \frac{xf(x)}{1-F(x)} = -x\{\log[1-F(x)]\}'$$

which is defined for $x \geqslant x_1$. Hence, for such x,

$$1 - F(x) = C \exp\left\{-\int_a^x \frac{g(y)}{y} dy\right\},$$

where $C > 0$ and $a > x_1$ are suitable constants. We thus have for $t > x_1$ and $x > 0$ for which $tx > x_1$

$$\frac{1-F(tx)}{1-F(t)} = \exp\left\{-\int_t^{tx} \frac{g(y)}{y} dy\right\} = \exp\left\{-\int_1^x \frac{g(ty)}{y} dy\right\}.$$

If we fix $x > 0$ and let $t \to +\infty$, (127) and the dominated convergence theorem yield

$$\lim_{t=+\infty} \frac{1-F(tx)}{1-F(t)} = \exp(-\gamma \log x) = x^{-\gamma},$$

which is exactly the criterion (10) for $F(x)$ to belong to the domain of attraction of $H_{1,\gamma}(x)$. The proof is completed. ▲

In view of Theorem 2.1.2, (127) can easily be transformed to a criterion for $F(x)$ with $\omega(F) < +\infty$ to belong to the domain of attraction of $H_{2,\gamma}(x)$. See Exercise 11. For the domain of attraction of $H_{3,0}(x)$ we prove the following result.

Theorem 2.7.2. *Let $F(x)$ be a distribution function. Let us assume that there is a real number x_1 such that, for all $x_1 \leqslant x < \omega(F)$, $f(x) = F'(x)$ and $F''(x)$ exist and $f(x) \neq 0$. Furthermore, let*

$$\lim_{x=\omega(F)} \frac{d}{dx}\left[\frac{1-F(x)}{f(x)}\right] = 0. \tag{128}$$

Then $F(x)$ is in the domain of attraction of $H_{3,0}(x)$.

Proof. We shall show that the conditions (15) and (17) of Theorem 2.1.3 are satisfied whenever (128) holds. Hence, the conclusion of Theorem 2.1.3 implies our theorem.

We first establish (15). Notice that if $\omega(F) < +\infty$, (15) holds. Hence,

2.7 FURTHER RESULTS ON DOMAINS OF ATTRACTION

only the case $\omega(F) = +\infty$ needs proof. For $x \geq x_1$, set

$$u(x) = \frac{1 - F(x)}{f(x)}.$$

By (128), we can choose a real number $z \geq x_1$ such that, for all $x \geq z$, $|u'(x)| < \frac{1}{2}$. Then, for $y > z$,

$$\int_z^y [1 - F(x)] dx = \int_z^y u(x) f(x) dx$$

$$= -u(x)[1 - F(x)]\Big|_{x=z}^y + \int_z^y u'(x)[1 - F(x)] dx$$

$$< u(z)[1 - F(z)] - u(y)[1 - F(y)] + \frac{1}{2} \int_z^y [1 - F(x)] dx.$$

Therefore, as $y \to +\infty$,

$$\limsup \int_z^y [1 - F(x)] dx < 2u(z)[1 - F(z)],$$

and thus (15) follows.

We now turn to (17). By definition, for $x_1 \leq t < y$,

$$\int_t^y \frac{1}{u(x)} dx = \log \frac{1 - F(t)}{1 - F(y)}.$$

On the other hand, since $u(x) > 0$ and continuous, the mean value theorem of integrals yields

$$\int_t^y \frac{1}{u(x)} dx = (y - t) \frac{1}{u(\xi)}, \qquad t < \xi < y.$$

Hence, with $y = t + sR(t)$, where $R(t)$ is the function defined in (16), we get

$$\log \frac{1 - F(t)}{1 - F[t + sR(t)]} = \frac{sR(t)}{u(\xi)}, \qquad t < \xi < t + sR(t).$$

(We use $s > 0$ in the notations, but all estimates remain the same if $s < 0$.) It remains only to show

$$\lim_{t = +\infty} \frac{R(t)}{u(\xi)} = 1, \qquad t < \xi < t + sR(t).$$

By writing

$$\frac{R(t)}{u(\xi)} = \frac{R(t)}{u(t)} \frac{u(t)}{u(\xi)},$$

it suffices to show that each fraction on the right hand side tends to one as $t \to +\infty$. Since, by Taylor's expansion,

$$u(\xi) = u(t) + (\xi - t)u'(\eta), \qquad t < \eta < \xi,$$

we get

$$\frac{u(\xi)}{u(t)} = 1 + \frac{\xi - t}{u(t)} u'(\eta).$$

But, for $t < \xi < t + sR(t)$,

$$\left| \frac{\xi - t}{u(t)} \right| < s \frac{R(t)}{u(t)},$$

which remains bounded for fixed s whenever $R(t)/u(t)$ is bounded. Therefore, in view of $u'(\eta) \to 0$ by the assumption (128), $u(\xi)/u(t) \to 1$ as $t \to \omega(F), t < \xi < t + sR(t)$, if we show that $R(t)/u(t) \to 1$ as $t \to \omega(F)$. This last limit relation can be obtained from the definition (16) of $R(t)$ and from (128) as follows. Since

$$\int_t^{\omega(F)} [1 - F(y)] dy = u(t)[1 - F(t)] + \int_t^{\omega(F)} u'(y)[1 - F(y)] dy,$$

we have

$$\frac{R(t)}{u(t)} = 1 + \int_t^{\omega(F)} u'(y)[1 - F(y)] dy.$$

Therefore, in view of (128), as $t \to \omega(F)$,

$$\left| \frac{R(t)}{u(t)} - 1 \right| < \varepsilon \int_t^{\omega(F)} [1 - F(y)] dy,$$

where $\varepsilon > 0$ is arbitrary. As was shown in the first part of the proof, the coefficient of ε is finite, and thus $R(t)/u(t) \to 1$ as $t \to \omega(F)$. This completes the proof. ▲

2.7 FURTHER RESULTS ON DOMAINS OF ATTRACTION

For the corresponding theorems on the minimum, see Exercise 14.

The rest of the present section is devoted to the analysis of moments of $F(x)$ when it belongs to one of the domains of attraction of limiting distributions for the extremes. We start with an $F(x)$ which belongs to one of the domains of attraction of $H_{1,\gamma}(x)$ and $H_{3,0}(x)$. Since, by Theorem 2.1.2, the case of $H_{2,\gamma}(x)$ is reduced to $H_{1,\gamma}(x)$ by a transformation, we exclude $H_{2,\gamma}(x)$ from our direct discussion.

Let X be a random variable with distribution function $F(x)$. Then, for $a > 0$, we define

$$m_a^+ = E\bigl[(X^+)^a\bigr] = \int_0^{+\infty} x^a dF(x),$$

where $X^+ = \max(0, X)$. Our aim is to analyze the sets

$$M_F = \{a : a > 0, m_a^+ < +\infty\}$$

for distributions $F(x)$ in terms of the domain of attraction they belong to.

It is evident that the problem would be meaningless if we attempted to discuss finite moments m_a of X itself, assuming only that its distribution $F(x)$ belongs to the domain of attraction of one of $H_{1,\gamma}(x)$ and $H_{3,0}(x)$. Indeed, the criteria for these inclusions do not impose any condition on $F(x)$ as $x \to -\infty$, and thus one could modify $F(x)$ so that no moment would exist.

We prove the following theorem.

Theorem 2.7.3. *If the distribution function $F(x)$ belongs to the domain of attraction of $H_{1,\gamma}(x)$, then M_F equals the open interval $(0, \gamma)$. On the other hand, for any member $F(x)$ of the domain of attraction of $H_{3,0}(x)$, M_F is the whole positive real line.*

Proof. Let us first record that integration by parts leads to the formula

$$m_a^+ = \int_0^1 \cdots + \int_1^{+\infty} \cdots$$

$$= \int_0^1 x^a dF(x) + 1 - F(1) + a \int_1^{+\infty} x^{a-1}[1 - F(x)]dx. \quad (129)$$

Therefore, $m_a^+ < +\infty$ if, and only if, the last integral in the preceding equation is finite.

Let now $F(x)$ be in the domain of attraction of $H_{1,\gamma}(x)$.

By Theorems 2.1.1 and 2.4.3(i), this assumption is equivalent to the validity of (10). Hence, for any $\varepsilon > 0$, we can find a real number t_0 such that, for all $t \geq t_0$,

$$(1-\varepsilon)2^{-\gamma}[1-F(t)] < 1 - F(2t) < (1+\varepsilon)2^{-\gamma}[1-F(t)]. \tag{130}$$

Let us put

$$M_k = \int_{2^k}^{2^{k+1}} x^{a-1}[1-F(x)]\,dx,$$

and let us write

$$\int_1^{+\infty} x^{a-1}[1-F(x)]\,dx = \int_1^{2^m} x^{a-1}[1-F(x)]\,dx + \sum_{k=m}^{+\infty} M_k, \tag{131}$$

where m is a fixed integer with $2^m \geq t_0$. With the substitution $x = 2y$, we get

$$M_k = 2^a \int_{2^{k-1}}^{2^k} y^{a-1}[1-F(2y)]\,dy, \tag{132}$$

and thus, by (130), for $k > m$,

$$(1-\varepsilon)2^{a-\gamma}M_{k-1} < M_k < (1+\varepsilon)2^{a-\gamma}M_{k-1}.$$

From these last inequalities we get by induction

$$(1-\varepsilon)^{k-m}2^{(a-\gamma)(k-m)}M_m < M_k < (1+\varepsilon)^{k-m}2^{(a-\gamma)(k-m)}M_m.$$

Since $\varepsilon > 0$ is arbitrary, these now yield that (131), and thus (129) as well, is convergent if, and only if $a - \gamma < 0$. This concludes the proof of the first part of the theorem.

Turning to the domain of attraction of $H_{3,0}(x)$, we first remark that, if $\omega(F) < +\infty$, then $m_a^+ < +\infty$ for all $a > 0$. We can thus assume that $\omega(F) = +\infty$. By an appeal to Theorem 2.4.3(iii), we get that (17) holds. In particular, we get that, for any fixed real number s, as $t \to +\infty$,

$$\lim[t + sR(t)] = +\infty,$$

which, with negative s, yields that $R(t)/t \to 0$. One more appeal to (17) now results in the inequalities

$$1 - F(2t) \leq 1 - F[t + uR(t)] < 2e^{-u}[1 - F(t)], \tag{133}$$

where t is sufficiently large and $u > 0$ is arbitrary. Therefore, if we start

2.7 FURTHER RESULTS ON DOMAINS OF ATTRACTION

with the decomposition (131), we get from (132)

$$M_k < 2^{a+1} e^{-u} M_{k-1} < 2^{(a+1)(k-m)} e^{-u(k-m)} M_m,$$

where $m = m(u)$ is chosen so that, for $t \geq 2^m$, (133) should hold. Since $u > 0$ is arbitrary, (131) is convergent, which in turn implies the convergence of (129). Theorem 2.7.3 is thus established. ▲

Theorem 2.7.3 is a very valuable tool for reducing the "guessing" part when we want to apply Theorems 2.1.1–2.1.3. Namely, let $F(x)$ be a distribution function with $\omega(F) = +\infty$. We want to decide that which of the three possible limit distributions may apply to the maximum Z_n if the distribution of the population is $F(x)$. By Theorems 2.4.1 and 2.4.3, one of Theorems 2.1.1–2.1.3 should be tried. Since $\omega(F) = +\infty$, Theorem 2.1.2 fails. Therefore, if $F(x)$ is such that, for arbitrary $a > 0$, the ath moment of X^+ is finite, then Theorem 2.1.3 is the only possibility. Otherwise, Theorem 2.1.1 should be tried. We may add that if, for all $a > 0$, $m_a^+ = +\infty$, then Z_n cannot be normalized to have a limiting distribution in the form of (6).

The theorem is also applicable when $\omega(F) < +\infty$. Namely, we argue with the function $F^*(x) = F[\omega(F) - 1/x]$, where $x > 0$, in the above manner.

Example 2.7.1. Let

$$F(x) = \frac{1}{\Gamma(\alpha)} \int_0^x t^{\alpha-1} e^{-t} dt, \quad x > 0,$$

where $\Gamma(\alpha)$ is the gamma function (which makes $F(x)$ equal to one at $x = +\infty$). Then $F(x)$ belongs to the domain of attraction of $H_{3,0}(x)$. As a matter of fact, $\omega(F) = +\infty$, and

$$m_a^+ = \frac{1}{\Gamma(\alpha)} \int_0^{+\infty} x^{a+\alpha-1} e^{-x} dx,$$

which is finite for all $a > 0$. Hence, by Theorem 2.7.3, $F(x)$ can belong only to the domain of attraction of $H_{3,0}(x)$. We now have two ways of showing that $F(x)$ indeed belongs to the domain of attraction of $H_{3,0}(x)$. For this exercise, we apply Theorem 2.7.2. We calculate

$$\frac{d}{dx}\left[\frac{1-F(x)}{f(x)}\right] = -1 - \frac{[1-F(x)]f'(x)}{f^2(x)}$$

$$= \frac{(x-\alpha+1)\int_x^{+\infty} t^{\alpha-1} e^{-t} dt}{x^\alpha e^{-x}} - 1 \to 0$$

as $x \to +\infty$, which easily follows by L'Hospital's rule. ▲

Example 2.7.2. Consider the Pareto distribution defined as

$$F(x) = 1 - x^{-\beta}, \quad x \geq 1, \quad \beta > 0.$$

With the formula (129), $m_a^+ < +\infty$ if, and only if, $a < \beta$. Thus, by Theorem 2.7.3, the possible limit distribution for Z_n, when normalized, is $H_{1,\beta}(x)$. The fact that $F(x)$ indeed belongs to the domain of attraction of $H_{1,\beta}(x)$ can again be shown either by Theorem 2.1.1 or by Theorem 2.7.1. We choose the latter, which requires formula (127). We get

$$\frac{xf(x)}{1-F(x)} = \frac{\beta x^{-\beta}}{x^{-\beta}} = \beta,$$

which value would have sufficed even in limit. ▲

Example 2.7.3. Let X be a discrete random variable which takes the integers $k \geq 2$ with distribution

$$p_k = P(X = k) = \frac{C}{k(\log k)^2}, \quad k \geq 2,$$

where $C > 0$ is a suitable constant. Then the distribution function $F(x)$ of X does not belong to the domain of attraction of any of $H_{1,\gamma}(x)$, $H_{2,\gamma}(x)$, and $H_{3,0}(x)$.

As a matter of fact,

$$m_a^+ = C \sum_{k=2}^{+\infty} k^{a-1}(\log k)^{-2} = +\infty$$

for all $a > 0$. Hence, Theorem 2.7.3 implies our claim. ▲

Example 2.7.4. Let

$$F(x) = 1 - \exp(x^{-3}), \quad x < 0.$$

Then $F(x)$ belongs to the domain of attraction of $H_{3,0}(x)$.

Now $\omega(F) = 0$, and thus $H_{2,\gamma}(x)$ and $H_{3,0}(x)$ are the possible limit laws for Z_n when normalized. We turn to the function

$$F^*(x) = F\left(-\frac{1}{x}\right) = 1 - \exp(-x^3), \quad x > 0.$$

2.7 FURTHER RESULTS ON DOMAINS OF ATTRACTION

Since, with this distribution, $m_a^+ < +\infty$ for all $a > 0$, only $H_{3,0}(x)$ remains by Theorem 2.7.3 as a possible limit law. For showing that $F(x)$, in fact, belongs to the domain of attraction of $H_{3,0}(x)$, we can use either Theorem 2.1.3 or Theorem 2.7.2. The choice and the corresponding calculations are left to the reader. ▲

Example 2.7.5 (The Standard Planck Distribution). Let $X_1, X_2, \ldots, X_n$ be i.i.d. random variables with density function

$$f(x) = \frac{Kx^3}{e^x - 1}, \qquad x > 0.$$

Then the distribution function $F(x)$ is in the domain of attraction of $L_{2,3}(x)$.

We first exclude the other possibilities for the minimum W_n by considering the existence of moments. Applying our standard method of expressing W_n as the maximum $Z_{n,1}$ of the sequence $(-X_j)$, $1 \le j \le n$, we can appeal to our criteria on maxima in terms of m_a^+. Since the distribution of $(-X_j)$ is $F_1(x) = 1 - F(-x)$, its density is

$$f_1(x) = \frac{Kx^3}{1 - e^{-x}}, \qquad x < 0.$$

Because $\omega(F_1) = 0$, we make one further transformation, namely $F^{**}(x) = F_1(-1/x)$, $x > 0$. Its density is

$$f^{**}(x) = \frac{K}{x^5(e^{1/x} - 1)}, \qquad x > 0$$

Hence, $M_{F^{**}} = (0, 3)$, which, by Theorem 2.7.3 yields that the only possibility for F^{**} is to belong to the domain of attraction of $H_{1,3}(x)$. This, in turn, relates $F_1(x)$ to $H_{2,3}(x)$, which finally means that $F(x)$ can only be in the domain of attraction of $L_{2,3}(x)$.

In order to conclude that $F(x)$ is, in fact, in the domain of attraction of $L_{2,3}(x)$, we appeal to Theorem 2.1.5. The condition (27) for $F^*(x)$ is equivalent to

$$\lim_{t=0} \frac{\int_0^{tx} f(y)\,dy}{\int_0^t f(y)\,dy} = x^3, \qquad x > 0,$$

which is immediately obtained if we substitute $y = ux$ in the numerator and observe that, as $t \to 0$,

$$\frac{e^{ux}-1}{e^u-1} \to x \quad \text{uniformly for } 0 < u \leq t.$$ ▲

This last example clearly describes the method of solving problems for W_n by first turning to the sequence $(-X_j)$, $1 \leq j \leq n$, and to their maximum. For this reason, we do not reformulate the results of this section for W_n (but see Exercise 14).

2.8. WEAK CONVERGENCE FOR THE kTH EXTREMES

Let us recall that the term "kth extreme" was introduced in a limiting sense (see Definition 1.4.2). That is, if $X_{r:n}$ denotes the rth order statistic (taken in increasing order), then, for fixed k, as $n \to +\infty$, $X_{k:n}$ and $X_{n-k+1:n}$ are called the kth extremes. The distribution functions of $X_{k:n}$ and of $X_{n-k+1:n}$ take very simple forms in the case of i.i.d. random variables. As a matter of fact, if $X_1, X_2, \ldots, X_n$ are i.i.d. with common distribution function $F(z)$, then, by the elementary formulas of the binomial distribution (see (34) and (35) of Section 1.4),

$$P(X_{k:n} \geq z) = \sum_{t=0}^{k-1} \binom{n}{t} [F(z)]^t [1-F(z)]^{n-t} \tag{134}$$

and

$$P(X_{n-k+1:n} < z) = \sum_{t=0}^{k-1} \binom{n}{t} [1-F(z)]^t [F(z)]^{n-t}. \tag{135}$$

Because k is fixed, the question of the existence, as well as the forms, of limiting distributions for the above random variables can easily be reduced to that for maximum and minimum. This fact is expressed in the following theorems.

Theorem 2.8.1. *For sequences a_n and $b_n > 0$ of real numbers, and for a fixed integer $k \geq 1$, as $n \to +\infty$,*

$$F_{n-k+1:n}(a_n + b_n x) = P(X_{n-k+1:n} < a_n + b_n x)$$

converges weakly to a nondegenerate distribution function $H^{(k)}(x)$ if, and only if,

$$H_n(a_n + b_n x) = F_{n:n}(a_n + b_n x)$$

converges weakly to a nondegenerate distribution function $H(x)$. If $H^{(k)}(x)$ exists, then for $\alpha(H) < x < \omega(H)$,

$$H^{(k)}(x) = H(x) \sum_{t=0}^{k-1} \frac{1}{t!} \left[\log \frac{1}{H(x)} \right]^t, \qquad (136)$$

where $H(x)$ is one of the three types $H_{1,\gamma}(x)$, $H_{2,\gamma}(x)$, and $H_{3,0}(x)$.

Theorem 2.8.2. *For sequences c_n and $d_n > 0$ of real numbers, and for a fixed integer $k > 1$, as $n \to +\infty$,*

$$F_{k:n}(c_n + d_n x) = P(X_{k:n} < c_n + d_n x)$$

converges weakly to a nondegenerate distribution function $L^{(k)}(x)$ if, and only if,

$$L_n(c_n + d_n x) = F_{1:n}(c_n + d_n x)$$

converges weakly to a nondegenerate distribution function $L(x)$. If $L^{(k)}(x)$ exists, then, for $\alpha(L) < x < \omega(L)$,

$$L^{(k)}(x) = 1 - [1 - L(x)] \sum_{t=0}^{k-1} \frac{1}{t!} \{ -\log[1 - L(x)] \}^t,$$

where $L(x)$ is one of the three types $L_{1,\gamma}(x)$, $L_{2,\gamma}(x)$, or $L_{3,0}(x)$.

The question of the weak convergence of the kth extremes, with normalization, therefore, does not present new mathematical problems. We can use the previous sections to determine a_n, b_n, c_n, and d_n as well as $H^{(k)}(x)$ and $L^{(k)}(x)$. Furthermore, for a given distribution function $F(x)$, we can decide if $H^{(k)}(x)$ and (or) $L^{(k)}(x)$ exist.

Proof of Theorem 2.8.1. Let $k > 1$ be a fixed integer. Let us first assume that the sequences a_n and $b_n > 0$ are such that, as $n \to +\infty$,

$$\lim F_{n-k+1:n}(a_n + b_n x) = H^{(k)}(x)$$

in the sense of weak convergence, where $H^{(k)}(x)$ is a nondegenerate distribution function. We want to show that $F^n(a_n + b_n x)$ also converges weakly to a nondegenerate distribution function $H(x)$. We shall apply (135) with $z = a_n + b_n x$. Since k is fixed, t is bounded, and thus the general term of (135) is asymptotically equal to

$$\frac{1}{t!} \{ n[1 - F(a_n + b_n x)] \}^t [F(a_n + b_n x)]^{n-t}. \qquad (137)$$

Hence, if x is such that, for a subsequence $n(s)$, $s \geq 1$, $F(a_{n(s)} + b_{n(s)} x) \leq q$

with some $q < 1$, then $H^{(k)}(x) = 0$. Therefore, for $x > \alpha(H^{(k)})$, $F(a_n + b_n x) \to 1$. Furthermore, as $n \to +\infty$,

$$\limsup n[1 - F(a_n + b_n x)] = u^*(x) < +\infty,$$

because, by the upper inequality of Lemma 1.3.1, the term in (137) is smaller than

$$\frac{1}{t!} n^t [1 - F(a_n + b_n x)]^t \exp\{-(n-t)[1 - F(a_n + b_n x)]\},$$

which would become zero in limit on a subsequence if $u^*(x) = +\infty$. Let us now take an arbitrary subsequence $n(s)$, $s \geq 1$, on which, as $s \to +\infty$,

$$\lim n(s)[1 - F(a_{n(s)} + b_{n(s)} x)] = u(x) \tag{138}$$

exists for a given x. Evidently, $u(x) \leq u^*(x) < +\infty$. By Corollary 1.3.1, the term in (137) tends to

$$\frac{1}{t!} u^t(x) \exp[-u(x)],$$

as $s \to +\infty$. Hence, by (135), when we put $z = a_n + b_n x$, we get the relation (136) with $H(x) = \exp[-u(x)]$. Consequently, if, for a single point x of continuity of $H^{(k)}(x)$, there were two subsequences for which the limit in (138) was different, $u_1(x)$ and $u_2(x)$, say, then

$$H^{(k)}(x) = \exp[-u_1(x)] \sum_{t=0}^{k-1} \frac{1}{t!} u_1^t(x)$$

$$= \exp[-u_2(x)] \sum_{t=0}^{k-1} \frac{1}{t!} u_2^t(x).$$

The last equality, however, cannot hold unless $u_1(x) = u_2(x)$. We thus obtain that (138) holds for $\{n(s)\} = \{1, 2, \ldots\}$, which, with one more appeal to Corollary 1.3.1, yields the relation (136). From (136) it follows that, if $H^{(k)}(x)$ is nondegenerate, then so is $H(x)$. This completes one part of the proof.

Turning to the converse situation, we assume that a_n and $b_n > 0$ are such that, as $n \to +\infty$, $H_n(a_n + b_n x)$ converges weakly to $H(x)$. We then know from Theorem 2.4.1 that $H(x)$ is one of the three types $H_{1,\gamma}(x)$, $H_{2,\gamma}(x)$, and $H_{3,0}(x)$. Let $x > \alpha(H)$. Then, for sufficiently large n, $F(a_n + b_n x) > 0$. We can therefore take its logarithm. We get

$$\lim_{n = +\infty} n \log F(a_n + b_n x) = \log H(x). \tag{139}$$

This implies that $F(a_n + b_n x) \to 1$ as $n \to +\infty$. By the Taylor expansion

$$\log s = \log[1 - (1-s)] = -(1-s) + o(1), \qquad s \to 1,$$

we thus get from (139) that, as $n \to +\infty$,

$$\lim n[1 - F(a_n + b_n x)] = \log \frac{1}{H(x)}. \tag{140}$$

Hence we can apply Corollary 1.3.1 again, which, in view of (135), yields (136). The proof is completed. ▲

The proof of Theorem 2.7.2 is identical to the above argument. Its details are therefore omitted.

Example 2.8.1 (The Normal Distribution). For the normal distribution

$$F(x) = \frac{1}{\sqrt{2\pi}} \int_{-\infty}^{x} e^{-y^2/2} dy$$

we have calculated a_n, b_n, c_n, and d_n for Z_n and W_n (Example 2.3.2). By Theorems 2.8.1 and 2.8.2, these same constants can be used for the kth extremes.

We thus have, for fixed $k \geq 1$ and for $n \to +\infty$,

$$\lim P(X_{n-k+1:n} < a_n + b_n x) = \exp(-e^{-x}) \sum_{t=0}^{k-1} \frac{e^{-tx}}{t!}$$

and

$$\lim P(X_{k:n} < c_n + d_n x) = 1 - \exp(-e^x) \sum_{t=0}^{k-1} \frac{e^{tx}}{t!}. \qquad \blacktriangle$$

2.9. THE RANGE AND MIDRANGE

In this section we will investigate the asymptotic distribution of the statistics

$$R_n = Z_n - W_n$$

and

$$M_n = \tfrac{1}{2}(Z_n + W_n),$$

which are called the range and the midrange, respectively. The method developed here can be applied to other functions of the extremes, too.

We first determine the asymptotic distribution of the vector (W_n, Z_n), when it exists after suitable normalization. More precisely, we first prove the following result.

Theorem 2.9.1. *Let $X_1, X_2, \ldots, X_n$ be i.i.d. random variables with distribution function $F(x)$. Assume that $F(x)$ is such that there are sequences $a_n, c_n, b_n > 0$ and $d_n > 0$ for which, as $n \to +\infty$,*

$$\lim F^n(a_n + b_n x) = H(x)$$

and

$$\lim \left[1 - F(c_n + d_n x)\right]^n = 1 - L(x)$$

exist and are nondegenerate. Then, as $n \to +\infty$,

$$\lim P(W_n < c_n + d_n x, Z_n < a_n + b_n y) = L(x) H(y).$$

In other words, if the asymptotic distribution of each of W_n and Z_n exists, when suitably normalized, then, with the same normalization, W_n and Z_n are asymptotically independent.

Proof. Consider the events

$$D_j = D_j(w, z) = \{w \leqslant X_j < z\},$$

where w and z are real numbers with $w < z$. Then

$$P(W_n \geqslant w, Z_n < z) = P(D_1 D_2 \cdots D_n) = \left[F(z) - F(w)\right]^n.$$

For $z > w$, for which $F(z) > F(w)$, let us write

$$\left[F(z) - F(w)\right]^n = \exp\{n \log[1 - (1 - F(z)) - F(w)]\},$$

which, by Taylor's expansion, becomes

$$\left[F(z) - F(w)\right]^n = \exp\{-n[1 - F(z)] - nF(w)$$
$$+ \nu_1 n[1 - F(z)]^2 + \nu_2 n[F(w)]^2\},$$

where $|\nu_1| \leqslant 1$ and $|\nu_2| \leqslant 1$ if $1 - F(z) < \frac{1}{2}$ and $F(w) < \frac{1}{2}$. We shall apply this formula with

$$z = a_n + b_n y, \qquad w = c_n + d_n x.$$

By our assumptions on the choice of a_n, b_n, c_n, d_n, for $y > \alpha(H)$ and for $x < \omega(L)$, respectively,

$$\lim_{n=+\infty} F(a_n + b_n y) = 1, \qquad \lim_{n=+\infty} F(c_n + d_n x) = 0.$$

The above expressions thus yield

$$P(W_n \geq c_n + d_n x, Z_n < a_n + b_n y)$$
$$= \exp\{-[1+o(1)]n[1-F(a_n+b_n y)] - [1+o(1)]nF(c_n+d_n x)\},$$

as $n \to +\infty$, where $\alpha(L) < x < \omega(L)$ and $\alpha(H) < y < \omega(H)$. As we have seen several times (see (140) and apply a similar method in terms of $1 - L(x)$), the right hand side tends to $H(y)[1 - L(x)]$. On the other hand, if $x > \omega(L)$ or $y < \alpha(H)$, the limit is evidently zero. Finally, for $x < \alpha(L)$ or $y > \omega(H)$, the limit is $H(y)$ or $1 - L(x)$, respectively. Since all of these cases can be written in the form of $H(y)[1 - L(x)]$, the proof is complete. ▲

For finding the limiting behavior of R_n and M_n, we shall need the following lemma.

Lemma 2.9.1. *Let the distribution function of the vector (Y_n, U_n) converge weakly to $T(y)E(u)$, where $T(y)$ and $E(u)$ are continuous distribution functions. Then, as $n \to +\infty$,*

$$\lim P(Y_n + U_n < x) = \int_{-\infty}^{+\infty} E(x-y) \, dT(y).$$

Proof. Let $\varepsilon > 0$ be an arbitrary real number. Let n_0, A, B, and C be such that, for all $n \geq n_0$,

$$P(Y_n < A) \leq \varepsilon, \qquad P(Y_n \geq C) \leq \varepsilon, \quad \text{and} \quad P(U_n \leq B) \leq \varepsilon.$$

Finally, let $A = y_0 < y_1 < \cdots < y_s = C$ and $B = u_0 < u_1 < \cdots < u_s = x - A$ be fixed numbers. We consider the sum

$$S(n; x) = \sum_{i,j}^{(x)} P(y_{i-1} \leq Y_n < y_i, u_{j-1} \leq U_n < u_j),$$

where $\sum_{i,j}^{(x)}$ signifies summation over all $1 \leq i \leq s$, $1 \leq j \leq s$ for which $y_i + u_j < x$. As the largest subdivisions $\Delta y = y_i - y_{i-1}$ and $\Delta u = u_j - u_{j-1}$ tend to zero,

$$\lim S(n; x) = P(Y_n + U_n < x, A \leq Y_n < C, U_n \geq B),$$

while, as $n \to +\infty$,

$$\lim S(n;x) = \sum_{i,j}^{(x)} [T(y_i) - T(y_{i-1})][E(u_j) - E(u_{j-1})],$$

which we denote by $S(x)$. Now, as the largest of Δy and Δu tends to zero,

$$\lim S(x) = \iint_{a(x)} dT(y) \, dE(u),$$

when $a(x) = \{(y,u): y + u < x, A \leq y < C, B \leq u\}$. If we combine the above limits and observe that

$$\iint_{a(x)} dT(y) \, dE(u) \to \int_{-\infty}^{+\infty} E(x-y) \, dT(y)$$

as $A, B \to -\infty$ and $C \to +\infty$, the claimed formula follows, if passage to the limit is taken in the following order: (i) the largest of Δy and Δu tends to zero, (ii) $n \to +\infty$, and (iii) $A, B \to -\infty$ and $C \to +\infty$. The lemma is established. ▲

The main result on the range and the midrange is contained in the following theorem.

Theorem 2.9.2. *Let $X_1, X_2, \ldots, X_n$ be i.i.d. random variables with common distribution function $F(x)$. Assume that $F(x)$ is such that there are sequences a_n, c_n, and $b_n > 0$ for which, as $n \to +\infty$,*

$$\lim F^n(a_n + b_n x) = H(x) \tag{141}$$

and

$$\lim [1 - F(c_n + b_n x)]^n = 1 - L(x) \tag{142}$$

exist and are nondegenerate. Then, as $n \to +\infty$,

$$\lim P(R_n < a_n - c_n + b_n x) = \int_{-\infty}^{+\infty} [1 - L(y-x)] \, dH(y)$$

and

$$\lim P(2M_n < a_n + c_n + b_n x) = \int_{-\infty}^{+\infty} L(x-y) \, dH(y).$$

Remark 2.9.1. The assumption in (141) and (142) that the same sequence $b_n > 0$ can be chosen may seem somewhat restrictive. However, this is the only interesting situation. Namely, with different values for the coefficient of x in (141) and (142), R_n and M_n reduce to one of Z_n and W_n in limit. Such a case will be illustrated by one of the examples which follow the proof.

Proof. By Theorem 2.9.1, the random variables $Y_n = (Z_n - a_n)/b_n$ and $U_n = (W_n - c_n)/b_n$ are asymptotically independent. Hence, Lemma 2.9.1 is applicable to both $(Y_n, -U_n)$ and (Y_n, U_n), from which the theorem follows. ▲

Let us look at some concrete distributions.

Example 2.9.1 (The Normal Distribution). If

$$F(x) = (2\pi)^{-1/2} \int_{-\infty}^{x} e^{-t^2/2} dt,$$

we know that (141) and (142) hold (Example 2.3.2). In fact

$$a_n = -c_n = (2\log n)^{1/2} - \frac{1}{2} \frac{\log\log n + \log 4\pi}{(2\log n)^{1/2}}$$

and

$$b_n = (2\log n)^{-1/2}.$$

In addition, $H(x) = \exp(-e^{-x})$ and $L(x) = 1 - \exp(-e^x)$. Thus, by Theorem 2.9.1, $(Z_n - a_n)/b_n$ and $(W_n + a_n)/b_n$ are asymptotically independent. Furthermore, by Theorem 2.9.2, as $n \to +\infty$,

$$\lim P(R_n < 2a_n + b_n x) = \int_{-\infty}^{+\infty} \exp(-e^{y-x}) d[\exp(-e^{-y})]$$

and

$$\lim P\left(M_n < \tfrac{1}{2} b_n x\right) = \int_{-\infty}^{+\infty} \exp(-e^{x-y}) d[\exp(-e^{-y})]$$

$$= \frac{1}{1 + e^{-x}}.$$

No explicit form is known for the integral which is obtained as the limiting distribution of R_n. It, of course, does not present any disadvantage, since numerical integration can give its value for any x. The asymptotic distribution of M_n, on the other hand, is the familiar logistic distribution. ▲

Example 2.9.2 (The Exponential Distribution). For $F(x)=1-e^{-x}$, as $n\to+\infty$,

$$P(Z_n < \log n + x) \to \exp(-e^{-x}) = H_{3,0}(x)$$

and

$$P(W_n < \frac{x}{n}) = 1 - e^{-x},$$

as was shown in Example 1.3.1. Thus, for any $\varepsilon > 0$,

$$P(W_n \geq \varepsilon) = e^{-\varepsilon n} \to 0, \qquad \text{as } n \to +\infty.$$

That is, $W_n \to 0$ in probability. Hence, by Lemma 2.2.1, as $n \to +\infty$,

$$\lim P(R_n < \log n + x) = \lim P(Z_n < \log n + x) = H_{3,0}(x)$$

and

$$\lim P(M_n < \tfrac{1}{2}\log n + \tfrac{1}{2}x) = \lim P(Z_n < \log n + x) = H_{3,0}(x). \qquad \blacktriangle$$

It should be noted that the fact of $W_n \to 0$ in probability was not the sole reason for R_n and M_n having been expressible by Z_n alone. The important part was the relation of Z_n and W_n. This is made clear in the next example.

Example 2.9.3 (The Uniform Distribution). Let $F(x) = x$ for $0 \leq x \leq 1$. Then, by the results in Example 2.3.1, (141) and (142) are satisfied with $a_n = 1, c_n = 0$, and $b_n = 1/n$. Furthermore, $H(x) = e^x$ for $x < 0$ and $L(x) = 1 - e^{-x}, x > 0$. Here, both W_n and $(Z_n - 1)$ tend to zero in probability, but the normalizing constant b_n is the same for both. Hence, they both contribute to R_n and M_n. Their asymptotic distribution can be obtained from Theorem 2.9.2. We get, for $n \to +\infty$,

$$\lim P\left(R_n < 1 + \frac{x}{n}\right) = \int_x^0 e^{-y+x} e^y \, dy + \int_{-\infty}^x e^y \, dy = (1-x)e^x$$

if $x < 0$, and the limit is one for $x > 0$. On the other hand, as $n \to +\infty$,

$$\lim P\left(M_n < \frac{1}{2} + \frac{x}{2n}\right) = \int_{-\infty}^{\min(x,0)} (1 - e^{y-x}) e^y \, dy,$$

which is $\tfrac{1}{2}e^x$ if $x < 0$ and $1 - \tfrac{1}{2}e^{-x}$ for $x \geq 0$. $\blacktriangle$

2.10. SPEED OF CONVERGENCE

We would like to estimate the error committed by the replacement of the exact distributions of the extremes by their limiting forms. Such estimates will help in determining the sample size required for the application of the asymptotic theory when observations are independent. We always assume that the distribution $F(x)$ of the population has correctly been chosen. Hence, the errors are due to the passage to the limit as the sample size increases indefinitely.

In mathematical terms, we want to estimate the following quantity. Let $X_1, X_2, \ldots, X_n$ be i.i.d. random variables and let E_n signify one of their extremes. We assume that the distribution $F(x)$ of the X_j is such that, with suitable normalizing constants β_n and $\gamma_n > 0$, as $n \to +\infty$,

$$\lim P(E_n < \beta_n + \gamma_n x) = R(x)$$

exists. Let us put

$$\Delta_n(x) = \Delta_n(x; E_n, \beta_n, \gamma_n) = P(E_n < \beta_n + \gamma_n x) - R(x).$$

Our aim is to investigate $\Delta_n(x)$ in terms of n. Since, for all extremes, $R(x)$ is continuous, the following result shows that $\Delta_n(x)$ can, in fact, be considered a function which does not depend on x.

Lemma 2.10.1. *Let a sequence $R_n(x)$ of distribution functions converge to a continuous distribution function $R(x)$. Then the convergence is uniform in x.*

Proof. What we have to prove is that, for arbitrary $\varepsilon > 0$, we can find a positive integer n_0, which depends on ε (but not on x) and which has the property that, for all $n \geq n_0$,

$$|R(x) - R_n(x)| < \varepsilon \quad \text{for all } x. \tag{143}$$

Let us first observe that, for any fixed real number a, there is an integer $n_1 = n_1(a)$, such that, for all $n \geq n_1$,

$$R_n(a) \leq 2R(a), \quad 1 - R_n(a) \leq 2(1 - R(a)).$$

Thus, since $R_n(x)$ and $R(x)$ are distribution functions, we can choose real numbers A and B and a positive integer n_2 such that, for all $n \geq n_2$,

$$R_n(x) \leq R_n(A) \leq 2R(A) < \frac{\varepsilon}{2} \quad \text{for all } x \leq A$$

and

$$1 - R_n(x) \leq 1 - R_n(B) \leq 2(1 - R(B)) < \frac{\varepsilon}{2} \quad \text{for all } x \geq B.$$

Thus, for all $n \geq n_2$,

$$|R(x) - R_n(x)| \leq R(x) + R_n(x) \leq R(A) + R_n(A) < \varepsilon, \quad x \leq A$$

and

$$|R(x) - R_n(x)| = |1 - R_n(x) - [1 - R(x)]| \leq 1 - R_n(x) + 1 - R(x)$$
$$\leq 1 - R_n(B) + 1 - R(B) < \varepsilon, \quad x \geq B.$$

Now let $A < x < B$. Since $R(x)$ is continuous, it is known to be uniformly continuous on the finite interval $[A, B]$. That is, we can divide $[A, B]$ into N equal parts by the points $A = x_0 < x_1 < \cdots < x_N = B$ and, for each k with $1 \leq k \leq N$,

$$R(x_k) - R(x_{k-1}) < \frac{\varepsilon}{6}.$$

Here, N depends on ε only. For a given $\varepsilon > 0$, let us fix the value of N. We can then find a positive integer n_3 such that, for each k with $1 \leq k \leq N$ and for all $n \geq n_3$,

$$|R_n(x_k) - R(x_k)| < \frac{\varepsilon}{6}.$$

Since $A < x < B$, there is one k with $x_{k-1} < x \leq x_k$. Then, by the choices above and on account of $R_n(x)$ being nondecreasing, we have, for all $n \geq n_3$,

$$|R_n(x) - R(x)| < |R_n(x_k) - R(x_k)| + |R_n(x_{k-1}) - R(x_{k-1})|$$
$$+ [R_n(x_k) - R_n(x_{k-1})] + [R(x_k) - R(x_{k-1})]$$
$$< \frac{3\varepsilon}{6} + |R_n(x_k) - R(x_k) + R(x_k) - R(x_{k-1}) + R(x_{k-1}) - R_n(x_{k-1})|,$$

which is smaller than ε if we apply the triangle inequality once again. Any positive integer n_0 which is larger than both n_2 and n_3 satisfies the requirements of n_0's being independent of x and, at the same time, (143) holds for all $n \geq n_0$. The proof is completed. ▲

Let us return to $\Delta_n(x; E_n, \beta_n, \gamma_n)$. We have just seen that we can estimate it uniformly in x. On the other hand, its dependence on β_n and γ_n is very significant. While, in view of Lemma 2.2.2, we have a large freedom in choosing β_n and γ_n without affecting the convergence of $\Delta_n(x)$ to zero, the speed of this convergence depends on their actual choice. The clearest example for such a claim can be the exponential distribution, when

$E_n = W_n$, the minimum of the observations. Then (Example 1.3.1) $\Delta_n(x; W_n, 0, 1/n) \equiv 0$, while for any other permissible values of β_n and γ_n, $\Delta_n(x) > 0$ for all $x > 0$. In fact, if $g(n) \to 0$ in an arbitrary manner, then, as $n \to +\infty$,

$$\lim \frac{\Delta_n(x; W_n, g(n)/n, 1/n)}{g(n)} = e^{-x}$$

(see Exercise 15). Therefore, no meaningful result can be expected as estimates on $\Delta_n(x)$ which do not involve β_n and γ_n. This explains the choices of the terms in the estimates of the next theorem.

Theorem 2.10.1. *Let $H(x)$ be one of the possible extreme value distributions for the maximum. Let $F(x)$ be in the domain of attraction of $H(x)$. For given sequences a_n and $b_n > 0$ of real numbers, we put*

$$z_n(x) = n[1 - F(a_n + b_n x)]$$

and, for x's for which $H(x) > 0$,

$$\rho_n(x) = z_n(x) + \log H(x).$$

Then, if x is such that $H(x) > 0$ and if $z_n(x)/n \leq \frac{1}{2}$,

$$|P(Z_n < a_n + b_n x) - H(x)| \leq H(x)[r_{1,n}(x) + r_{2,n}(x) + r_{1,n}(x) r_{2,n}(x)],$$

(144)

where

$$r_{1,n}(x) = \frac{2z_n^2(x)}{n} + \frac{2z_n^4(x)}{n^2} \cdot \frac{1}{1-q},$$

$$r_{2,n}(x) = |\rho_n(x)| + \frac{\rho_n^2(x)}{2} \cdot \frac{1}{1-s},$$

with $q < 1$ and $s < 1$ such that $\frac{2}{3} z_n^2(x)/n \leq q$ and $\frac{1}{3}|\rho_n(x)| \leq s$, respectively.

Remark 2.10.1. Notice that the statement did not impose any direct condition on the choices of a_n and b_n. Therefore, the estimate (144) is applicable for arbitrary choices, whenever $z_n(x)/n$, $z_n^2(x)/n$, and $\rho_n(x)$ satisfy the stated inequalities. On the other hand, if a_n and b_n are chosen so that $(Z_n - a_n)/b_n$ converges weakly to $H(x)$, then the assumed bounds on $z_n(x)/n$, $z_n^2(x)/n$, and $|\rho_n(x)|$ are valid for very moderate values of n (usually for n as small as 2 or 3). Namely, if $H(x) > 0$, we can take logarithm in (6) and we see that $z_n(x) \to -\log H(x)$ (see (101a) and the formula preceding it). Hence, $z_n(x)$ is bounded and $\rho_n(x) \to 0$, as $n \to +\infty$.

Remark 2.10.2. Referring once again to the fact that, with the proper choice of a_n and b_n, $z_n(x) \to -\log H(x)$ and $\rho_n(x) \to 0$ as $n \to +\infty$, we can see that $r_{1,n}(x)$ and $r_{2,n}(x)$ represent two essentially different kinds of error. Since

$$z_n(x) = \rho_n(x) - \log H(x),$$

the major contribution to $z_n(x)$ is $-\log H(x)$, hence $r_{1,n}(x)$ is of the order of $1/n$ for all $F(x)$ and for all proper choices of a_n and b_n. On the other hand, $r_{2,n}(x)$ expresses the error due to the choices of a_n and b_n and it also depends on the population distribution $F(x)$.

Proof. By the basic formula (4) and by the triangle inequality

$$|P(Z_n < a_n + b_n x) - H(x)| \leq |F^n(a_n + b_n x) - e^{-z_n(x)}| + |e^{-z_n(x)} - H(x)|. \tag{145}$$

If we write $F = 1 - (1 - F)$, Lemma 1.3.1 yields

$$|F^n(a_n + b_n x) - e^{-z_n(x)}| \leq e^{-z_n(x)} \left\{ \exp\left[\frac{2z_n^2(x)}{n}\right] - 1 \right\},$$

whenever $z_n(x)/n \leq \frac{1}{2}$. Hence, by the inequality (which is a consequence of the Taylor expansion)

$$|e^y - 1| \leq |y| + \frac{y^2}{2} \frac{1}{1-q} \qquad \text{uniformly for } \tfrac{1}{3}|y| \leq q < 1 \tag{146}$$

we have

$$|F^n(a_n + b_n x) - e^{-z_n(x)}| \leq e^{-z_n(x)} \left\{ \frac{2z_n^2(x)}{n} + \frac{2z_n^4(x)}{n^2} \frac{1}{1-q} \right\}, \tag{147}$$

where $z_n(x)/n \leq \frac{1}{2}$ and $\frac{2}{3} z_n^2(x)/n \leq q < 1$.

Let us now turn to the second term on the right hand side of (145). We assume that $H(x) > 0$. Since

$$e^{-z_n(x)} - H(x) = H(x)[e^{-\rho_n(x)} - 1],$$

the inequality (146) leads to the estimate

$$|e^{-z_n(x)} - H(x)| \leq H(x)\left[|\rho_n(x)| + \frac{\rho_n^2(x)}{2} \frac{1}{1-s}\right], \tag{148}$$

where $\frac{1}{3}|\rho_n(x)| \leq s < 1$. If we now write

$$e^{-z_n(x)} = e^{-z_n(x)} - H(x) + H(x)$$

on the right hand side of (147), the inequality (144) follows from (145), (147), and (148). The theorem is established. ▲

A similar estimate applies for the case of minima, too.

Theorem 2.10.2. *Let $L(x)$ be one of the possible extreme value distributions for the minimum. Let $F(x)$ be in the domain of attraction of $L(x)$. For given sequences c_n and $d_n > 0$ of real numbers, we put*

$$z_n(x) = nF(c_n + d_n x)$$

and, for x's for which $L(x) < 1$,

$$\rho_n(x) = z_n(x) + \log[1 - L(x)].$$

If x is such that $L(x) < 1$ and if $z_n(x)/n \leq \frac{1}{2}$, then

$$|P(W_n < c_n + d_n x) - L(x)| \leq [1 - L(x)][r_{1,n}(x) + r_{2,n}(x) + r_{1,n}(x)r_{2,n}(x)],$$

where $r_{1,n}(x)$ and $r_{2,n}(x)$ are defined as in Theorem 2.10.1.

Proof. Notice that

$$|P(W_n < w) - L(w)| = |[1 - P(W_n < w)] - [1 - L(w)]|.$$

Hence, the proof of Theorem 2.10.1 can be repeated by changing the roles of $P(Z_n < a_n + b_n x)$ and $H(x)$ to $1 - P(W_n < c_n + d_n x)$ and $1 - L(x)$, respectively. The details are therefore omitted. ▲

We illustrate the results of Theorems 2.10.1 and 2.10.2 in the following examples.

Example 2.10.1. Let $F(x) = 1 - e^{-x}$, $x \geq 0$. We know that $F(x)$ is in the domain of attraction of $H_{3,0}(x)$. Let us choose $a_n = \log n$ and $b_n = 1$. Then

$$z_n(x) = e^{-x}, \qquad \rho_n(x) = 0.$$

Consequently,

$$r_{1,n}(x) = \frac{2e^{-2x}}{n} + \frac{2e^{-4x}}{n^2(1-q)},$$

where $q < 1$ and $\frac{2}{3}e^{-2x}/n \leq q$. Furthermore, $r_{2,n}(x) = 0$. Thus, by Theorem 2.10.1,

$$|P(Z_n < \log n + x) - H_{3,0}(x)| \leq H_{3,0}(x)r_{1,n}(x).$$

For example, if $x=2$ and $n=10$,

$$P(Z_{10} < \log 10 + 2) = .8726, \qquad H_{3,0}(2) = .8734$$

while our estimate above gives the error term

$$H_{3,0}(2) r_{1,10}(2) = 0.0032.$$

For this last value, we took from a table $e^{-4} = 0.018316$ and $e^{-8} = 0.000335$. Hence, any $q < 1$ can be taken which is not smaller than 0.00123. If we choose $q = 0.002$, we get $r_{1,10}(2) = 0.00366$. ▲

While the error estimate 0.0032 is large compared with the actual error 0.0008, it should be noted that our method is applicable to any distribution $F(x)$ and for all values of x and n.

Example 2.10.2. In Example 2.6.3, we approximated $P(Z_{50} < 2.6)$ by $H_{3,0}(1.3961)$, assuming that the population has standard normal distribution (the experimenter's assumption in the quoted example). Let us now estimate the error term in this approximation. In the example, we used a_n and b_n as obtained in Section 2.3.2. With these choices, we got $a_{50} = 2.1009$ and $b_{50} = 0.3575$. Hence, if we write $2.6 = a_{50} + b_{50} x$, we get $x = 1.3961$. From a table for the standard normal distribution, we have

$$z_{50}(1.3961) = 50[1 - F(2.6)] = 0.235,$$

and thus $\rho_{50}(1.3961) = -0.01256$. For computing the error terms, we first observe that we can choose $q = s = 0.005$. Therefore, by definitions,

$$r_{1,50}(1.3961) = 0.00221, \qquad r_{2,50}(1.3961) = 0.01264.$$

We can now compute the estimate (144) of the error term. We get

$$|P(Z_n < 2.6) - H_{3,0}(1.3961)| \leq 0.01162,$$

while $F^{50}(2.6) = .7901$ and $H_{3,0}(1.3961) = .7807$. We can therefore claim that the estimate is quite good. It may also be added here that the convergence of the distribution of Z_n to its limit is somewhat slow for the normal distribution, since an error occurs in the second digit even for $n = 50$ (compare this with the result of the preceding example, where n was only 10). ▲

Estimates on the speed of convergence to limiting distributions for the kth extremes can be reduced to the case of maxima and minima. See Exercise 16.

2.11. SURVEY OF THE LITERATURE

The interest in the distribution of extremes goes back as far as applications of laws of chance to actuarial and insurance problems. Gumbel (1958) remarks that the problem did indeed come up in the works of N. Bernoulli in the early eighteenth century. Historians will certainly discover scattered solutions ever since Bernoulli's time. What is clear now is that accurate and general solutions are implicitly contained in the works of Poisson, whose influence on the theory of sums and the distribution of rare events actually led to a basic change of thinking in the theory of probability. This change could not come in his own time but only several decades later.

Even the basic work of M. Fréchet (1927) was not appropriately recognized at that time because of his departure from the assumption of normality of the population distribution. Statistics in the first decades of the present century was associated with normal populations; a systematic theory therefore received attention only if it was for normal variates. The early works along this line were by L. von Bortkiewicz (1922), R. von Mises (1923), and L. H. C. Tippett (1925). Tippett produced extensive tables for the distribution of the maximum of normal variates for different sample sizes; these tables have been in use up to date. E. L. Dodd (1923) was the first to deviate from normality, but like Fréchet's more detailed work, his was also neglected. The theoretical work was continued by R. A. Fisher and L. H. C. Tippett (1928), who found the three possible limiting distributions of the extremes, and by R. von Mises (1936), who classified absolutely continuous population distributions according to their limits for maxima (domains of attraction) (Theorems 2.7.1 and 2.7.2). It was pointed out by Fisher and Tippett that the speed of convergence to the asymptotic distribution of the maxima for normal populations is slow, a fact which further delayed practitioners' acceptance of the asymptotic theory.

B. V. Gnedenko (1943) developed the theory to a high level by establishing practically all results contained in Sections 2.1–2.4, except that he obtained Theorem 2.4.3(iii) in an abstract form, which can be stated in our approach as follows. The existence of a function $h(t)$ such that (99) holds is a necessary and sufficient condition for $F(x)$ to belong to the domain of attraction of $H_{3,0}(x)$. The final neat step leading to Theorem 2.4.3(iii) was made by L. de Haan (1970 or 1971), whose major contribution is that he made $h(t)$ specific ($R(t)$ in our nation). The proof adopted here is a combination of methods of Gnedenko, de Haan, and D. G. Mejzler (1949), whose result is Step 3 in Section 2.5.

The approach of de Haan made also possible to obtain several equivalent forms of Theorem 2.4.3, some of which are listed among the exercises. This same technique led to the interesting result of A. A. Balkema and L. de Haan (1972), which roughly says that functions in the domain of attraction of $H_{3,0}(x)$ are "comparable" with functions which satisfy the criterion of von Mises (see Exercise 20). Comparable simply means here the concept of tail equivalence, which was introduced by S. Resnick (1971a) (see Exercise 21). This concept is also basic in de Haan's work. As a matter of fact, Step 5 is essential in Section 2.5. If the population distribution is assumed smooth enough, then already the result of von Mises leads to $h(t) = R(t)$ (we have reversed the procedure here). Some of the results of de Haan are also contained in the paper by M. Marcus and M. Pinsky (1969). We should add that some steps of Gnedenko are used in the present book in forms which are well known in the theory of slowly varying functions (although we did not use the theory itself). Some of the reformulations of Gnedenko's work into the set-up mentioned are due to W. Feller (1966). See also L. de Haan (1974a).

The relation of moment convergence to weak convergence is settled in J. Pickands, III (1968). Some special cases were discussed earlier by von Mises (see selected works), J. Geffroy (1958), P. K. Sen (1961), and J. R. McCord (1964). See also L. K. Chan and G. A. Jarvis (1970).

The theory of the kth extremes was developed by T. Kawata (1951) and N. V. Smirnov (1949 and 1952). The asymptotic independence of the maximum and minimum was observed by Smirnov (1949) and T. Homma (1951). Tiago de Oliveira (1961) established the asymptotic independence of the sample mean and the extremes. A more systematic study of this subject matter is given by H. J. Rossberg (1965a,b); see also J. E. Walsh (1969a and 1970), S. Ikeda and T. Matsunawa (1970), and S. Smid and A. J. Stam (1975). A. Rényi (1953) developed a general method to treat order statistics. His method is based on Sukhatme's theorem that we presented as an introduction to Theorem 1.6.3.

Theorem 2.7.3 is a collection of scattered results in the literature. The special distributions of Section 2.3, although one can treat them as examples for the theory, received special attention in the literature. The case of the normal distribution has been emphasized. The lognormal is receiving more and more attention in different branches of applied science. In connection with Example 2.6.3, see S. Kotz (1973). With reference to air pollution studies, N. D. Singpurwalla (1972) investigated Z_n for lognormal populations. Some other special distributions, including lognormal, are discussed by K. V. Bury (1974) and J. Villasenor (1976). Villasenor also discusses extensively tail properties of convolutions, which are far-reaching extensions of Feller's (1966) initial results. S. Resnick (1972a) gives conditions under which the product of a finite set of distributions belongs to the domain of attraction of a limit law. Because of the lack of asymptotic laws for most popular discrete distributions, C. W. Anderson (1970) suggests asymptotic bounds for the distribution of normalized extremes of these populations.

As seen in Section 2.4, a central fact of the theory of extremes is that a limit law should satisfy the functional equation (69). In its solution, we used the validity of (69) for all $m \geq 1$. It is, however, not necessary. J. Sethuraman (1965) has shown that if (69) holds for two different values of m such that the ratio of log B_m for the two values is irrational, then the solution of (69) is unique (if $B_m = 1$ for all m, then the irrationality of the ratio of the two A_m is to be assumed). Another extension was mentioned in Chapter 1, by which a moment assumption can replace (69) (the theorem of Huang (1974)).

The speed of convergence estimate is due to J. Galambos (1978). The uniformity of convergence is essentially due to G. Pólya (1920). The range and midrange is extensively studied by Gumbel (1958) and de Haan (1974b). The strange fact expressed by Example 2.6.5 and Exercise 17 was discovered by R. F. Green (1976b). Tables for random numbers for the three extreme value distributions are presented by N. Goldstein (1963).

It is pointed out by S. B. Weinstein (1973) that nonlinear normalization can result in more accurate approximation of distributions for a given sample size than the linear normalization adapted in the present chapter.

The error estimates by B. Grigelionis (1962 and 1970) in the Poisson approximation of the distribution of sums of indicator variables can be used to estimate the number of sample elements which exceed a predetermined number.

For statistical methods of estimating the parameters of the asymptotic distributions, when the sample elements are i.i.d., the major work is still the book by Gambel (1958) (see also J. Tiago de Oliveira (1972a)). But since the basic assumption of independence is frequently (if not mostly) questioned, caution is to be applied. A preferable approach would be the method of J. Pickands, III (1975), which perhaps can be developed to dependent samples and its asymptotic character simplified.

The increasing popularity of the asymptotic theory of extremes can be measured by the fact that more and more books, scientific or educational, include this theory. The most popular is the method of Cramér (1946); the book by W. A. Thompson, Jr. (1969) gives a simpler approach. The reader can find newer methods of estimation and testing for the extreme value distributions in N. R. Mann, R. E. Schafer, and N. D. Singpurwalla (1974). For tables for the distribution of extremes, see those by the National Bureau of Standards of 1953, W. Q. Meeker and W. B. Nelson (1975), and M. W. Mahmoud and A. Ragab (1975).

A large deviation result is formulated in Exercise 26 which is due to L. de Haan and A. Hordijk (1972). Other types can be deduced from results of S. A. Book (1972) and J. Steinebach (1976) in view of the representation of Z_n as in Theorem 1.6.3 for the exponential distribution. This can then be transformed to continuous distributions which are sufficiently smooth.

The titles of the following papers clearly indicate their content. They are relevant to our subject matter and they stress some points which were raised in other contexts: C. Singh (1967), J. E. Walsh (1969b), Barlow et al. (1969), L. Weiss (1969), E. G. Enns (1970), and C. E. Antle and F. Rademaker (1972).

2.12. EXERCISES

(In all exercises the basic random variables are i.i.d.)

1. Let $F(x)$ be the uniform distribution on the interval $(-2, 5)$. Find a_n and $b_n > 0$ such that $(Z_n - a_n)/b_n$ converges weakly. Determine the limit. Use your result to approximate $P(Z_{50} < 4.5)$. What does the approximation give for $P(X_{46:50} < 4.5)$?

2. Find $P(Z_{100} < 8)$ by the appropriate approximation if the population is normal with mean 2 and variance 4. Determine also the asymptotic value of $P(X_{6:100} < -8)$.

3. Let the distribution function of X be $H_{3,0}(x)$. Find $E(X)$ and $V(X)$.

4. If the distribution of X is $H_{1,\gamma}(x)$, what is the condition on γ if X has finite variance?

5. Let X be a random variable with distribution function $H_{3,0}(x)$. Show that the distribution of the random variable $Y = \exp(X/\gamma)$, where $\gamma > 0$, is $H_{1,\gamma}(x)$, and that of $-1/Y$ is $H_{2,\gamma}(x)$.

6. Show that the maxima of samples from the geometric distribution cannot be normalized to have a nondegenerate asymptotic distribution.

7. Show the same fact about the Poisson distribution that was stated in Exercise 6.

8. Extend Exercises 6 and 7 and prove the following result. If X_1 is discrete and takes the nonnegative integers only, and if the limit

$$\lim_{n = +\infty} \frac{P(X_1 \geq n)}{P(X_1 \geq n+1)} = 1$$

fails, then, whatever be the constants a_n and $b_n > 0$, $(Z_n - a_n)/b_n$ does not converge to a nondegenerate distribution.

9. Let $X > 0$ and let $\log X$ be standard normal variate (i.e., X is a lognormal variate). Show that, as $\sigma \to 0$, the distribution of $\sigma^{-1}(X^\sigma - 1)$ converges to the standard normal distribution.

10. Using the results in Section 2.1, reobtain the conclusions of Examples 1.3.1 and 1.3.2.

11. Find a criterion from (127) which guarantees that $F(x)$ belongs to the domain of attraction of $H_{2,\gamma}(x)$.

12. Show that under the conditions of Theorem 2.7.2 the normalizing constants a_n and $b_n > 0$, for which $(Z_n - a_n)/b_n$ converges weakly, satisfy the limit relation

$$\lim_{n=+\infty} nb_n f(a_n + b_n z) = e^{-z}.$$

13. Let $X_1, X_2, \ldots$ be i.i.d. unit exponential variates. The observations X_j are collected in blocks of n, and mn blocks are evaluated. For the kth block $X_{(k-1)n+1}, X_{(k-1)n+2}, \ldots$, let $W_{k,n}$ be the minimum. Find constants $a(m,n)$ and $b(m,n) > 0$ such that as $mn \to +\infty$, the largest Z_{mn} of $W_{k,n}$, $1 \leq k \leq mn$, when normalized $[Z_{mn} - a(m,n)]/b(m,n)$, converges weakly to a nondegenerate limit. What is the limiting distribution?

14. Restate the results of Section 2.7 for W_n.

15. With the notation of Section 2.10, show that if $g(n) \to 0$ as $n \to +\infty$, then, for the exponential distribution,

$$\lim_{n=+\infty} \frac{\Delta_n(x; W_n, g(n)/n, 1/n)}{g(n)} = e^{-x}.$$

16. Use formula (137) to conclude that the speed of convergence of the distribution of the kth extremes is of the same order of magnitude as the speed for maxima or minima according as the kth extremes are upper extremes or lower extremes, respectively.

17. By using the method of construction in Example 2.6.5 show, for an arbitrary distribution function $T(x)$, that there is an increasing sequence n_k of natural numbers and two sequences a_k and $b_k > 0$ of numbers such that

$$\frac{Z_{n_k} - a_k}{b_k} \to T(x) \quad \text{weakly.}$$

[R. F. Green (1976b)]

18. Show that if $F(x)$ is in the domain of attraction of $H_{3,0}(x)$ and if $F_1(x)$ is a distribution function such that, as $x \to \omega(F)$,

$$\lim \frac{1 - F(x)}{1 - F_1(x)} = 1, \qquad (149)$$

then $F_1(x)$ is also in the domain of attraction of $H_{3,0}(x)$. Furthermore, the same normalizing constants can be used for the maxima in the case of $F_1(x)$ as for $F(x)$.

19. Show that if $F(x)$ is in the domain of attraction of $H_{3,0}(x)$, then so is the function

$$F_1(x) = 1 - \int_x^{\omega(F)} [1 - F(t)] dt, \quad x \geq x_0.$$

[L. de Haan (1970)]

20. Show that if $F(x)$ is in the domain of attraction of $H_{3,0}(x)$, then there is a distribution function $F_1(x)$ which satisfies the conditions of Theorem 2.7.2 and, as $x \to \omega(F)$,

$$\lim \frac{1 - F(x)}{1 - F_1(x)} = 1.$$

[A. A. Balkema and L. de Haan (1972)]

21. Show the converse of Exercise 18: if, for some sequences a_n and $b_n > 0$, both $F^n(a_n + b_n x)$ and $F_1^n(a_n + b_n x)$ converge to $H_{3,0}(x)$, then (149) holds. (If (149) holds, then F and F_1 are said to be tail equivalent.)

[S. I. Resnick (1971)]

22. Let $X_1, X_2, \ldots, X_n$ be i.i.d. random variables with common distribution function $F(x)$. Let $\omega = \omega(F) < +\infty$. Show that $(Z_n - a_n)/b_n$ converges weakly with some constants a_n and $b_n > 0$ if, and only if, there are sequences a_n^* and b_n^* such that $(Z_n^* - a_n^*)/b_n^*$ converges weakly, where $Z_n^* = \max\{1/(\omega - X_j) : 1 \leq j \leq n\}$ (the limits are nondegenerate).

[L. de Haan (1970)]

23. Let $X_1, X_2, \ldots, X_n$ be i.i.d. with common discrete distribution $F(x)$. Assume that the jumps of $F(x)$ occur at positive integers and that, for all large integers, $F(x)$ has a positive jump. Show that there is a sequence a_n such that

$$\limsup P(Z_n < a_n + x) \leq \exp(-e^{-bx})$$

and

$$\liminf P(Z_n < a_n + x) \geq \exp(-e^{-b(x-1)})$$

if, and only if, as $n \to +\infty$,

$$\lim \frac{1 - F(n)}{1 - F(n+1)} = e^b.$$

Apply the result to the geometric distribution.

[C. W. Anderson (1970)]

24. Show that if $\omega(F) = +\infty$ and if, for some u and $v > 0$, as $x \to +\infty$,

$$\lim x^u [\exp(x^v)][1 - F(x)] = c, \quad 0 < c < +\infty,$$

then $(Z_n - a_n)/b_n$ converges weakly to $H_{3,0}(x)$. The constants a_n and $b_n > 0$ can be chosen as

$$b_n = (1/v)(\log nc)^{(1-v)/v}$$

and

$$a_n = (\log nc)^{1/v} - \frac{u \log \log nc}{v^2 (\log nc)^{(v-1)/v}}.$$

[J. Villasenor (1976)]

25. Let $F(x)$ be the gamma distribution with parameters (u, v), where u is an integer. Let $T(x)$ be an absolutely continuous distribution such that $\omega(T) = +\infty$ and

$$\int_{-\infty}^{+\infty} x^{u-1} e^{x/v} dT(x) < +\infty.$$

Then the convolution of F and T belongs to the domain of attraction of $H_{3,0}(x)$.

[J. Villasenor (1976)]

26. Let $F(x)$ be absolutely continuous and put $T(x) = [1 - F(x)]/f(x)$, where $f(x) = F'(x)$. Let $\omega(F) = +\infty$. Let $s(x)$ be nondecreasing and $s(x) \to +\infty$ with x. Show that if $T'(t) s^2 [1/(1 - F(t))] \to 0$ as $t \to +\infty$ and if $x_n = O(s(n))$, then as $n \to +\infty$,

$$\lim \frac{1 - F^n[a_n + x_n T(a_n)]}{1 - \exp[-\exp(-x_n)]} = 1,$$

where a_n is defined by $F(a_n) = 1 - 1/n$.

[L. de Haan and A. Hordijk (1972)]

CHAPTER 3

Weak Convergence of Extremes in the General Case

Our aim is to get rid of the restrictive assumptions of the preceding chapter, that the basic random variables $X_1, X_2, \ldots, X_n$ are i.i.d., and to replace these assumptions by less restrictive ones. These new models will cover most fields of interest to the applied scientist, when the solution to a problem can be expressed by the asymptotic distribution of extremes.

We shall first give a very general result for the exact distribution of extremes for a fixed n. We shall then deduce several limit theorems from it. Finally, in a number of sections, we shall specialize these results for specific structures of the X_j, $1 \leq j \leq n$. In these latter models we shall focus on two problems: (i) what restrictions on the structure of the X_j will lead to the same types of limiting distribution for the extremes which we obtained in the i.i.d. case, and (ii) by dropping the restrictions of (i), what new types of limiting distribution are obtained for the extremes? After completing the theoretical investigations, we shall discuss several applied models.

Throughout the chapter we shall use the notations introduced in Section 1.2.

3.1. A NEW LOOK AT FAILURE MODELS

Let us call a component of an equipment essential if its failure causes the failure of the equipment. Let X_j denote the random life length of the jth essential component and let n be the number of such components. Then the time to failure of the equipment is evidently $W_n = \min(X_1, X_2, \ldots, X_n)$. Assume that our use of the components is what they were produced for; that is, X_j is not affected by our particular equipment, but by its production procedure alone. Therefore, W_n for a particular equipment is determined at the factory (or factories) where the components were produced and by the selection procedure of purchasing the n (essential) components. In order to simplify the discussion (and this will turn out not to be a restriction at all, see Section 3.2), we consider the following case. Let all components be similar in nature; thus a particular equipment can be assembled by purchasing n items out of a large lot of N products. If there are $M = M(N,x)$ products in the whole lot which would fail in the time interval $(0,x)$, then $\{W_n \geq x\}$ if all components were purchased from the $N - M$ items, each of which lasts for at least x time units. Now, if at any time a fixed number N of items is available in a store but the unknown M is random, depending on the time of purchase, and if the n items we buy are selected at random, then by the total probability rule and by the basic formula for random selection from a finite population (the hypergeometric distribution)

$$P(W_n \geq x) = \sum_{t=0}^{N-n} \frac{\binom{N-t}{n}}{\binom{N}{n}} P(M=t), \qquad (1)$$

where the distribution $P(M=t)$ depends on x. The formula (1) is, in fact, a special form of the following result. If $\nu_n(N,x)$ denotes the number of components among those n which we purchased and which fail in less than x time units, then

$$P(\nu_n(N,x) = k) = \sum_{t=k}^{N-n} \frac{\binom{t}{k}\binom{N-t}{n-k}}{\binom{N}{n}} P(M=t). \qquad (2)$$

The surprising result of the next section is that the model we have just described is the most general one we can get for W_n. In other words, the distribution of W_n is always of the form (1), and (2) also represents the most general form of the distribution of the number of occurrences in a given sequence of events. Hence, the distribution of all order statistics can be reduced to (2) (see (34) and (35) of Section 1.4).

Since our model is very significant, let us look at it more closely.

First, why is our approach a new look at failure models? In all published works known to the present author the manufacturer of the components of the equipment is never considered. Instead, assumptions are made about the interrelation of those n components which are used in the equipment. We reverse this approach and assign all blame (or credit) to the manufacturer of the components. The interrelation of the X_j does not have any direct role in (1); the sole influencing factor is the distribution of M. Of course, indirectly, we also impose a structure on the X_j, since by (2) their joint distribution is determined by the distributions of $M = M(x)$ as x varies over the real numbers.

The following example illustrates this point. In the example we denote the life length of the N original components by $Y_1, Y_2, \ldots, Y_N$. They are assumed to have the same distribution function $F(x)$. As before, $X_1, X_2, \ldots, X_n$ are those Y's that we purchased for assembling the equipment, and $M = M(x)$ denotes the number of Y_j, $1 \leq j \leq N$, for which $\{Y_j < x\}$. (In the example, we shall use an identity for binomial coefficients. For this, see Exercise 3 of Chapter 1.)

Example 3.1.1. Let the multivariate distribution of the Y_j, $1 \leq j \leq N$, be such that, for all choices of the subscripts $1 \leq i_1 < i_2 < \ldots < i_k \leq N$,

$$P(Y_{i_1} < x, Y_{i_2} < x, \ldots, Y_{i_k} < x) = H_k(x)$$

depend only on k and x. (Notice that it is not required that the distribution of the vector $(Y_{i_1}, Y_{i_2}, \ldots, Y_{i_k})$ be independent of the subscripts $(i_1, i_2, \ldots, i_k)$. Such an assumption is made only when, for each t, $\{Y_{i_t} < x\}$ is considered with the same x.) Since $M(x)$ is the number of the events $\{Y_j < x\}$, $1 \leq j \leq N$, which occur, Theorem 1.4.1 yields

$$P(M(x) = t) = \sum_{k=0}^{N-t} (-1)^k \binom{k+t}{t} \binom{N}{k+t} H_{k+t}(x)$$

$$= \binom{N}{t} \sum_{k=0}^{N-t} (-1)^k \binom{N-t}{k} H_{k+t}(x).$$

A substitution into (1) thus leads to

$$P(W_n \geq x) = \sum_{t=0}^{N-n} \frac{\binom{N-t}{n}}{\binom{N}{n}} \binom{N}{t} \sum_{k=0}^{N-t} (-1)^k \binom{N-t}{k} H_{k+t}(x)$$

$$= \sum_{t=0}^{N-n} \binom{N-n}{t} \sum_{k=0}^{N-t} (-1)^k \binom{N-t}{k} H_{k+t}(x). \quad (3)$$

The case $H_k(x) = F^k(x)$, of course, yields independence, but other functions $H_k(x)$ result in dependent distributions, for the X_j, $1 \leq j \leq n$. ▲

3.2. THE SPECIAL ROLE OF EXCHANGEABLE VARIABLES

One widely used class of dependent random variables in probability theory is the exchangeable one.

Definition 3.2.1. We say that the random variables $U_1, U_2, \ldots, U_n$ are exchangeable if the distribution of the vector $(U_{i_1}, U_{i_2}, \ldots, U_{i_n})$ is identical to that of $(U_1, U_2, \ldots, U_n)$ for all permutations $(i_1, i_2, \ldots, i_n)$ of the subscripts $(1, 2, \ldots, n)$. Furthermore, an infinite sequence $U_1, U_2, \ldots$, of random variables is called exchangeable if each finite segment $U_1, U_2, \ldots, U_n$ constitutes exchangeable variables.

Notice that a subset of exchangeable variables is also exchangeable. It suffices to show this for a finite number of variables, since the concept of exchangeability for an infinite sequence is reduced to that for a finite sequence. In order to avoid complicated notations, let us first show this claim for a fixed n, $n=5$, say. We assume that $U_1, U_2, \ldots, U_5$ are exchangeable and we show that so are U_2, U_3, and U_5, say. Let j_1, j_2, and j_3 be an arbitrary permutation of 2, 3, and 5. By definition

$$P(U_1 < x_1, U_4 < x_2, U_{j_1} < x_3, U_{j_2} < x_4, U_{j_3} < x_5)$$
$$= P(U_1 < x_1, U_4 < x_2, U_2 < x_3, U_3 < x_4, U_5 < x_5).$$

Let x_1 and x_2 tend to $+\infty$. We get

$$P(U_{j_1} < x_3, U_{j_2} < x_4, U_{j_3} < x_5) = P(U_2 < x_3, U_3 < x_4, U_5 < x_5)$$

for all permutations (j_1, j_2, j_3) of $(2, 3, 5)$. This is, however, the definition of U_2, U_3, U_5 being exchangeable. The proof for arbitrary n and for an arbitrary subset is similar to the one above. One simply lets those x_j tend to $+\infty$ the subscripts of which are not contained in the subset in question.

It is evident that if $U_1, U_2, \ldots, U_n$ are i.i.d., then they are exchangeable. Hence, the concept of exchangeability extends that of independence if the random variables are known to be identically distributed.

The assumptions that a finite number of random variables are exchangeable or that they are a finite segment of an infinite sequence of exchangeable random variables are significantly different. The following random variables are exchangeable but they cannot be extended to more than six variables without violating exchangeability. Let U_1, U_2, and U_3 be random variables which take the values 0 and 1 only. Let $P(U_j = 1) = .5$, $1 \leq j \leq 3$, and $P(U_1 = U_2 = 1) = P(U_1 = U_3 = 1) = P(U_2 = U_3 = 1) = .2$. These are indeed exchangeable, since, for any real numbers x_1, x_2, and x_3, and for any permutation (i_1, i_2, i_3) of $(1,2,3)$,

$$P(U_{i_1} < x_1, U_{i_2} < x_2, U_{i_3} < x_3)$$

is either empty (whenever one of the x_j is negative), or the sure event (if all $x_j > 1$), or one of the following forms: $P(U_{i_t} = 0)$ or $P(U_{i_t} = 0, U_{i_s} = 0)$ or $P(U_1 = U_2 = U_3 = 0)$. But, by assumption, $P(U_{i_t} = 0) = 1 - P(U_{i_t} = 1) = .5$, whatever be i_t, and for any values of i_t and i_s,

$$P(U_{i_t} = 0, U_{i_s} = 0)$$

$$= P(U_{i_t} = 0) - P(U_{i_t} = 0, U_{i_s} = 1)$$

$$= P(U_{i_t} = 0) - [P(U_{i_s} = 1) - P(U_{i_t} = 1, U_{i_s} = 1)]$$

$$= .2.$$

Let us now assume that U_1, U_2, and U_3 were the first three in a sequence $U_1, U_2, \ldots, U_n$ of exchangeable variables. Then, in particular, $P(U_j = 1) = .5$ and $P(U_j = 1, U_t = 1) = .2$ for all $1 \leq j < t \leq n$. Let ν_n be the number of ones among $U_1, U_2, \ldots, U_n$. We can calculate the variance V of ν_n by Lemma 1.4.1. We get

$$0 \leq V = E(\nu_n^2) - E^2(\nu_n)$$

$$= E[\nu_n(\nu_n - 1)] + E(\nu_n) - E^2(\nu_n)$$

$$= 2S_{2,n} + S_{1,n} - S_{1,n}^2$$

$$= 2\binom{n}{2} \times 0.2 + 0.5n - (0.5n)^2$$

$$= 0.3n - 0.05n^2$$

—that is, $0.05n^2 \leq 0.3n$, and thus $n \leq 6$, as stated.

In the preceding paragraph we have seen for a special case that if $U_1, U_2, \ldots, U_n$ are indicator variables (that is, they take the values 0 and 1 only), then the concept of exchangeability is equivalent to the following property. For all $k \geq 1$ and for arbitrary subscripts $1 \leq j_1 < j_2 < \ldots < j_k \leq n$,

$$P(U_{j_1} = U_{j_2} = \ldots = U_{j_k} = 1) = P(U_1 = U_2 = \ldots = U_k = 1).$$

It is indeed true in the general case, and its proof is exactly the same as the one presented in the preceding paragraph. Hence, the details are omitted.

3.2 THE SPECIAL ROLE OF EXCHANGEABLE VARIABLES

Definition 3.2.2. A finite or infinite sequence of events is called exchangeable if their indicator variables are exchangeable. Therefore, Definition 3.2.1 and the remark in the preceding paragraph imply: the events $C_1, C_2, \ldots, C_n$ are exchangeable if, for all $k \geq 1$ and for arbitrary subscripts $1 \leq j_1 < j_2 < \ldots < j_k \leq n$,

$$P(C_{j_1} C_{j_2} \ldots C_{j_k}) = P(C_1 C_2 \ldots C_k).$$

Furthermore, an infinite sequence $C_1, C_2, \ldots$, of events is exchangeable if so are all segments $C_1, C_2, \ldots, C_n, n \geq 2$.

We now state our somewhat surprising result.

Theorem 3.2.1. *Let $A_1, A_2, \ldots, A_n$ be an arbitrary sequence of events. Let $\nu_n(A)$ denote the number of the events A_i which occur. Then there is a sequence $C_1, C_2, \ldots, C_n$ of exchangeable events such that, for $t \geq 0$,*

$$P(\nu_n(A) = t) = P(\nu_n(C) = t),$$

where $\nu_n(C)$ is the number of the C_j which occur.

Recall Section 1.4, where the significance of the distribution $P(\nu_n(A)=t)$ in the theory of order statistics was pointed out. In particular, $P(\nu_n(A)=0) = P(W_n \geq x)$ for the special choice $A_j = A_j(x) = \{X_j < x\}$, where $X_1, X_2, \ldots, X_n$ are random variables. Our theorem above thus implies that, instead of arbitrary random variables, we can always consider sequences when $\{X_j < x\}$ are exchangeable. A similar remark applies to Z_n as well as to arbitrary order statistics.

For proving Theorem 3.2.1, we first present a lemma.

Lemma 3.2.1. *A decreasing sequence $1 = \alpha_0 \geq \alpha_1 \geq \ldots \geq \alpha_n \geq 0$ of nonnegative real numbers can be associated with a sequence $C_1, C_2, \ldots, C_n$ of exchangeable events as*

$$\alpha_k = P(C_1 C_2 \ldots C_k) \tag{4}$$

if, and only if, the differences $\delta^r \alpha_k$ satisfy

$$\delta^r \alpha_{n-r} \geq 0, \quad 0 \leq r \leq n, \tag{5}$$

and

$$\sum_{r=0}^{n} \binom{n}{r} \delta^r \alpha_{n-r} = 1. \tag{6}$$

Here, $\delta^r \alpha_k$ is defined recursively by $\delta^0 \alpha_k = \alpha_k$, and

$$\delta \alpha_k = \alpha_k - \alpha_{k+1}, \quad \delta^r \alpha_k = \delta(\delta^{r-1} \alpha_k), \quad r \geq 1. \tag{7}$$

Proof. Let us first assume that the sequence α_k is defined by (4), where $C_1, C_2, \ldots, C_n$ are exchangeable. Then, evidently, $\alpha_k \geq \alpha_{k+1}$, and thus $\delta\alpha_k \geq 0$, $0 \leq k \leq n-1$. For showing (5) for $r > 1$, notice that

$$\delta^r \alpha_k = P(C_1 C_2 \cdots C_k C'_{k+1} \cdots C'_{k+r}), \quad k \leq n-r, \quad r \geq 1. \tag{8}$$

Indeed,

$$\delta\alpha_k = P(C_1 C_2 \cdots C_k) - P(C_1 C_2 \cdots C_{k+1}) = P(C_1 C_2 \cdots C_k C'_{k+1}),$$

which gives (8) for $r = 1$. Assume that (8) holds for a given r and for all $k \leq n - r$. Then

$$\delta^{r+1}\alpha_k = \delta(\delta^r \alpha_k)$$

$$= P(C_1 C_2 \cdots C_k C'_{k+1} \cdots C'_{k+r})$$

$$- P(C_1 C_2 \cdots C_{k+1} C'_{k+2} \cdots C'_{k+r+1}). \tag{9}$$

But, by exchangeability,

$$P(C_1 C_2 \cdots C_{k+1} C'_{k+2} \cdots C'_{k+r+1}) = P(C_1 C_2 \cdots C_k C'_{k+1} \cdots C'_{k+r} C_{k+r+1}),$$

and thus (9) yields

$$\delta^{r+1}\alpha_k = P(C_1 C_2 \cdots C_k C'_{k+1} \cdots C'_{k+r} C'_{k+r+1}).$$

Induction over r thus establishes (8), which, in turn, implies (5). Equation (6) also follows from (8) in view of the exchangeability assumption. By exchangeability, for an arbitrary permutation $(i_1, i_2, \ldots, i_n)$ of the integers $(1, 2, \ldots, n)$, (8) implies

$$\delta^r \alpha_{n-r} = P(C_{i_1} C_{i_2} \cdots C_{i_{n-r}} C'_{i_{n-r+1}} \cdots C'_{i_n}). \tag{10}$$

The events on the right hand side are mutually exclusive, and their union over all permutations and over all r equals the whole sample space. Thus summation in (10) leads to (6), which completes the proof of one part of the lemma.

Let us now assume that (5) and (6) hold. We shall construct a probability space and a sequence $C_1, C_2, \ldots, C_n$ of exchangeable events on it for which (4) is valid.

Let the sample space be the set $(0, 1, 2, \ldots, 2^n - 1)$ of consecutive integers. Write

$$x = \sum_{k=0}^{n-1} a_k 2^k, \quad a_k = 0 \text{ or } 1, \quad 0 \leq x \leq 2^n - 1. \tag{11}$$

3.2 THE SPECIAL ROLE OF EXCHANGEABLE VARIABLES

Define the probability measure on the set of all subsets of the sample space by assigning the value $\delta^r \alpha_{n-r}$ to $P(\{x\})$, where r equals the number of zeros among the a_k for x in (11). Let now C_j be the set of those x's, $0 \leq x \leq 2^n - 1$, for which $a_j = 1$. Then, for $1 \leq i_1 < \cdots < i_k \leq n$, the event $C_{i_1} C_{i_2} \cdots C_{i_k}$ is the set of x's for which $a_{i_1} = a_{i_2} = \cdots = a_{i_k} = 1$. The remaining $(n-k)$ a_t's can be either 0 or 1 and, since P is defined through the number of zeros among the a_t's,

$$P(C_{i_1} C_{i_2} \cdots C_{i_k}) = \sum_{r=0}^{n-k} \binom{n-k}{r} \delta^r \alpha_{n-r}. \tag{12}$$

Since the right hand side of (12) does not depend on $(i_1, i_2, \ldots, i_k)$, the events $C_1 C_2, \ldots, C_n$ are exchangeable. In addition, we can deduce from (7)

$$\sum_{r=0}^{n-k} \binom{n-k}{r} \delta^r \alpha_{n-r} = \alpha_k, \tag{13}$$

which, together with (12), yields (4). Therefore, the proof is completed if we show the validity of (13). We first note that induction over r and (7) yield

$$\delta^r \alpha_k = \sum_{j=0}^{r} (-1)^j \binom{r}{j} \alpha_{k+j}, \quad k + r \leq n. \tag{14}$$

(details are left as Exercise 4). Thus

$$\sum_{r=0}^{n-k} \binom{n-k}{r} \delta^r \alpha_{n-r} = \sum_{r=0}^{n-k} \sum_{j=0}^{r} (-1)^j \binom{n-k}{r} \binom{r}{j} \alpha_{n-r+j}$$

$$= \sum_{t=k}^{n} \alpha_t \sum\nolimits^{*} (-1)^j \binom{n-k}{r} \binom{r}{j},$$

where $\sum^*$ signifies summation over r and j such that $n - r + j = t$, $0 \leq j \leq r$ and $0 \leq r \leq n - k$. Hence, if $t = k$, $\sum^*$ contains a single term, namely, $j = 0$, and $r = n - k$; that is, the coefficient of α_k equals one. On the other hand, for $t > k$, $r = n - t + j \leq n - k$ results in $j \leq t - k$, and thus the coefficient of α_t now equals

$$\sum\nolimits^{*} (-1)^j \binom{n-k}{r} \binom{r}{j} = \sum_{j=0}^{t-k} (-1)^j \binom{n-k}{n-t+j} \binom{n-t+j}{j}$$

$$= \binom{n-k}{t-k} \sum_{j=0}^{t-k} (-1)^j \binom{t-k}{j} = 0.$$

Therefore, (13) follows, and the proof is completed. ▲

Proof of Theorem 3.2.1 We appeal to Theorem 1.4.1, which says that the binomial moments determine the distribution of the number of occurrences in a sequence of events. Let $S_{k,n}(A)$ and $S_{k,n}(C)$ denote the binomial moments of $\nu_n(A)$ and $\nu_n(C)$, respectively. For proving Theorem 3.2.1, we thus have to show that, for a given sequence $A_1, A_2, \ldots, A_n$ of events, we can find exchangeable events $C_1, C_2, \ldots, C_n$ such that, for all $k \geq 1, S_{k,n}(A) = S_{k,n}(C)$. But, for exchangeable events,

$$S_{k,n}(C) = \binom{n}{k}\alpha_k, \qquad k \geq 1,$$

where α_k is defined as at (4). Therefore, we have to show that the sequence $\alpha_0 = 1$ and

$$\alpha_k = \frac{S_{k,n}(A)}{\binom{n}{k}}, \qquad 1 \leq k \leq n, \tag{15}$$

is decreasing and that it can be associated with a sequence $C_1, C_2, \ldots, C_n$ of exchangeable events as at (4). The fact that α_k is decreasing can be shown by several methods of Chapter 1 (see Exercise 2 of Chapter 1). On the other hand, for showing that the sequence (15) can be associated with a sequence of exchangeable events as at (4), we apply Lemma 3.2.1. We got that the theorem is equivalent to the validity of (5) and (6) with the numbers defined in (15). Notice that (13) immediately yields (6), since $\alpha_0 = 1$. Hence, only the validity of (5) needs proof. The case $r = 0$ is, of course, trivial. Let $r \geq 1$. Then, by (14), (5) becomes

$$\sum_{j=0}^{r} (-1)^j \binom{r}{j} \frac{S_{n-r+j,n}(A)}{\binom{n}{n-r+j}} \geq 0, \qquad 1 \leq r \leq n. \tag{16}$$

In view of Lemma 1.4.1,

$$\frac{S_{n-r+j,n}(A)}{\binom{n}{n-r+j}} = E\left\{\frac{\nu_n(A)[\nu_n(A)-1]\cdots[\nu_n(A)-n+r-j+1]}{n(n-1)\cdots(r-j+1)}\right\}.$$

Hence, putting

$$\alpha_k^* = \frac{\nu_n(A)[\nu_n(A)-1]\cdots[\nu_n(A)-k+1]}{n(n-1)\cdots(n-k+1)}, \tag{17}$$

3.2 THE SPECIAL ROLE OF EXCHANGEABLE VARIABLES

by another appeal to (14) we can rewrite (16) as

$$E(\delta^r \alpha^*_{n-r}) \geq 0, \quad 1 \leq r \leq n. \tag{18}$$

For proving (18), consider the following simple selection problem for each point ω of the sample space. Let ω be fixed and thus $\nu_n(A)$ a well-defined integer. Let an urn contain n balls, out of which $\nu_n(A)$ are white. We select balls without replacement and let C_j^* be the event that the jth selection results in a white ball. Then the events C_j^* are exchangeable and the corresponding value as at (4) is α_k^* of (17) (see Exercise 5). Thus, by Lemma 3.2.1, $\delta^r \alpha^*_{n-r} \geq 0$ for each ω, which evidently implies (18). The theorem is established. ▲

We now give a general formula for the distribution of $\nu_n(C)$ for exchangeable events $C_1, C_2, \ldots, C_n$. This formula, when combined with Theorem 3.2.1, will be similar to (2). This will, therefore, justify the remark which followed (2).

Theorem 3.2.2. *Let $C_1, C_2, \ldots, C_n$ be exchangeable events. Then, with the previous meaning for $\nu_n(C)$, for $t \geq 0$,*

$$P(\nu_n(C) = t) = \sum_{T=t}^{N} \frac{\binom{T}{t} \binom{N-T}{n-t}}{\binom{N}{n}} P_T,$$

where $N \geq n$ is a fixed integer and P_T, $0 \leq T \leq N$, is a probability distribution.

Remark 3.2.1. With $N = n$, the statement is, of course, trivial, but it is significant whenever $N > n$. In particular, this representation is of great value in limit theorems when N is large compared with n. The proof, which follows, is constructive, and the actual sequence P_T, $1 \leq T \leq N$, is given in (19). To give lower estimates on possible values of N, one can use Lemma 3.2.1, although the actual calculations can be very demanding without a high-speed computer. The aim in presenting Theorem 3.2.2 is, however, not to gain computational advantages; rather, it is included for theoretical purposes: to prove our claim in Section 3.1 that general models are equivalent to the simple failure model discussed there. This is, in fact, expressed in Corollary 3.2.1. In addition, Theorem 3.2.2 will serve as the basis for the limit theorems of Section 3.4 as well as for the model of Section 3.5.

Before the proof, let us combine Theorems 3.2.1 and 3.2.2 into a corollary.

Corollary 3.2.1. Let $A_1, A_2, \ldots, A_n$ be arbitrary events. Let $\nu_n(A)$ be the number of those A's which occur. Then there is a distribution P_T on the integers $0 \leq T \leq N$ with some integer $N \geq n$ such that

$$P(\nu_n(A) = t) = \sum_{T=t}^{N} \frac{\binom{T}{t}\binom{N-T}{n-t}}{\binom{N}{n}} P_T.$$

This corollary is immediate from the mentioned theorems. Consequently, only Theorem 3.2.2 needs proof.

Proof of Theorem 3.2.2. Let $N \geq n$ be an integer such that $C_1, C_2, \ldots, C_n$ can be enlarged to a set $C_1, C_2, \ldots, C_N$ of exchangeable events. Then, by (10),

$$P(\nu_N(C) = T) = \binom{N}{T} \delta^{N-T} \alpha_T = P_T, \text{ say.} \tag{19}$$

We now recalculate $P(C_{i_1} C_{i_2} \cdots C_{i_k})$ for $1 \leq i_1 < i_2 < \cdots < i_k$ in terms of P_T. By the total probability rule

$$P(C_{i_1} \cdots C_{i_k}) = \sum_{T=k}^{N} P(C_{i_1} \cdots C_{i_k} | \nu_N(C) = T) P_T. \tag{20}$$

On the other hand, by another appeal to (10) and by (19),

$$P(C_{i_1} \cdots C_{i_k} | \nu_N(C) = T) = \frac{\binom{N-k}{T-k} \delta^{N-T} \alpha_T}{\binom{N}{T} \delta^{N-T} \alpha_T} = \frac{\binom{N-k}{T-k}}{\binom{N}{T}}$$

$$= \frac{T(T-1) \cdots (T-k+1)}{N(N-1) \cdots (N-k+1)}, \quad T \geq k. \tag{21}$$

This is the familiar formula in connection with sampling without replacement from a population of N items which contain T marked ones (see Exercise 5). Hence, (20) and (21) yield the claimed formula of Theorem 3.2.2. The proof is complete. ▲

3.3. PREPARATIONS FOR LIMIT THEOREMS

We would like to determine the limiting distribution of $(Z_n - a_n)/b_n$ and $(W_n - c_n)/d_n$ with suitable constants a_n, $b_n > 0$, c_n, $d_n > 0$, when the random variables $X_1, X_2, \ldots, X_n$ are not i.i.d. We would like to impose on their

3.3 PREPARATIONS FOR LIMIT THEOREMS

interdependence as little restriction as possible. However, in order to achieve meaningful limit theorems, restrictions are necessary. The following examples will indicate the kind of restrictions needed for nontrivial limit theorems.

Example 3.3.1. Let $X_1, X_2, \ldots, X_n$ be n identical repetitions of a random variable X with distribution function $F(x)$. Then $Z_n = W_n = X$, and thus the arbitrary $F(x)$ would serve as limiting distribution for both Z_n and W_n. Hence, some assumptions are definitely needed. ▲

Example 3.3.2. Let $Y_1, Y_2, \ldots, Y_n$ be i.i.d. standard normal variates. Let U be an arbitrary random variable with distribution function $F(x)$. Let $X_j = U + Y_j$, $1 \le j \le n$. Then

$$Z_n = U + \max(Y_1, Y_2, \ldots, Y_n).$$

However, by Section 2.3.2, for any $\varepsilon > 0$, as $n \to +\infty$,

$$\lim P\left(|\max(Y_1, Y_2, \ldots, Y_n) - (2\log n)^{1/2}| \ge \varepsilon\right) = 0.$$

Therefore, by Lemma 2.2.1, as $n \to +\infty$,

$$\lim P\left(Z_n - (2\log n)^{1/2} < x\right) = P(U < x) = F(x).$$

Since $F(x)$ is arbitrary, a meaningful model for the asymptotic theory of extremes should not combine such X_j's with other structures. ▲

Example 3.3.3. Let $F(x)$ be an arbitrary distribution function. Let $r(t)$, $t = 1, 2, \ldots$, be a probability distribution with $r(t) > 0$ for all t. Let $X_1, X_2, \ldots$, be independent random variables and let the distribution function of X_j be $F^{r(j)}(x)$. Then, by the basic formula (5) of Chapter 1,

$$P(Z_n < x) = F^{r(1)}(x) F^{r(2)}(x) \cdots F^{r(n)}(x) = F^{s(n)}(x),$$

where $s(n) = r(1) + r(2) + \cdots + r(n)$. Hence, as $n \to +\infty$,

$$\lim P(Z_n < x) = F(x),$$

which was again arbitrary. ▲

Each example can suitably be modified to lead to warnings concerning W_n.

Let us analyze the reasons why arbitrary distributions were obtained as limiting distribution of Z_n. More importantly, what conditions would exclude these structures?

The trivial model of the first example is evidently excluded if it is guaranteed that W_n and Z_n are substantially different. This is achieved by the simplest assumptions.

In order to understand the structure of the second example, let us look at all extremes of $X_1, X_2, \ldots, X_n$. Evidently, all order statistics of the X_j can be obtained from the order statistics of the Y_j by adding U to the latter. Since, by Theorem 2.8.1 and by Section 2.3.2, as $n \to +\infty$,

$$P\left(|Y_{n-k:n} - (2\log n)^{1/2}| \geq \varepsilon\right) \to 0, \qquad (22)$$

where $\varepsilon > 0$ is arbitrary and k is a fixed integer, again by Lemma 2.2.1, as $n \to +\infty$,

$$\lim P\left(X_{n-k:n} < (2\log n)^{1/2} + x\right) = P(U < x) = F(x). \qquad (23)$$

That is, for all upper extremes, both the normalizing constants $a_n = (2\log n)^{1/2}$, $b_n = 1$, and the limiting distribution $F(x)$ are independent of k. We can go even further to see that (22), and thus (23) as well, holds for $k = k(n)$, which tends to infinity very slowly with n (we do not go into detail, but the interested reader can easily deduce this fact from the formulas of Section 2.8). Therefore, we exclude this structure from a model by assuming that there are at least two upper (lower) extremes which are significantly different. We can also exclude it by assuming that $X_{n-k:n}$ has different asymptotic properties according as k is fixed or k goes to infinity with n.

Turning to the last example, we note that the contribution of the distribution of X_j to the distribution of Z_n did not change with n. Hence, the effect of each X_j on Z_n was permanent as n increased to infinity. Such a situation will be excluded by the following type of assumption. In some models, we shall assume that the normalizing constants a_n and b_n in $(Z_n - a_n)/b_n$ are such that $F_j(a_n + b_n x)$ tends to one uniformly in j as $n \to +\infty$. Here, $F_j(x)$, as in general, denotes the distribution function of X_j.

The assumptions, specified in the preceding paragraphs, will of course exclude several models, not just the ones given in the three examples. For example, assuming that all upper extremes have limiting distributions with suitable normalizing constants automatically excludes models where $(Z_n - a_n)/b_n$ converges weakly, but with no normalizing constants a_n^* and $b_n^* > 0$ would $(X_{n-1:n} - a_n^*)/b_n^*$ converge weakly. Therefore, we shall not aim at obtaining a single general model. Rather, we shall specify general structures, which may provide overlapping models, for which both the mathematical theory and its applicability are interesting and important.

3.4. A LIMIT THEOREM FOR MIXTURES

Let $f_k(n, M, N)$ signify the hypergeometric distribution

$$\frac{\binom{M}{k}\binom{N-M}{n-k}}{\binom{N}{n}}, \qquad k=0,1,\ldots,\min(M,n).$$

Then Corollary 3.2.1 can be restated as follows: the distribution of the number of occurrences in a given sequence of events is always a mixture of $f_k(n, M, N)$ and an arbitrary discrete distribution for M. In an integral form, Corollary 3.2.1 becomes

$$P(\nu_n(A)=t) = \int_0^N f_t(n,y,N)\,dU_n(y), \qquad (24)$$

where $U_n(y)$, for each $n=1,2,\ldots$, is a distribution function with positive increamments over such intervals only which contain nonnegative integers and $U_n(N)=1$.

We shall prove the following important result.

Theorem 3.4.1. *With the notations of (24), for each t, as both n and N/n tend to infinity,*

$$\lim P(\nu_n(A)=t) = g_t$$

exists and $\{g_t\}$ is a distribution if, and only if, $U_n(Ny/n)$ converges weakly to a distribution function $U(y)$. The limits g_t satisfy

$$g_t = \frac{1}{t!}\int_0^{+\infty} y^t e^{-y}\,dU(y). \qquad (25)$$

In order to shorten the proof, we separate one part of it as a lemma.

Lemma 3.4.1. *Let g_t, $t=0,1,\ldots$, be a distribution and let $U(y)$ be a distribution function. Assume that (25) holds. Then the sequence $\{g_t\}$ uniquely determines $U(y)$.*

Proof. Let z be a real number with $z \leq 1$. Then by (25)

$$M(z) = \sum_{t=0}^{+\infty} g_t z^t = \int_0^{+\infty}\left[e^{-y}\sum_{t=0}^{+\infty}\frac{(zy)^t}{t!}\right]dU(y)$$

$$= \int_0^{+\infty} e^{-y(1-z)}\,dU(y), \qquad z \leq 1,$$

where the interchange of the integration and summation is justified by the dominated convergence theorem (Appendix I). In the final integral, let us substitute $u=e^{-y}$ and define $V(u)=1-U(y)$ at points of continuity of $U(y)$. Then

$$M(1-s)=\int_0^1 u^s dV(u), \quad s\geq 0.$$

That is, $M(1-s)$ is the sth moment of $V(u)$. Since the moments uniquely determine $V(u)$ ($V(u)$ is the distribution function of a bounded random variable; see Appendix II), we now get our lemma by the following chain. The sequence $\{g_t\}$ determines $M(z)$, $M(1-s)$ determines $V(u)$, which finally uniquely determines $U(y)$, which was to be proved. ▲

The following lemma is well known in elementary probability theory.

Lemma 3.4.2. *As* $n\to+\infty$, $M\to+\infty$, *and* $nM/N\to a>0$,

$$\lim f_k(n,M,N) = \frac{a^k e^{-a}}{k!}, \quad k=0,1,2,\ldots .$$

Proof. Let us introduce the notation $b_m(x) = x(x-1)\cdots(x-m+1)$. Then

$$f_k(n,M,N) = \frac{b_k(n)b_k(M)b_{n-k}(N-M)}{k! b_n(N)}.$$

Let us write in the denominator $b_n(N) = b_k(N) b_{n-k}(N-k)$. Next we observe that, under the assumptions on the passage to limit, for fixed k,

$$\lim \frac{b_k(n)b_k(M)}{b_k(N)} = a^k.$$

Hence, it remains to establish

$$\lim \frac{b_{n-k}(N-M)}{b_{n-k}(N-k)} = e^{-a}. \tag{26}$$

We start with

$$\frac{b_{n-k}(N-M)}{b_{n-k}(N-k)} = \left(\frac{N-M}{N-k}\right)^{n-k} \prod_{j=1}^{n-k-1} \frac{1-j/(N-M)}{1-j/(N-k)}.$$

3.4 A LIMIT THEOREM FOR MIXTURES

From elementary calculus, under our assumptions (k is fixed)

$$\lim\left(\frac{N-M}{N-k}\right)^{n-k} = \lim\left(\frac{N-M}{N-k}\right)^{n}$$

$$= \lim\left(\frac{N}{N-k}\right)^{n}\left(1-\frac{M}{N}\right)^{n}$$

$$= \lim\left(1-\frac{k}{N}\right)^{-n}\left(1-\frac{a}{n}\right)^{n}$$

$$= e^{-a}\lim\left(1-\frac{M}{N}\frac{k}{M}\right)^{-n} = e^{-a}.$$

On the other hand, by the Taylor expansion,

$$\log(1-y) = -y + \tau y^2, \qquad |\tau| \leq 1 \text{ for } |y| \leq \tfrac{1}{2},$$

one can easily deduce that

$$\prod_{j=1}^{n-k-1}\frac{1-j/(N-M)}{1-j/(N-k)} \sim \exp\left(-\frac{cMn^2}{N^2}\right) \to 1.$$

Hence, (26) now follows, and the proof is completed. ▲

We can now turn to the proof of Theorem 3.4.1.

Proof of Theorem 3.4.1. Let us first assume that $U_n(Ny/n)$ converges weakly to a proper distribution function $U(y)$. Let B be a continuity point of $U(y)$ such that $1 - U(B) < \varepsilon$. Then

$$\int_B^{+\infty}\frac{1}{t!}y^t e^{-y}dU(y) \leq 1 - U(B) < \varepsilon \tag{27a}$$

and, for n sufficiently large,

$$\int_B^{+\infty} f_t\left(n,\frac{Ny}{n},N\right)dU_n\left(\frac{Ny}{n}\right) \leq 1 - U_n\left(\frac{NB}{n}\right) < 2(1-U(B)) < 2\varepsilon. \tag{27b}$$

Let us replace in (24) y by Ny/n. We get

$$P(\nu_n(A) = t) = \int_0^n f_t\left(n,\frac{Ny}{n},N\right)dU_n\left(\frac{Ny}{n}\right). \tag{28}$$

For estimating the difference $P(\nu_n(A)=t)-g_t$, where g_t is defined by (25), we first estimate

$$\int_0^B f_t\left(n,\frac{Ny}{n},N\right)dU_n\left(\frac{Ny}{n}\right)-\int_0^B \frac{1}{t!}y^t e^{-y}dU(y).$$

By the triangle inequality

$$\left|\int_0^B f_t\left(n,\frac{Ny}{n},N\right)dU_n\left(\frac{Ny}{n}\right)-\int_0^B \frac{1}{t!}y^t e^{-y}dU(y)\right|$$

$$\leq \left|\int_0^B f_t\left(n,\frac{Ny}{n},N\right)dU_n\left(\frac{Ny}{n}\right)-\int_0^B \frac{1}{t!}y^t e^{-y}dU_n\left(\frac{Ny}{n}\right)\right|$$

$$+\left|\int_0^B \frac{1}{t!}y^t e^{-y}dU_n\left(\frac{Ny}{n}\right)-\int_0^B \frac{1}{t!}y^t e^{-y}dU(y)\right|. \quad (29)$$

Lemma 3.4.2 guarantees that, as n and N/n tend to infinity,

$$f_t\left(n,\frac{Ny}{n},N\right)\to \frac{1}{t!}y^t e^{-y},$$

and, in fact, this convergence is uniform over the finite interval, $0\leq y\leq B$ (see Exercise 6). Therefore, for arbitrary $\varepsilon>0$ and for sufficiently large n,

$$\left|\int_0^B \left[f_t\left(n,\frac{Ny}{n},N\right)-\frac{1}{t!}y^t e^{-y}\right]dU_n\left(\frac{Ny}{n}\right)\right|\leq \varepsilon U_n\left(\frac{NB}{n}\right)\leq \varepsilon. \quad (30)$$

In order to estimate the second difference on the right hand side of (29), we construct Riemann sums which are close to the integrals there. Let T be a fixed number. Let $0=y_0<y_1<\cdots<y_T=B$ be continuity points of $U(y)$. Furthermore, let T and the y_j be such that

$$\left|\int_0^B \frac{1}{t!}y^t e^{-y}dU_n\left(\frac{Ny}{n}\right)-\sum_{j=1}^T \frac{1}{t!}y_j^t e^{-y_j}\left[U_n\left(\frac{Ny_j}{n}\right)-U_n\left(\frac{Ny_{j-1}}{n}\right)\right]\right|<\varepsilon$$

and

$$\left|\int_0^B \frac{1}{t!}y^t e^{-y}dU(y)-\sum_{j=1}^T \frac{1}{t!}y_j^t e^{-y_j}\left[U(y_j)-U(y_{j-1})\right]\right|<\varepsilon.$$

Since, by assumption, $U_n(Ny_j/n) \to U(y_j)$, $0 \leq j \leq T$, as $n \to +\infty$, the two Riemann sums are closer to each other than ε for all n sufficiently large. Thus, once again by the triangle inequality, the absolute value of the difference of the integrals is smaller than 3ε. Combining this fact with (30), the left hand side of (29) becomes smaller than 4ε for all large n. Therefore, in view of (27) and (28),

$$|P(\nu_n(A)=t)-g_t| < \left| \int_0^B f_t dU_n - \int_0^B \frac{1}{t!} y^t e^{-y} dU(y) \right|$$

$$+ \int_B^{+\infty} f_t dU_n + \int_B^{+\infty} \frac{1}{t!} y^t e^{-y} dU(y) < 7\varepsilon,$$

where the variables of f_t and U_n are as in (28). This completes the proof of the sufficiency of the theorem.

We now turn to the converse. We assume that, as $n \to +\infty$, $P(\nu_n(A)=t)$ converges to g_t, $t=0,1,2,\ldots$, which sequence forms a distribution. Then, starting with (28), we select a subsequence $n(m)$, $m=1,2,\ldots$, of the integers for which $U_{n(m)}(Ny/n(m))$ converges weakly to an extended distribution function $U(y)$ (such a subsequence exists by the compactness of distribution functions, see Appendix II). Then, by repeating the first part of the theorem for the subsequence $n(m)$, with the exception that we choose B so that $U(+\infty) - U(B) < \varepsilon$, we get

$$g_t = \frac{1}{t!} \int_0^{+\infty} y^t e^{-y} dU(y). \tag{31}$$

The sequence g_t is a distribution. Thus, summing the two sides above with respect to t, we get $U(+\infty)=1$, that is, $U(y)$ is a distribution function. Now, if $U_n(Ny/n)$ did not converge weakly, then we could select two subsequences $n(m)$ and $n(s)$ such that $U_{n(m)}(Ny/n(m))$ would converge weakly to $U(y)$ and $U_{n(s)}(Ny/n(s))$ to another distribution function $U^*(y)$. But formula (31) would then hold both with $U(y)$ and $U^*(y)$, which contradicts Lemma 3.4.1. This completes the proof of Theorem 3.4.1. ▲

In the proof, it was not significant that f_t represented the hypergeometric distribution. Several distributions could have been chosen for f_t. In particular, the replacement of f_t by the binomial distribution does not require any change in the proof. We thus have the following results.

Theorem 3.4.2.a. *Let $U_n(y)$ be a sequence of distribution functions with $U_n(0)=0$ and $U_n(1+0)=1$. Then, for each t, as $n \to +\infty$,*

$$\lim \int_0^1 \binom{n}{t} p^t (1-p)^{n-t} dU_n(p) = g_t$$

exists, and $\{g_t\}$ is a distribution if, and only if, $U_n(p/n)$ converges weakly to a distribution function $U(y)$. The limits g_t satisfy (25).

Theorem 3.4.2.b. *Let $U_p(y)$, for each p with $0 \leq p \leq 1$, be a distribution function and such that $U_p(y)$ can have positive increments over only those intervals which contain nonnegative integers. Let $p(s)$ be a sequence of numbers with $0 \leq p(s) \leq 1$ and $p(s) \to 0$ as $s \to +\infty$. Then, as $s \to +\infty$,*

$$\lim \int_0^{+\infty} \binom{y}{t} p(s)^t [1-p(s)]^{y-t} dU_{p(s)}(y) = g_t$$

exists for each t, and $\{g_t\}$ is a distribution if, and only if, $U_{p(s)}(y/p(s))$ converges weakly to a distribution function $U(y)$. The limits g_t satisfy (25).

Notice that in each of the three theorems, when the limits g_t exist, they satisfy the formula (25). Therefore, Lemma 3.4.1 implies the following interesting fact. Since the sequence g_t uniquely determines $U(y)$ in (25), a specific distribution g_t can be obtained only by a well-defined $U(y)$. We illustrate this in the following two corollaries.

Corollary 3.4.1. *With the notations of (24), as n and N/n tend to infinity,*

$$\lim P(\nu_n(A) = t) = \frac{a^t e^{-a}}{t!}, \qquad a > 0, \quad t = 0, 1, 2, \ldots, \tag{32}$$

if, and only if, $U_n(Ny/n)$ converges to one for $y > a$ and to zero for $y \leq a$.

Proof. By Theorem 3.4.1, the left hand side of (32) has a limit for each t, and the limiting sequence is a distribution if, and only if, $U_n(Ny/n)$ converges weakly to a distribution function $U(y)$. Furthermore, (25) holds. Since, with

$$U(y) = \begin{cases} 1 & \text{if } y > a, \\ 0 & \text{otherwise,} \end{cases} \tag{33}$$

g_t of (25) does become the right hand side of (32), Lemma 3.4.1 says that there is no other $U(y)$ with which this same relation would hold. The proof is complete. ▲

Corollaries of the above nature can be produced by working backward. We start with a function $U(y)$, we compute g_t by (25), and these limits g_t can be obtained only by the weak convergence of $U_n(Ny/n)$ to the function $U(y)$ that we started with. For future reference we record one more specific case. This is obtained by working with $U(y) = 1 - e^{-ay}$, $a > 0$, $y \geq 0$.

Corollary 3.4.2. *With the notations of (24), as n and N/n tend to infinity,*

$$\lim P(\nu_n(A)=t) = \frac{a}{(1+a)^{t+1}}, \qquad a>0, \quad t=0,1,2,\ldots,$$

if, and only if, $U_n(Ny/n)$ converges to $U(y)=1-e^{-ay}$, $y \geq 0$.

3.5. A THEORETICAL MODEL

Exploiting the results of the previous section, we shall describe a model of great theoretical value. In particular, we shall obtain a large class of possible limiting distributions for the extremes in nontrivial situations (see Section 3.3). Therefore, if the assumptions of the special models of the forthcoming sections (Sections 3.6–3.10) are not justified in a given situation, we can appeal to the general model of the present section and try to fit the data to one of the extreme value distributions to be obtained.

As remarked earlier, we use the notations of Section 1.2. Hence $X_1, X_2, \ldots, X_n$ denote the basic random variables and $F_j(x) = P(X_j < x)$. Furthermore, $H_n(x)$ and $L_n(x)$ denote the distribution of the maximum Z_n and the minimum W_n, respectively. Finally, a frequently used notation will be

$$F_{i(k)}(x_1, x_2, \ldots, x_k) = P(X_{i_s} < x_s, 1 \leq s \leq k) \tag{34}$$

and

$$G_{i(k)}(x_1, x_2, \ldots, x_k) = P(X_{i_s} \geq x_s, 1 \leq s \leq k), \tag{35}$$

where $1 \leq i_1 < i_2 < \cdots < i_k \leq n$ are given integers. When all variables $x_1 = x_2 = \cdots = x_k = x$, then we abbreviate

$$F_{i(k)}(x_1, x_2, \ldots, x_k) = F^*_{i(k)}(x), \qquad G_{i(k)}(x_1, x_2, \ldots, x_k) = G^*_{i(k)}(x). \tag{36}$$

Evidently, $F_{i(1)}(x_1) = F^*_{i(1)}(x_1) = F_{i_1}(x_1)$ and $G_{i(1)}(x_1) = G^*_{i(1)}(x_1) = 1 - F_{i_1}(x_1)$. Since the vector $i(n) = (1, 2, \ldots, n)$ is unique, we also use $F^*_{i(n)} = F^*_n$ and $G^*_{i(n)} = G^*_n$. Notice that $H_n(x) = F^*_n(x)$ and $1 - L_n(x) = G^*_n(x)$.

We now describe the model for the upper extremes.

With the above notations, define $S_{0,n}(x) = 1$ and

$$S_{k,n}(x) = \sum_{1 \leq i_1 < \cdots < i_k \leq n} G^*_{i(k)}(x), \qquad k \geq 1. \tag{37}$$

Thus, $S_{k,n}(x)$ is the kth binomial moment of the number $\nu_n(x)$ of those

$X_j, 1 \leqslant j \leqslant n$, which satisfy $X_j \geqslant x$. We introduce additional theoretical random variables $X_{n+1}, X_{n+2}, \ldots, X_N$ with the property that, with this enlarged set X_j, $1 \leqslant j \leqslant N$, of random variables, for $k \geqslant 1$,

$$S_{k,N}(x) = \binom{N}{k} \frac{S_{k,n}(x)}{\binom{n}{k}}. \tag{38}$$

This formula actually imposes conditions on the multivariate distribution on cubes of the enlarged set X_j, $1 \leqslant j \leqslant N$—that is, when all variables in the distribution functions equal x (see Remark 3.5.1). Finally, in order to avoid situations like the one in Example 3.3.2 (see (23) and the discussion thereafter), we make the following assumption on the normalizing constants a_n and $b_n > 0$.

Let a_n and $b_n > 0$ be such that, as $n \to +\infty$,

$$\lim P(X_{n-k+1:n} < a_n + b_n x) = E_k(x) \tag{39}$$

exists for each fixed k, and such that, for all $x \geqslant x_0$, there is at least one k with $0 < E_k(x) < 1$. Furthermore, if $k = k(n) \to +\infty$ with n,

$$\lim_{n=+\infty} P(X_{n-k(n):n} < a_n + b_n x) = 1, \qquad x \geqslant x_0. \tag{40}$$

We shall refer to this condition as the normalizing constants a_n and $b_n > 0$ being characteristic to the upper extremes.

Notice that we did not require that the functions $E_k(x)$ be distribution functions. They are, however, distribution functions in the extended sense; that is, $0 \leqslant E_k(x) \leqslant 1$, but the equalities are not necessarily achieved. It does not cause any problem or inconvenience later, since, in all statements where they occur, actual formulas will be given for $E_k(x)$. Hence, it is always clear when they become proper distribution functions.

We now have the following result.

Theorem 3.5.1. *Let $X_1, X_2, \ldots, X_n$ be such that they admit additional random variables $X_{n+1}, X_{n+2}, \ldots, X_N$ with distributions which satisfy (38); furthermore, $N/n \to +\infty$ with n. Then there are normalizing constants a_n and $b_n > 0$, which are characteristic to the upper extremes (so that (39) and (40) hold) if, and only if, as $n \to +\infty$, $P[\nu_N(a_n + b_n x) < (N/n)y]$ converges weakly to a distribution function $U(y) = U(y; x)$. The limiting distributions $E_k(x)$ of (39) satisfy the relation*

$$E_k(x) = \sum_{t=0}^{k-1} \frac{1}{t!} \int_0^{+\infty} y^t e^{-y} dU(y; x), \qquad k \geqslant 1. \tag{41}$$

Remark 3.5.1. The theorem is theoretical only in the sense that it requires that the original variables X_j, $1 \leq j \leq n$, be extendible into a larger set X_j, $1 \leq j \leq N$, with some distributional requirements (38), and such that N be large compared with n ($N/n \to +\infty$ with n). Results which would guarantee such a possibility are not readily available in the literature (of course, if the original X_j's are i.i.d., or if they form a segment of a large set of exhangeable random variables, then (38) evidently holds). One can construct very complicated sufficient conditions for it, by applying Lemma 3.2.1, but they would not simplify the model. We therefore adopt the present approach, leaving the condition of extendibility untouched and labeling the model as theoretical. The fact that the choice of a_n and $b_n > 0$ is given in terms of the distribution of $\nu_N(x)$ further justifies the label "theoretical," although, with the binomial moments $S_{k,N}(x)$ specified in (38), there is a fast developing theory on $\nu_N(x)$ and its distribution. In addition, we shall have several cases when a_n and $b_n > 0$ can easily be computed. In spite of these remarks, the theorem is a valuable contribution even from the applied scientist's point of view. Namely, (41) provides a general class of possible limit distributions for the upper extremes.

Remark 3.5.2. If we replace $G_{i(k)}^*(x)$ in (37) by $F_{i(k)}^*(x)$, then $S_{k,n}(x)$ is the kth binomial moment of the number $\nu_n(x)$ of those X_j, $1 \leq j \leq n$, which are less than x. Furthermore, if we replace $X_{n-k+1:n}$ and $X_{n-k(n):n}$ by $X_{k:n}$ and $X_{k(n):n}$, respectively, then (39) and (40) lead to the definition of normalizing constants characteristic to the lower extremes. With these new terms, but with no other modification, Theorem 3.5.1 gives a general limit theorem for the distribution of the lower extremes. In particular, formally (41) remains unchanged (the actual distribution functions $U(y) = U(y;x)$ will, of course, be different).

Proof of Theorem 3.5.1. We shall apply Theorem 3.4.1 with the events $A_j = A_j(x) = \{X_j \geq x\}$, $1 \leq j \leq n$. With this choice, $\nu_n(A)$ of Section 3.4 is the present $\nu_n(x)$. We next observe

$$P(X_{n-k+1:n} < x) = \sum_{t=0}^{k-1} P(\nu_n(x) = t).$$

Therefore, the convergence of $P(X_{n-k+1:n} < a_n + b_n x)$ for each $k \geq 1$ is equivalent to the convergence of $P(\nu_n(a_n + b_n x) = t)$ for each $t \geq 0$. If these latter limits are denoted by $g_t = g_t(x)$, then the fact that $\{g_t\}$ be a distribution is equivalent to (40). Therefore, by Theorem 3.4.1, our theorem is proved by recalling that $U_n(y)$ in (24) is in fact $P(\nu_N(a_n + b_n x) < y)$. Indeed, (24) is a reformulation of Corollary 3.2.1, where $U_n(y)$ represents the discrete distribution $\{P_T\}$. Thus, by (19), $U_n(y)$ is the distribution function of $\nu_N(a_n + b_n x)$. This completes the proof. ▲

As remarked, the difficulty of appying Theorem 3.5.1 lies in the extendibility requirement of $X_1, X_2, \ldots, X_n$ into a larger set of N variables for which (38) holds. This, however, evidently holds if the X_j are i.i.d., or if they are a segment of an infinite sequence of exchangeable variables. Let us work out the other conditions of Theorem 3.5.1 for these special cases.

Example 3.5.1. Let $X_1, X_2, \ldots, X_n$ be i.i.d. with common distribution function $F(x)$. Then, it is well known in the foundations of probability theory that, for any N, the original set, can be extended to a larger set X_j, $1 \leq j \leq N$, which are still i.i.d. Hence, (38) holds, and the assumption $N/n \to +\infty$ with n can also be satisfied. Since

$$P(\nu_N(x)=t) = \binom{N}{t}[1-F(x)]^t F^{N-t}(x),$$

by the Chebishev inequality, for any $\varepsilon > 0$,

$$P\left(|\nu_N(a_n + b_n x) - N[1 - F(a_n + b_n x)]| \geq \frac{N\varepsilon}{n}\right) \leq \frac{n^2[1 - F(a_n + b_n x)]}{N\varepsilon^2}.$$

Therefore, if a_n and $b_n > 0$ are such that, as $n \to +\infty$,

$$\lim n[1 - F(a_n + b_n x)] = -\log H(x), \quad \text{say}, \tag{42}$$

is finite, then $P[\nu_N(a_n + b_n x) < (N/n)y]$ converges weakly to $U(y)$, which is degenerate at $y = -\log H(x)$. Hence, by Theorem 3.5.1, the upper extremes, when normalized by a_n and b_n, converge weakly. Their asymptotic distributions are $E_k(x), k \geq 1$, given in (41). With the special $U(y) = U(y; x)$ obtained above

$$E_k(x) = H(x) \sum_{t=0}^{k-1} \frac{1}{t!} [-\log H(x)]^t, \quad k \geq 1. \tag{43}$$

Condition (42) and formula (43) are, of course, the familiar expressions from Chapter 2. ▲

Example 3.5.2. Let $X_1, X_2, \ldots, X_n$ be a segment from an infinite sequence of exchangeable random variables. Then, by definition, we can extend them to X_j, $1 \leq j \leq N$, for any N, without violating exchangeability. In particular, (38) holds for all x and N. For this example, let us further assume

$$G^*_{i(k)}(x) = \int_0^{+\infty} (1 - e^{-1/\lambda x})^k dV(\lambda), \quad k \geq 1, \quad x > 0,$$

where $V(\lambda)$ is a continuous distribution function. Then, by Theorem 1.4.1,

$$P(\nu_N(x)=t)=\binom{N}{t}\int_0^{+\infty}(1-e^{-1/\lambda x})^t e^{-(N-t)/\lambda x}dV(\lambda), \qquad t\geq 0,$$

which, by the substitution $s=n(1-e^{-1/\lambda x})$, becomes

$$P(\nu_N(x)=t)=\binom{N}{t}\int_0^n \left(\frac{s}{n}\right)^t \left(1-\frac{s}{n}\right)^{N-t} dV^*(s;x) \qquad (44a)$$

with

$$V^*(s;x) = 1 - V\left[-\frac{1}{x\log(1-s/n)}\right]. \qquad (44b)$$

Now, since $\log(1-s/n)\sim -s/n$, for fixed $s>0$ and $x>0$, as $n\to +\infty$,

$$V^*(s;nx)\to 1 - V\left(\frac{1}{sx}\right). \qquad (45)$$

With one more appeal to the Chebishev inequality for the binomial distribution, we thus get from (44) and (45),

$$P\left(\nu_N(nx) < \frac{N}{n}y\right) \to 1 - V\left(\frac{1}{yx}\right),$$

as n and N/n tend to infinity. Therefore, in view of Theorem 3.5.1, (45) implies that the upper extremes, if divided by n, converge weakly. We can again compute the limiting distributions by (41). In particular,

$$\lim_{n=+\infty} P(Z_n < nx) = \int_0^{+\infty} e^{-y} d\left[1 - V\left(\frac{1}{yx}\right)\right]$$

$$= \int_0^{+\infty} (e^{-1/x})^y d\left[1 - V\left(\frac{1}{y}\right)\right]. \qquad \blacktriangle$$

We shall see that the general asymptotic theory of extremes for segments of infinite sequences of exchangeable variables goes along the line of Example 3.5.2. However, we first have to prove that $G^*_{i(k)}(x)$ can always be represented as the kth moment of a bounded random variable. Because of its significance in Bayesian statistics, we devote the next section to this theory.

We conclude the present section with the remark that, although our model was labeled as theoretical, there are several other possibilities for its direct application. In particular, if $X_1, X_2, \ldots, X_n$ are known to have come from a larger set of N exchangeable random variables, then (38) is automatically satisfied. Hence, Theorem 3.5.1 is directly applicable. This is always the case for the failure model of Section 3.1, for which some results are contained in Exercises 1 and 2.

3.6. SEGMENTS OF INFINITE SEQUENCES OF EXCHANGEABLE VARIABLES

In order to incorporate prior knowledge on the observed variables, the Bayesian statistician always assumes that the parameter of the distribution considered is itself a random variable. Hence, in this view, the observations $X_1, X_2, \ldots, X_n$ are i.i.d. for a given value of the parameter λ of the common distribution function $F(x, \lambda)$. Consequently, the joint distribution

$$P(X_1 < x_1, X_2 < x_2, \ldots, X_n < x_n) = \int_{-\infty}^{+\infty} F(x_1, y) F(x_2, y) \cdots F(x_n, y) dV(y),$$

where $V(y)$ denotes the distribution function of λ. Since i.i.d. variables can always be extended to an infinite sequence, $X_1, X_2, \ldots, X_n$ are in fact a segment of an infinite sequence of exchangeable variables. In particular,

$$G^*_{i(k)}(x) = \int_{-\infty}^{+\infty} [1 - F(x, y)]^k dV(y) \tag{46a}$$

and

$$F^*_{i(k)}(x) = \int_{-\infty}^{+\infty} F^k(x, y) dV(y). \tag{46b}$$

Hence, Example 3.5.2 was a typical example of this viewpoint. In order to find the limiting distributions of the extremes of $X_1, X_2, \ldots, X_n$, we can thus proceed as in Example 3.5.2. Before doing this, however, let us record an important result for arbitrary infinite sequences of exchangeable variables.

Theorem 3.6.1 (the deFinetti representation). *Let $X_1, X_2, \ldots$, form an infinite sequence of exchangeable random variables. Then for each real number x, there is a random variable $Y(x) = Y(x, \omega)$ with $0 \leq Y(x) \leq 1$ and such that*

$$F_{i(k)}(x_1, x_2, \ldots, x_k) = E(Y(x_1) Y(x_2) \cdots Y(x_k)). \tag{47}$$

Furthermore, $Y(x)$, for every random point ω, is a distribution function in x.

Remark 3.6.1. Those familiar with the concept of conditional expectation will realize from the proof that the variables $X_1, X_2, \ldots,$ are conditionally independent given the process $\{Y(x), x \text{ real}\}$.

Remark 3.6.2. If $x_1 = x_2 = \cdots = x_k$, (47) differs only in emphasis from (46b). Both express $F^*_{i(k)}(x)$ as the kth moment of a random variable which is bounded by zero and one, and which is a distribution function in x.

Proof. For a fixed x, let $I_j(x)$ be the indicator variable of the event $\{X_j < x\}$. That is, $I_j(x) = 1$ or 0 according as $X_j < x$ or not. Define

$$Y_n(x) = \frac{1}{n} \sum_{j=1}^{n} I_j(x).$$

We prove that $Y_n(x)$ converges in probability to a random variable $Y(x)$ which satisfies the claimed properties of the theorem.

First observe that, by exchangeability, for $m > n$,

$$E\{[Y_n(x) - Y_m(x)]^2\} = E\left\{\left[\frac{m-n}{mn} \sum_{j=1}^{n} I_j(x) - \frac{1}{m} \sum_{j=n+1}^{m} I_j(x)\right]^2\right\}$$

$$= \frac{m-n}{mn}\left[P(X_1 < x) - P(X_1 < x, X_2 < x)\right].$$

Therefore, as n and m tend to infinity, $E\{[Y_n(x) - Y_m(x)]^2\}$ tends to zero. But then by the completeness theorem (Appendix II), $Y_n(x)$ converges in probability to a random variable $Y(x)$. Evidently, $0 \leq Y(x) \leq 1$. Thus, by the dominated convergence theorem (Appendix I), for any fixed real numbers $x_1, x_2, \ldots, x_k$,

$$\lim_{n = +\infty} E(Y_n(x_1) Y_n(x_2) \cdots Y_n(x_k)) = E(Y(x_1) Y(x_2) \cdots Y(x_k)).$$

On the other hand, by exchangeability,

$$E(Y_n(x_1) Y_n(x_2) \cdots Y_n(x_k)) = \frac{1}{n^k}\binom{n}{k} F_{i(k)}(x_1, x_2, \ldots, x_k) + 0\left(\frac{1}{n}\right),$$

and thus (47) follows.

Finally, since for each n, $Y_n(x_1) \leq Y_n(x_2)$ for $x_1 \leq x_2$, $Y(x)$ is also nondecreasing in x. We also easily get that $Y(-\infty) = 0$ and $Y(+\infty) = 1$. Of course, all statements can be made only for almost all ω-points. However, if $Y(x) = Y(x, \omega)$ has the above properties for almost all ω, then it can be modified on a null set so that it has these properties for all ω.

Furthermore, the left continuity requirement can also be achieved. Details are left to the reader. The theorem is established. ▲

Before formulating the result for the asymptotic distribution of extremes, let us recall Example 3.3.2. The sequence X_j, $j \geq 1$, of this example is exchangeable, and in fact (46) applies with $F(x,y) = \Phi(x+y)$, where $\Phi(x)$ is the standard normal distribution function, and with $V(y)$ arbitrary, signifying the distribution of U. Hence, by the discussion of Section 3.3, we should impose restrictions on the normalizing constants in order to have meaningful results. We shall use our concept of a_n and $b_n > 0$ being characteristic to the upper extremes, a concept introduced for the specific purpose of excluding the structures represented by Example 3.3.2. (see (39) and (40) in Section 3.5). We first formulate Theorem 3.5.1 for exchangeable variables.

Theorem 3.6.2. *Let $X_1, X_2, \ldots$ be an infinite sequence of exchangeable random variables. Let $Y(x)$, x real, be the set of random variables occurring in the deFinetti representation (47). Then there are normalizing constants a_n and $b_n > 0$, which are characteristic to the upper extremes, if, and only if,*

$$\lim_{n=+\infty} P\big(n[1-Y(a_n+b_n x)] < y\big) = U(y) = U(y;x) \qquad (48)$$

for all continuity points of $U(y)$, where $U(y)$ is a distribution function. For the limiting distribution of the extremes, formula (41) applies.

Proof. We apply Theorem 3.5.1. Let N be a sequence of integers such that $N/n \to +\infty$ with n. Then, if $\nu_N(x)$ denotes the number of X_j, $1 \leq j \leq N$, which satisfy $X_j \geq x$, Theorem 3.6.1 yields

$$P(\nu_N(a_n+b_n x) = t) = \binom{N}{t} E\left\{\left[1 - Y(a_n+b_n x)\right]^t Y^{N-t}(a_n+b_n x)\right\}.$$

By Theorem 3.5.1, we have to show that (48) is equivalent to the weak convergence of $n\nu_N(a_n+b_n x)/N$ to $U(y)$. That is, we have to show that, as n and N/n tend to infinity,

$$\lim E\left\{\sum_{t=0}^{yN/n} \binom{N}{t}\left[1-Y(a_n+b_n x)\right]^t Y^{N-t}(a_n+b_n x)\right\} = U(y) \qquad (49)$$

at continuity points of $U(y)$ if, and only if, (48) holds. Let us put

$$K_{n,N}(y;x) = \sum_{t=0}^{yN/n} \binom{N}{t}\left[1-Y(a_n+b_n x)\right]^t Y^{N-t}(a_n+b_n x).$$

The function $K_{n,N}(y;x)$ is, of course, a random variable, but, for each ω-point, it represents in y the distribution function of a binomial variate (normalized by n/N). Therefore, the Chebishev inequality yields (the computation is similar to the one in Example 3.5.1) that, as $N/n \to +\infty$ with n,

$$K_{n,N}(y;x) \to \begin{cases} 1 & \text{if } n[1 - Y(a_n + b_n x)] < y \text{ ultimately} \\ 0 & \text{if } n[1 - Y(a_n + b_n x)] \geq y \text{ ultimately.} \end{cases}$$

It now follows that (48) implies (49). Since this argument can be repeated for subsequences of n, we get that, on any sequences on which (48) applies, the same limit is obtained in (49). Consequently, if (49) applies, so does (48). In view of Theorem 3.5.1, the theorem is established. ▲

Corollary 3.6.1. *Let $X_1, X_2, \ldots$ be an infinite sequence of exchangeable random variables and let $Y(x)$ be as in Theorem 3.6.2. Let us assume that there is a distribution function $D(x)$ with the following two properties*: (i) *there are sequences a_n and $b_n > 0$ such that, as $n \to +\infty$, $D^n(a_n + b_n x)$ converges to a distribution function $H_D(x)$, and* (ii) *as $x \to \omega(D)$*,

$$\lim P\left(\frac{1 - Y(x)}{1 - D(x)} < y\right) = U^*(y) \tag{50}$$

exists for all continuity points y of $U^*(y)$, which is continuous at zero. Then the normalized upper extremes $(X_{n-k+1:n} - a_n)/b_n$ converge weakly. Their limiting distributions $E_k(x)$ are of the form

$$E_k(x) = \sum_{t=0}^{k-1} \frac{1}{t!} [-\log H_D(x)]^t \int_0^{+\infty} z^t H_D^z(x) dU^*(z). \tag{51}$$

Proof. The corollary is an easy consequence of Theorem 3.6.2 and of the fact that condition (i) is equivalent to

$$\lim_{n = +\infty} n[1 - D(a_n + b_n x)] = -\log H_D(x), \quad H_D(x) > 0. \tag{52}$$

(Although we have dealt with functions satisfying condition (i) extensively in Chapter 2, the reader can easily reproduce this fact by taking logarithms and using the first term in the Taylor expansion of $\log z = \log[1 - (1 - z)]$, $|z| < 1$). As a matter of fact, we deduce (48) from our assumptions (50) and

(52). As $n \to +\infty$,

$$\lim P\{n[1-Y(a_n+b_nx)]<y\}$$

$$=\lim P\left\{n[1-D(a_n+b_nx)]\frac{1-Y(a_n+b_nx)}{1-D(a_n+b_nx)}<y\right\}$$

$$=\lim P\left\{\frac{1-Y(a_n+b_nx)}{1-D(a_n+b_nx)}<\frac{y}{-\log H_D(x)}\right\}$$

$$=U^*\left(\frac{y}{-\log H_D(x)}\right),$$

where the result is formally true even if $H_D(x)=0$ or 1, since $U^*(0)=0$ and is continuous at zero, while, for $H_D(x)=1$, one should write $U^*(+\infty)=1$ in limit. Notice that we applied the fact that $n\to+\infty$ implies $a_n+b_nx\to\omega(D)$, which is now evident by the experience of Chapter 2 (and can easily be reproduced). In addition, we in fact applied Lemma 2.2.2 at the second step, when we replaced $n[1-D(a_n+b_nx)]$ by its limit. We thus proved (48) with $U(y;x)=U^*\{y/[-\log H_D(x)]\}$. Hence, (41) applies for $E_k(x)$, which can be rewritten as (51). From this specific form, one immediately gets that the $E_k(x)$ are proper distribution functions. This completes the proof. ▲

Although Corollary 3.6.1 expresses only a sufficient condition for the existence of $E_k(x)$, $k \geq 1$, it has the convenience of reducing the choice of the normalizing constants to the case of Chapter 2, where condition (i) was extensively investigated.

We restate Corollary 3.6.1, where $Y(x)$ does not occur explicitly.

Corollary 3.6.2. *Let $X_1, X_2,\ldots$ be an infinite sequence of exchangeable random variables. Let $D(x)$ be a distribution function which satisfies condition (i) of Corollary 3.6.1. If $U^*(x;y)$ is a distribution function for each real number $x<\omega(D)$ such that $U^*(x;0)=0$ and*

$$\int_0^{+\infty} y^k dU^*(x;y) = \frac{G^*_{i(k)}(x)}{[1-D(x)]^k}, \qquad k \geq 1, \tag{53}$$

and if $U^*(x;y)$ converges weakly to a distribution function $U^*(y)$ as $x\to\omega(D)$, then the conclusion of Corollary 3.6.1 is valid.

Proof. We first show that (53) uniquely determines $U^*(x;y)$ for each

$x < \omega(D)$. As a matter of fact, we show

$$U^*(x;y) = P\left(\frac{1-Y(x)}{1-D(x)} < y\right), \tag{54}$$

where $Y(x)$ is defined as in Theorem 3.6.2. Namely, on account of Theorem 3.6.1, (53) holds if the function $U^*(x;y)$ is defined by (54). But the random variable $[1-Y(x)]/[1-D(x)]$ is bounded for each $x < \omega(D)$, and thus its moment sequence uniquely determines its distribution (Appendix II). Hence, the assumption of weak convergence of $U^*(x;y)$ is exactly the assumption (50), from which the conclusion follows. This completes the proof. ▲

The following statement gives a case when, for the asymptotic distribution of extremes, infinite sequences of exchangeable variables can be approximated by i.i.d. ones.

Corollary 3.6.3. *With the notations of Corollary 3.6.1, the asymptotic distribution of $(Z_n - a_n)/b_n$ is one of the three possible types $H_{1,\gamma}(x)$, $H_{2,\gamma}(x)$, and $H_{3,0}(x)$ for i.i.d. variables, if $U^*(y)$ of (50) is degenerate at some positive constant. In particular, if, as $n \to +\infty$,*

$$\lim nP(X_1 \geq a_n + b_n x) = -\log H_D(x) \tag{55}$$

and

$$\lim n^2 P(X_1 \geq a_n + b_n x, X_2 \geq a_n + b_n x) = [\log H_D(x)]^2, \tag{56}$$

where a_n, b_n, and $H_D(x)$ are defined in condition (i) of Corollary 3.6.1, then the limit distribution of $(X_{n-k+1:n} - a_n)/b_n$ is of the same type as for i.i.d. variables.

Proof. The statement is immediate from Corollaries 3.6.1 and 3.6.2 by evaluating $E_k(x)$ of (51) and by noting that, for each of the three types mentioned in the Corollary, $H^c(x)$ is of the same type as $H(x)$ for any constant $c > 0$. For the particular case, one has to observe that, by the Chebishev inequality, (55) and (56) imply that $U^*(x;y)$ of Corollary 3.6.2 tends to one for $y > 1$ and to zero otherwise. The proof is completed. ▲

Notice that, in (46b), the deFinetti representation is readily available. Therefore, all preceding statements are applicable in connection with it. There are, however, situations when only the existence of $Y(x)$ is guaranteed, but it is not known explicitly. In this case, Corollary 3.6.2 and the particular case of Corollary 3.6.3 are applicable.

We now work out three examples. Let us, however, first remark that the theory of lower extremes can be obtained by a change of sign from the previous statements. These results are collected as Exercise 7 and 8.

Example 3.6.1. Let $X_1, X_2, \ldots, X_n$ be unit exponential variables with random location parameter α. We assume that, for a given value of $\alpha, X_1, X_2, \ldots, X_n$ are i.i.d. Hence

$$G^*_{i(k)}(x) = \int_{-\infty}^{x} e^{-k(x-y)} dV(y) + \int_{x}^{+\infty} dV(y), \qquad (57)$$

where $V(y)$ denotes the distribution function of α. Let $V(y) = 1 - \exp(-e^y)$. We now deduce from Theorem 3.6.2

$$\lim_{n=+\infty} P(Z_n < \log n + x) = \frac{1}{1 + e^{-x}}, \qquad -\infty < x < +\infty. \qquad (58)$$

Indeed, (57) already shows that $Y(x)$ in the deFinetti representation is $1 - e^{-(x-\alpha)}$ if $x \geq \alpha$ and zero otherwise. Thus, for $y > 0$,

$$\lim P(n[1 - Y(\log n + x)] < y) = P(e^\alpha < ye^x) = V(x + \log y).$$

Consequently, Theorem 3.6.2 is applicable with

$$U(y; x) = V(x + \log y) = 1 - \exp(-ye^x),$$

and with $a_n = \log n$ and $b_n = 1$. The limiting distributions $E_k(x)$ for the kth extremes can be computed by (41). In particular, $k = 1$ yields

$$\int_0^{+\infty} e^{-y} dU(y; x) = \int_0^{+\infty} e^{-y} d[1 - \exp(-ye^x)]$$

$$= \int_0^{+\infty} \exp(-ze^{-x}) d(1 - e^{-z})$$

$$= \frac{1}{1 + e^{-x}},$$

where x is arbitrary, as stated in (58). ▲

Example 3.6.2. Let $X_1, X_2, \ldots$ be an infinite sequence of exchangeable variables and let

$$G^*_{i(k)}(x) = k! \left[1 - \exp\left(-\frac{1}{x^2}\right) \right]^k, \qquad x > 0, \quad k \geq 1.$$

We do not know $Y(x)$ and we cannot determine it from the information

3.6 INFINITE SEQUENCES OF EXCHANGEABLE VARIABLES

given above. Hence, the only applicable result is Corollary 3.6.2. Since the ratios

$$\frac{G^*_{i(k)}(x)}{[1-\exp(-1/x^2)]^k} = k!, \qquad k \geq 1, \quad x > 0,$$

are the moment sequence of the exponential distribution $U^*(x;y) = 1 - e^{-y}, y > 0$, the conditions of Corollary 3.6.2 are satisfied. Namely, $D(x) = \exp(-1/x^2)$ satisfies $D(x) = D^n(\sqrt{n}\,x)$; thus $H_D(x) = D(x)$ and $a_n = 0$, $b_n = \sqrt{n}$. Furthermore, the requirement on $U^*(x;y)$ evidently holds. Thus, for each fixed $k \geq 1, n^{-1/2} X_{n-k+1:n}$ converges weakly. The limiting distributions can be computed by (51). For example,

$$E_1(x) = \int_0^{+\infty} D^z(x)\, d(1-e^{-z}) = \left(1 + \frac{1}{x^2}\right)^{-1}, \qquad x > 0.$$

▲

Example 3.6.3. Let $X_1, X_2, \ldots$ be an infinite sequence of exchangeable variables. Let the common distribution be $F(x) = 1 - e^{-x}, x > 0$. Let further

$$F^*_{i(2)}(x) = P(X_1 < x, X_2 < x) = (1-e^{-x})^2(1+ce^{-3x}),$$

where c is a given number. Then, with $a_n = \log n$ and $b_n = 1$,

$$n[1 - F(a_n + b_n x)] = e^{-x},$$

and, since

$$G^*_{i(2)}(x) = F^*_{i(2)}(x) + 2e^{-x} - 1$$
$$= e^{-2x} + ce^{-3x}(1-e^{-x})^2,$$

$$n^2 G^*_{i(2)}(a_n + b_n x) = e^{-2x} + O\left(\frac{1}{n}\right).$$

Without specifying higher-dimensional distributions, we can apply Corollary 3.6.3. It yields that, for each $k \geq 1$, $X_{n-k+1:n} - \log n$ has an asymptotic distribution, and these distributions are of the same type as if the X_j were i.i.d. In particular, as $n \to +\infty$,

$$P(Z_n < \log n + x) \to \exp(-e^{-x}).$$

▲

3.7. STATIONARY SEQUENCES

In the present section we shall deal with sequences $X_1, X_2,\ldots$ of random variables which satisfy the properties (i) $P(X_j < x) = F(x)$ for each j and (ii) for any positive integer s,

$$F_{i(k)}(x_1, x_2, \ldots, x_k) = F_{i(k)+s}(x_1, x_2, \ldots, x_k),$$

where $i(k) + s$ signifies the vector $(i_1 + s, i_2 + s, \ldots, i_k + s)$. We shall refer to such a sequence of random variables as stationary (the usual term in the mathematical literature is "stationary in the strict sense," but we do not need this distinction here). Evidently, exchangeable variables are stationary, and thus so are i.i.d. variables. On the other hand, important classes of stationary sequences of random variables are not exchangeable. One class is exemplified by the following construction.

Let $Y_1, Y_2, \ldots$ be i.i.d. random variables. Let $m > 0$ be a fixed integer. Let us now define $X_j = g(Y_j, Y_{j+1}, \ldots, Y_{j+m-1})$, $j \geq 1$, where g is a (measurable) function of m variables. For example, $g(u_1, u_2, \ldots, u_m) = u_1 + u_2 + \cdots + u_m$ or $g(u_1, u_2, \ldots, u_m) = u_1 u_2 \cdots u_m$, etc. This special sequence X_j, $j \geq 1$, has the following additional property. For any integers $1 \leq i_1 < i_2 < \cdots < i_t < j_1 < j_2 < \cdots < j_k$, the vectors $(X_{i_1}, X_{i_2}, \ldots, X_{i_t})$ and $(X_{j_1}, X_{j_2}, \ldots, X_{j_k})$ are independent whenever $i_t + m \leq j_1$. A sequence of random variables with this last property is called m-dependent. Hence, the above constructed sequence $X_1, X_2, \ldots$, for any function g, is an m-dependent stationary sequence.

A more general important class of sequences to be investigated is the so-called mixing random variables. This concept gives accurate mathematical meaning to the requirement that the terms in a sequence of random variables be less and less dependent as they are further and further apart. Since, for extreme value theory, not the whole sequence $X_1, X_2, \ldots$ of random variables, but only the events $\{X_j \geq x\}$ or $\{X_j < x\}$ are significant, we adopt the following definition for mixing.

Definition 3.7.1. A stationary sequence $X_1, X_2, \ldots$ of random variables is called mixing (in the upper tail) if the following condition holds. For vectors $i(k) = (i_1, i_2, \ldots, i_k)$ and $j(t) = (j_1, j_2, \ldots, j_t)$ with $1 \leq i_1 < i_2 < \cdots < i_k$, $i_k + s \leq j_1 < \cdots < j_t$,

$$|F^*_{i(k),j(t)}(u) - F^*_{i(k)}(u) F^*_{j(t)}(u)| \leq \tau(s, u), \tag{59}$$

where $i(k), j(t)$ signifies the combined vector $(i_1, \ldots, i_k, j_1, \ldots, j_t)$, and where $\tau(s, u)$ is nonincreasing in s and is such that, for at least one sequence $u_n \to \omega(F)$ as $n \to +\infty$, there is a sequence $s_n \to +\infty$ with n and $\tau(s_n, u_n) \to 0$ as $n \to +\infty$. If (59) is replaced by

$$|G^*_{i(k),j(t)}(u) - G^*_{i(k)}(u) G^*_{j(t)}(u)| \leq \tau(s, u) \tag{60}$$

3.7 STATIONARY SEQUENCES

and if the condition on u_n is changed to $u_n \to \alpha(F)$, then we speak of sequences mixing in the lower tail.

It is evident that if a stationary sequence is m-dependent, then it is mixing (in both tails). Namely, $\tau(s,u)=0$ for $s \geq m$, identically in u.

Our aim is to find additional conditions to mixing which guarantee that Z_n, W_n, and other extremes would have the same types of limiting distribution as those obtained in the i.i.d. case. Exchangeable sequences (Section 3.6), Gaussian sequences (Section 3.8), and others (Section 3.9) provide ample situations when the distribution of extremes for stationary sequences cannot be approximated by i.i.d. variables.

Theorem 3.7.1. *Let $X_1, X_2, \ldots$ be a stationary sequence of random variables with common distribution function $F(x)$. Let a_n and $b_n > 0$ be sequences of real numbers such that, for each real number x,*

$$\lim_{n = +\infty} n[1 - F(a_n + b_n x)] = u(x) \tag{61}$$

exists and $0 < u(x) < +\infty$ on an interval of positive length. Set $H(x) = e^{-u(x)}$, where $e^{-\infty} = 0$. Assume that (59) holds and $\tau(s_n, u_n) \to 0$ as $n \to +\infty$, where $u_n = a_n + b_n x$ and s_n is a sequence of integers such that $s_n/n \to 0$ as $n \to +\infty$. Finally, let us assume that, with the above u_n,

$$\limsup_{n = +\infty} n \sum_{j=2}^{n} P(X_1 \geq u_{nM}, X_j \geq u_{nM}) = o\left(\frac{1}{M}\right) \tag{62}$$

as $M \to +\infty$. Then, as $n \to +\infty$,

$$P(Z_n < a_n + b_n x) \to H(x). \tag{63}$$

Remark 3.7.1. If $X_1, X_2, \ldots, X_n$ are i.i.d., then the only assumption is (61). Hence, in this case, the theorem reduces to Corollary 1.3.1. This fact implies that the limit distribution $H(x)$ is of the type in the general case as it was for i.i.d. variables.

Remark 3.7.2. If the stationary sequence $X_1, X_2, \ldots$ is m-dependent, then (59) holds with $\tau(s,u)$ such that $\tau(s,u) = 0$ for $s \geq m$, identically in u. Therefore, for m-dependent variables, the only assumptions are (61) and (62). Notice that (62) follows from the simpler assumption

$$\lim_{u = \omega(F)} \frac{\max_{1 \leq j \leq m} P(X_1 \geq u, X_j \geq u)}{1 - F(u)} = 0. \tag{64}$$

As a matter of fact, from the estimates

$$\sum_{j=2}^{n} P(X_1 \geq u, X_j \geq u) = \sum_{j=2}^{m} P(X_1 \geq u, X_j \geq u) + (n-m-1)[1-F(u)]^2$$

$$\leq m \left[\max_{1 \leq j \leq m} P(X_1 \geq u, X_j \geq u) \right] + n[1-F(u)]^2$$

(62) is immediate.

Let us turn to the proof of Theorem 3.7.1. Instead of trying to deduce it from the result of Section 3.5, we give a direct proof.

Proof. Throughout the proof, we put $u_n = a_n + b_n x$, where a_n and b_n satisfy (61). We first prove that from the assumptions it follows that, for any fixed integer $M > 0$, as $n \to +\infty$,

$$P(Z_{nM} < u_{nM}) - P^M(Z_n < u_{nM}) \to 0. \tag{65}$$

The proof of (65) is based on the following simple observation. If we remove a finite number of blocks of s_n (or s_{nM}) from the original sequence $X_1, X_2, \ldots, X_{nM}$, it does not have any effect on the asymptotic distribution of the maximum. On the other hand, this procedure makes (59) applicable. The details are as follows.

Let $i(k_1) = (1, 2, \ldots, n), i(k_2) = (n+s+1, n+s+2, \ldots, 2n+s), \ldots, i(k_M) = ((M-1)(n+s)+1, (M-1)(n+s)+2, \ldots, (M-1)(n+s)+n)$, and let $i(k_1), i(k_2), \ldots, i(k_M)$ signify the vector which combines the components of $i(k_1), i(k_2), \ldots, i(k_M)$. We shall later choose s as s_{nM}. Now, using the triangular inequality, by induction and by stationarity, we get from (59)

$$|F_{i(k_1), i(k_2), \ldots, i(k_M)}(u) - F_{i(k_1)}^M(u)| \leq (M-1)\tau(s, u). \tag{66}$$

Since $F_{i(k_1)}(u) = P(Z_n < u)$, in order to obtain (65) from (66), we have to estimate the difference

$$P(Z_{nM} < u) - F_{i(k_1), i(k_2), \ldots, i(k_M)}(u).$$

One more appeal to the triangular inequality yields

$$|P(Z_{nM} < u) - F_{i(k_1), i(k_2), \ldots, i(k_M)}(u)|$$

$$\leq P(Z_{nM} < u) - P(Z_{nM+s(M-1)} < u)$$

$$+ |P(Z_{nM+s(M-1)} < u) - F_{i(k_1), \ldots, i(k_M)}(u)|. \tag{67}$$

We can further simplify the second term in the above estimate. If we denote by A_j the event that each $X_t < u$ with $(j-1)(n+s) < t \leq j(n+s) - s$ and $X_t \geq u$ for $jn + (j-1)s < t \leq j(n+s)$, then

$$0 \leq F_{i(k_1),\ldots,i(k_M)}(u) - P(Z_{nM+sM} < u) = P\left(\bigcup_{j=1}^{M} A_j\right)$$

$$\leq \sum_{j=1}^{M} P(A_j) = MP(A_1) = M[P(Z_n < u) - P(Z_{n+s} < u)],$$

where we used the first term in the inequality of Theorem 1.4.1 and then the assumption of stationarity. Hence, (66) and (67) can be combined to read

$$|P(Z_{nM} < u) - P^M(Z_n < u)| \leq (M-1)\tau(s,u)$$
$$+ [P(Z_{nM} < u) - P(Z_{nM+s(M-1)} < u)]$$
$$+ M[P(Z_n < u) - P(Z_{n+s} < u)]. \quad (68)$$

The last two terms in (68) are similar in nature, for which we show

$$0 \leq P(Z_T < u) - P(Z_{T+t} < u) \leq \frac{1}{R} + 2R\tau(s,u), \quad (69)$$

where R is an arbitrary integer with $0 < R < T/(t+s)$. In the proof of (69), we follow the argument which led to (66). We divide the integers $1, 2, 3, \ldots, T+t$ into R blocks of length t as follows. We work backward, and thus the first block $B_1 = (T+1, T+2, \ldots, T+t)$. We then neglect at least s numbers, and B_2 is the last t integers which remain. We then delete at least s terms again, construct B_3, etc. Thus

$$P(Z_T < u) - P(Z_{T+t} < u) = P(Z_T < u, Z_{T+t} \geq u)$$

$$\leq P\left[\left(\bigcap_{r=2}^{R} \bigcap_{j \in B_r} \{X_j < u\}\right) \cap \left\{\bigcup_{j=T+1}^{T+t} (X_j \geq u)\right\}\right]$$

$$= P\left[\bigcap_{r=2}^{R} \bigcap_{j \in B_r} \{X_j < u\}\right] - P\left[\bigcap_{r=1}^{R} \bigcap_{j \in B_r} \{X_j < u\}\right],$$

which, by the inequality (66) and by stationarity, becomes

$$0 \leq P(Z_T < u) - P(Z_{T+t} < u) \leq P^R(Z_t < u) - P^{R+1}(Z_t < u) + 2R\tau(s,u),$$

from which (69) evidently follows. Notice that when we apply (69) in (68), we can choose R as an arbitrary positive integer with $R < n/2s$. Therefore, with $s = s_{nM}$, R can be fixed arbitrarily large as $n \to +\infty$ and M is fixed, on account of the assumption of $s_n/n \to 0$ as $n \to +\infty$. Consequently, with $u = u_{nM}$, (68) and (69) imply (65).

We can now complete the proof as follows. By Theorem 1.4.1 and by stationarity

$$1 - nP(X_1 \geq u_{nM}) \leq P(Z_n < u_{nM}) \leq 1 - nP(X_1 \geq u_{nM}) + S_{2,n},$$

where

$$S_{2,n} = \sum_{1 \leq i < j \leq n} P(X_i \geq u_{nM}, X_j \geq u_{nM}) \leq n \sum_{j=2}^{n} P(X_1 \geq u_{nM}, X_j \geq u_{nM}).$$

Thus, by (61), (62), and (65), for any fixed $M > 0$, as $n \to +\infty$,

$$\left(1 - \frac{u(x)}{M}\right)^M \leq \liminf P(Z_{nM} < u_{nM})$$

$$\leq \limsup P(Z_{nM} < u_{nM})$$

$$\leq \left(1 - \frac{u(x)}{M} + o\left(\frac{1}{M}\right)\right)^M. \tag{70}$$

Now let N be an arbitrary integer. Let us write $N = nM + t$, where $0 \leq t < M$. Then

$$P(Z_N < u_N) \leq P(Z_{nM} < u_N) \leq P(Z_{nM} < u_{nM+M}). \tag{71}$$

Since, by (69), as $n \to +\infty$,

$$\limsup P(Z_{nM} < u_{nM+M}) = \limsup P(Z_{nM+M} < u_{nM+M}),$$

(70) and (71) imply that, for arbitrary $M > 0$,

$$\limsup_{N = +\infty} P(Z_N < u_N) \leq \left(1 - \frac{u(x)}{M} + o\left(\frac{1}{M}\right)\right)^M.$$

This in turn, by M's being arbitrary, results in

$$\limsup_{N = +\infty} P(Z_N < u_N) \leq e^{-u(x)} = H(x).$$

In the same manner, the inequalities

$$P(Z_N < u_N) \geq P(Z_{nM+M} < u_N) \geq P(Z_{nM+M} < u_{nM}),$$

on account of (69) and (70), yield

$$\liminf_{N=+\infty} P(Z_N < u_N) \geq H(x).$$

This completes the proof. ▲

The preceding proof can be analyzed to obtain necessary conditions for the conclusion of Theorem 3.7.1. Let us record one of these conditions.

Theorem 3.7.2. *Let us assume that all conditions, except perhaps (62), of Theorem 3.7.1 hold. Furthermore, let (63) be also valid. Then, for any fixed i, as $N \to +\infty$,*

$$\lim NP(X_1 \geq a_N + b_N x, X_i \geq a_N + b_N x) = 0, \tag{72}$$

whenever $H(x)$ of (63) is positive.

Proof. We again use the abbreviation $u_m = a_m + b_m x$. Since (62) was not applied in the proof of (65), the estimates (68) and (69) remain to hold under the present assumptions. Therefore, if we choose a sequence $M \to +\infty$ and such that, as $N \to +\infty$, $M\tau(s_N, u_N) \to 0$ and $N/M \to +\infty$, then with n the integer part of N/M, (65) follows again. But then by (63)

$$P^M(Z_n < u_{nM}) \to H(x). \tag{73}$$

Since $M \to +\infty$, the limit relation just obtained can hold only if $P(Z_n < u_{nM}) \to 1$. Therefore, by the frequently applied Taylor expansion

$$-\log u \sim 1 - u, \quad 0 < u < 1, \quad u \to 1,$$

we can rewrite (73) as

$$\lim M[1 - P(Z_n < u_{nM})] = -\log H(x) = u(x), \tag{74}$$

whenever $H(x) > 0$, where n and M are specified as in (73).

We now estimate $1 - P(Z_n < u_{nM}) = P(Z_n \geq u_{nM})$ by terms which involve $P(X_1 \geq u_{nM}, X_i \geq u_{nM})$ and $1 - F(u_{nM})$, an estimate that will lead to (72). For fixed i, combine some of the random variables $X_1, X_2, \ldots, X_n$ into pairs (X_t, X_{t+i}) in such a way that no X_j would occur twice among those selected into the pairs. Let the number of pairs (X_t, X_{t+i}) be at least $\frac{1}{4}n$ and, for easier reference, let the set of t's be T. Now, by stationarity,

$$P(X_t \geq u, X_{t+i} \geq u) = P(X_1 \geq u, X_{i+1} \geq u)$$

and thus, since for any events A and B, $P(A \cup B) = P(A) + P(B) -$

$P(AB)$,

$$P(X_t \geq u \text{ or } X_{t+i} \geq u) = 2[1 - F(u)] - P(X_1 \geq u, X_{i+1} \geq u).$$

Hence

$$P(Z_n \geq u_{nM}) = P(X_j \geq u_{nM} \text{ for at least one } j)$$

$$\leq \sum_{t \in T} P(X_t \geq u_{nM} \text{ or } X_{t+i} \geq u_{nM}) + \sum_{j \notin T} P(X_j \geq u_{nM})$$

$$= n[1 - F(u_{nM})] - \sum_{t \in T} P(X_t \geq u_{nM}, X_{t+i} \geq u_{nM})$$

$$\leq n[1 - F(u_{nM})] - \tfrac{1}{4} n P(X_1 \geq u_{nM}, X_{i+1} \geq u_{nM})$$

$$\leq n[1 - F(u_{nM})].$$

Let us multiply these inequalities by M and let $M \to +\infty$. By (61) and (74), the extreme sides tend to $u(x)$ and thus so do all terms in between. In particular, as $M \to +\infty$,

$$\lim \{ Mn[1 - F(u_{nM})] - \tfrac{1}{4} MnP(X_1 \geq u_{nM}, X_{i+1} \geq u_{nM}) \} = u(x),$$

from which, by one more appeal to (61), (72) follows with nM for N, where n and M both tend to infinity with N, and n is the integer part of N/M. Therefore, $nM \leq N$, and thus by monotonicity

$$MnP(X_1 \geq u_N, X_i \geq u_N) \to 0.$$

Finally, $nM \leq N < (n+1)M < 2nM$, from which (72) is now evident. The theorem is established. ▲

Remark 3.7.2 and Theorem 3.7.2 can be combined into a necessary and sufficient condition for m-dependent sequences which guarantees (63).

Corollary 3.7.1. *Let $X_1, X_2, \ldots$ be an m-dependent stationary sequence with common distribution function $F(x)$. Assume that (61) holds. Then (63) is valid if, and only if, for each $0 < i \leq m$, as $n \to +\infty$,*

$$\lim nP(X_1 \geq a_n + b_n x, X_i \geq a_n + b_n x) = 0,$$

or, equivalently, as $u \to \omega(F)$,

$$\lim \frac{P(X_1 \geq u, X_i \geq u)}{1 - F(u)} = 0, \qquad 1 \leq i < m.$$

This easily follows from the above mentioned statements; hence we omit details.

All the results of the present section can be restated for the minimum by our usual change of the sequence $\{X_j\}$ to $\{-X_j\}$, which changes the assumptions which were in terms of $1-F$ and $\{X_j \geq u\}$ to F and $\{X_j < u\}$, respectively. In addition, in all limits, the argument of F should tend to $\alpha(F)$ rather than to $\omega(F)$. The reader is invited to carry out the details. The proofs, of course, do not have to be repeated.

We delay the discussion of the asymptotic distribution of the kth extremes to Section 3.11, where the relevant result will be obtained as a corollary to general Poisson limit theorems.

In the next section we deal with special stationary sequences, when the finite dimensional distributions are normal. Since the early development of statistics was based on the assumption of normality, normal sequences received special attention. This, of course, led to finer results in its basic theory, permitting us to obtain specific and concrete conclusions with simple assumptions. Some of the results could be deduced as corollaries to theorems of the present section, while others would follow from general statements of Section 3.9. We shall, however, give direct proofs which are very specific to normal sequences.

3.8. STATIONARY GAUSSIAN SEQUENCES

Let $X_1, X_2, \ldots$ be a stationary sequence of random variables. In addition, we assume that, for all $n \geq 1$, the distribution of the vector $(X_1, X_2, \ldots, X_n)$ is normal with $E(X_j) = 0$ and $V(X_j) = 1, j \geq 1$. This means that $(X_1, X_2, \ldots, X_n)$ has a density function of the form

$$f_n(\mathbf{x}) = f_n(\mathbf{x}; R) = \frac{|R|^{1/2}}{(2\pi)^{n/2}} \exp\left(-\frac{1}{2} \mathbf{x} R^{-1} \mathbf{x}'\right), \tag{75}$$

where R is a positive definite $n \times n$ matrix whose (i,j)th entry is $r(i,j) = E(X_i X_j)$, $|R|$ is its determinant, and $\mathbf{x} = (x_1, x_2, \ldots, x_n)$, while $\mathbf{x}'$ is the same vector written in a column. The assumption of stationarity results in $r(i,j) = r_m$, where $m = |i-j|$, and the assumption on the first two moments yields $r(j,j) = 1, j \geq 1$.

A sequence $X_1, X_2, \ldots, X_n$ of random variables with $E(X_j) = 0$ and $V(X_j) = 1, j \geq 1$, is called Gaussian if their joint density function is given by (75). An infinite sequence $X_1, X_2, \ldots$ is Gaussian if, for all $n \geq 1, X_1, X_2, \ldots, X_n$ is Gaussian. Notice that a subsequence of a Gaussian sequence is also Gaussian. This easily follows from the definition if one turns to the distribution function and lets $x_j \to +\infty$ if j does not occur in the subsequence in question.

We shall again give details about the asymptotic distribution of Z_n under different assumptions on the sequence $r_m, m \geq 1$. The results can then be changed into results on W_n, which, by symmetry about the origin of $f_n(\mathbf{x})$, does not even require additional computations.

One of our basic tools of proof will be the following lemma.

Lemma 3.8.1. *The n-variate normal distribution function $\Phi_n(\mathbf{x}; R)$ is an increasing function of each element $r(i,j)$ of R.*

Proof. We do not give a detailed proof, which is quite simple in principle. Simply differentiate $\Phi_n(\mathbf{x}; R)$ with respect to $r(i,j)$, which will be the integral of the density $f_n(\mathbf{y}; R)$, $\mathbf{y}=(y_1, y_2, \ldots, y_n)$, over the following region. If $t \neq i, j$, then the integration with respect to y_t is from $-\infty$ to x_t. On the other hand, $y_i = x_i$ and $y_j = x_j$. Therefore, this derivative is positive, which implies the monotonicity as stated. ▲

In fact, the mentioned explicit formula for the derivative of $\Phi_n(\mathbf{x}; R)$ with respect to $r(i,j)$ helps us to arrive at an important inequality. The integral quoted can be rewritten as

$$\frac{\partial \Phi_n(\mathbf{x}; R)}{\partial r(i,j)} = f_2(x_i, x_j) \Phi_{n-2}(\mathbf{s}; V), \qquad i \neq j, \tag{76}$$

where $f_2(x_i, x_j)$ is the density of (X_i, X_j) and $\mathbf{s}$ and V in Φ_{n-2} are defined as follows. The vector $\mathbf{s}$ contains $n-2$ components which will be labeled as s_t, $1 \leq t \leq n$, but $t \neq i$ or j. Then $s_t = x_t - z_t$ with

$$z_t = [1 - r(i,j)]^{-1/2} \{[r(i,t) - r(j,t)r(i,j)] x_i + [r(j,t) - r(i,t)r(i,j)] x_j\}.$$

Finally, V is the conditional variance-covariance matrix of X_t, $1 \leq t \leq n$, $t \neq i$ or j, given (X_i, X_j).

We can now prove the following result.

Lemma 3.8.2 *Let $\{X_{1,i}\}$ and $\{X_{2,i}\}$, $1 \leq i \leq n$, be Gaussian sequences and assume that both $r_1(i,j) = E(X_{1,i} X_{1,j})$ and $r_2(i,j) = E(X_{2,i} X_{2,j})$ depend on $i - j$ only. For $i < j$, $m = j - i$, we put $r_{1,m}$ and $r_{2,m}$ for $r_1(i,j)$ and $r_2(i,j)$, respectively. Then*

$$|P(Z_{1,n} < x) - P(Z_{2,n} < x)| \leq n \sum_{k=1}^{n} |r_{1,k} - r_{2,k}|(1 - m_k^2)^{-1/2} \exp\left(-\frac{x^2}{1+m_k}\right),$$

where $m_k = \max(|r_{1,k}|, |r_{2,k}|)$.

Proof. By the definition of Gaussian sequences and by the basic formula (5) of Chapter 1,

$$P(Z_{t,n} < x) = \Phi_n(\mathbf{x}; R_t), \qquad t = 1, 2,$$

where $\mathbf{x} = (x, x, \ldots, x)$. We assumed that the (i,j)th entry of R_t, $t = 1, 2$, is $r_{1,m}$ and $r_{2,m}$, respectively, where $m = |i - j|$. Therefore, if R denotes the $n \times n$ matrix, the (i,j)th entry of which is $r_{|i-j|}$, the difference

$$P(Z_{1,n} < x) - P(Z_{2,n} < x) = \Phi_n(\mathbf{x}; R_1) - \Phi_n(\mathbf{x}; R_2)$$

is an "increment" of the function $\Phi_n(\mathbf{x}; R)$, where the variables are the entries of R. Hence, from the elements of calculus (one of the so-called mean value theorems),

$$P(Z_{1,n} < x) - P(Z_{2,n} < x) = \sum_{k=1}^{n} (r_{1,k} - r_{2,k}) \frac{\partial \Phi_n(\mathbf{x}; R)}{\partial r_k} \bigg|_{R=R^*}, \qquad (77)$$

where R^* is also of the structure that its (i,j)th entry $r^*_{|i-j|}$ depends on $|i-j|$ only, and r_k^* is a number between $r_{1,k}$ and $r_{2,k}$. We now estimate the partial derivatives on the right hand side of (77) by an appeal to (76). First note that, for general symmetric matrices R, and for $\mathbf{x} = (x, x, \ldots, x)$, (76) implies

$$\left| \frac{\partial \Phi_n(\mathbf{x}; R)}{\partial r(i,j)} \right| \leq f_2(x, x) = \frac{1}{2\pi} [1 - r^2(i,j)]^{-1/2} \exp\left[-\frac{x^2}{1 + r(i,j)} \right],$$

which is further increased if we replace $r(i,j)$ by a number m such that $|r(i,j)| \leq m < 1$. That is, for any number m with $|r(i,j)| \leq m < 1$,

$$\left| \frac{\partial \Phi_n(\mathbf{x}; R)}{\partial r(i,j)} \right| \leq \frac{1}{2\pi} (1 - m^2)^{-1/2} \exp\left(-\frac{x^2}{1 + m} \right). \qquad (78)$$

Next, we get by the chain rule that, for the special matrices R with $r(i,j) = r_{|i-j|}$,

$$\frac{\partial \Phi_n(\mathbf{x}; R)}{\partial r_k} = \sum \frac{\partial \Phi_n(\mathbf{x}; R)}{\partial r(i,j)},$$

where the summation is for all $i < j$ for which $j - i = k$. Since the number of terms here is smaller than n and each term can be estimated by the formula (78), we immediately get from (77) the inequality that was to be proved. ▲

We can now prove asymptotic results for Z_n by the following method. We first investigate sequences for which the sequence r_m has very simple properties. We then use Lemmas 3.8.1 and 3.8.2 to show that a deviation from these "neat" sequences r_m will not affect the limiting form of the distribution of Z_n, when properly normalized.

We start with the following simple structure. Let $X_1, X_2, \ldots, X_n$ be a Gaussian sequence with zero expectation, unit variance, and constant correlation $r = r(n) = E(X_i X_j)$, $i \neq j$. This sequence, for $r \geq 0$, can be exemplified, and for distributional results can also be replaced, by the following sequence. Let $Y_0, Y_1, Y_2, \ldots, Y_n$ be i.i.d. standard normal variates and let

$$X_j = r^{1/2} Y_0 + (1-r)^{1/2} Y_j, \qquad 1 \leq j \leq n, \quad r \geq 0. \tag{79}$$

The case $r = 0$, of course, reduces to $X_j = Y_j$—that is, to i.i.d. standard normal variates, which, in view of Lemma 3.8.1, will give a lower bound for the distribution of the maximum when $r > 0$. We now prove the following result.

Theorem 3.8.1. *Let $Z_n(r)$ be the maximum of a Gaussian sequence $X_1, X_2, \ldots, X_n$ with zero expectation, unit variance, and constant correlation $r = r(n)$. Let*

$$a_n = \frac{1}{b_n} - \frac{1}{2} b_n (\log \log n + \log 4\pi), \qquad b_n = (2 \log n)^{-1/2}. \tag{80}$$

If, as $n \to +\infty$, $r(n) \log n$ converges to a finite value τ, then $(Z_n(r) - a_n)/b_n$ has a limiting distribution $H(x)$. For $\tau = 0$, $H(x) = H_{3,0}(x) = \exp(-e^{-x})$, while, for $\tau > 0$, $H(x)$ is the convolution of $H_{3,0}(x+\tau)$ and $\Phi[x(2\tau)^{-1/2}]$, where $\Phi(x)$ is the standard normal distribution function.

On the other hand, if $r(n) \log n \to +\infty$, then, as $n \to +\infty$,

$$\lim P\left[Z_n(r) < a_n (1-r)^{1/2} + x r^{1/2}\right] = \Phi(x).$$

Proof. By the representation (79)

$$Z_n(r) = r^{1/2} Y_0 + (1-r)^{1/2} Z_n^*, \tag{81}$$

where $Z_n^* = \text{mas}(Y_1, Y_2, \ldots, Y_n)$, and Y_0 and Z_n^* are independent. Hence, by (80),

$$\frac{Z_n(r) - a_n}{b_n} = U_n + V_n,$$

where

$$U_n = (2r\log n)^{1/2} Y_0, \qquad V_n = (1-r)^{1/2} \frac{Z_n^* - (1-r)^{-1/2} a_n}{b_n},$$

and U_n and V_n are independent. We shall show that, if $r\log n \to \tau$, which is finite, both U_n and V_n have a limiting distribution. In fact, if $\tau=0$, then, for any $\varepsilon>0$, as $n\to+\infty$,

$$\lim P(|U_n| \geq \varepsilon) = 0.$$

Therefore, by Lemma 2.2.1,

$$\lim P(Z_n(r) < a_n + b_n x) = \lim P(V_n < x),$$

which will be shown to equal $H_{3,0}(x)$. On the other hand, for $0 < \tau < +\infty$, as $n \to +\infty$,

$$\lim P(U_n < x) = \Phi\left[x(2\tau)^{-1/2}\right].$$

Consequently, if we show

$$P(V_n < x) \to H_{3,0}(x+\tau), \tag{82}$$

Lemma 2.9.1 will yield our claim. It remains now to investigate the distribution of V_n. Notice that it suffices to prove (82), assuming that $0 \leq \tau < +\infty$, since the limit reduces to $H_{3,0}(x)$ for $\tau = 0$.

We know from Section 2.3.2 that, as $n \to +\infty$,

$$P(Z_n^* < a_n + b_n x) \to H_{3,0}(x). \tag{83}$$

Now, let us write

$$P(V_n < x) = P(Z_n^* < A_n + B_n x),$$

where

$$A_n = (1-r)^{-1/2} a_n, \qquad B_n = (1-r)^{-1/2} b_n.$$

In view of Lemma 2.2.2, we immediately get (82) if we show that, as $n \to +\infty$

$$\frac{A_n - a_n}{b_n} \to \tau \quad \text{and} \quad \frac{B_n}{b_n} = 1.$$

The latter is evident from the assumption $r\log n\to\tau$, and thus $r\to 0$. Hence, only the first relation needs proof. Applying that, as $r\to 0$,

$$(1-r)^{-1/2} = 1 + \tfrac{1}{2}r + O(r^2),$$

we get by (80)

$$\frac{A_n - a_n}{b_n} = \frac{a_n\left[(1-r)^{-1/2} - 1\right]}{b_n}$$

$$= \left[\tfrac{1}{2}r + O(r^2)\right]\left[2\log n + o(\log n)\right]$$

$$= \left[1 + o(1)\right]r\log n \to \tau, \quad \text{as } n\to +\infty.$$

This completes the first part of the proof.

Turning to the case of $r\log n\to +\infty$ with n, we use the new normalizing constants

$$r^{-1/2}\left[Z_n(r) - a_n(1-r)^{1/2}\right],$$

which, by (81), reduces to

$$Y_0 + (1-r)^{1/2}r^{-1/2}(Z_n^* - a_n) = Y_0 + T_n, \text{ say.}$$

If we show that, for any $\varepsilon > 0$, as $n\to +\infty$,

$$P(|T_n| \geq \varepsilon) \to 0, \tag{84}$$

then Lemma 2.2.1 would imply that the limiting distribution of $Z_n(r)$, normalized as above, is the actual distribution of Y_0, which is $\Phi(x)$. We thus have to prove (84). It, however, immediately follows from (83) by the estimate

$$P(|T_n| \geq \varepsilon) \leq P(r^{-1/2}|Z_n^* - a_n| \geq \varepsilon) = P\left[\frac{|Z_n^* - a_n|}{b_n} \geq \varepsilon(2r\log n)^{1/2}\right]$$

and by $r\log n \to +\infty$ with n. The proof is complete. ▲

Theorem 3.8.1 is interesting in that it shows that the limiting form of $Z_n(r)$, when suitably normalized, depends on the relation of $r(n)$ to $\log n$. We now show that the assumption of constant correlation is not essential in this regard.

In order to simplify proofs, we separate an important step as a lemma.

Lemma 3.8.3. Let $X_{1,1}, X_{1,2}, \ldots, X_{1,n}$ and $X_{2,1}, X_{2,2}, \ldots, X_{2,n}$ be two Gaussian sequences, each term with zero expectation and unit variance. Let $r_{1,k} = E(X_{1,j} X_{1,j+k})$ and $r_{2,k} = E(X_{2,j} X_{2,j+k})$ for all j. Let us assume that $r_{1,k} = r_{2,k}$ for all $k \geq s(n) = n^t$ with some $0 < t \leq \frac{1}{3}$. Finally, let $|r_{j,k}| \leq M < 1$ and $r_{j,k} \to 0$, as $k \to +\infty$, for $j = 1, 2$. Then, as $n \to +\infty$,

$$\lim \left[P(Z_{1,n} < a_n + b_n x) - P(Z_{2,n} < a_n + b_n x) \right] = 0,$$

where a_n and b_n are defined in (80).

Proof. By Lemma 3.8.2

$$|P(Z_{1,n} < c_n) - P(Z_{2,n} < c_n)| \leq n \sum_{k=1}^{s(n)} (1 - m_k^2)^{-1/2} \exp\left(-\frac{c_n^2}{1 + m_k}\right), \quad (85)$$

where $m_k = \max(|r_{1,k}|, |r_{2,k}|)$. By assumption, $m_k \to 0$ as $k \to +\infty$. Therefore, there is a fixed N such that, for all $k \geq N$, $1 + m_k \leq 2/(1 + 2t)$. Furthermore, $m_k \leq M$ and $(1 - m_k^2)^{-1/2}$ is bounded. We now estimate the terms of (85) for $c_n = a_n + b_n x$, which we use in the form $c_n^2 = 2 \log n + o(\log n)$. Now, if $k < N$, each term of (85) is smaller than

$$n(1 - M^2)^{-1/2} \exp\left[-\frac{2}{1 + M} \log n + o(\log n) \right] \to 0$$

as $n \to +\infty$. Since N is fixed, their total contribution to (85) also tends to zero. Next, for $k \geq N$, we estimate the number of terms by $s(n)$ itself, and, in the individual terms, the choice of N will be used. Furthermore, $ns(n) = n^{1+t} = \exp[(1+t)\log n]$. Hence, the total contribution of these terms to (85) does not exceed

$$\exp\left[-(1 + 2t)\log n + (1 + t)\log n + o(\log n) \right],$$

which also tends to zero. This completes the proof. ▲

Theorem 3.8.2. Let $X_1, X_2, \cdots$ be a stationary Gaussian sequence with zero expectation and unit variance. Assume that the correlations $r_m = E(X_j X_{j+m})$ satisfy $r_m \log m \to 0$ as $m \to +\infty$. Then, as $n \to +\infty$,

$$P(Z_n < a_n + b_n x) \to H_{3,0}(x),$$

where a_n and b_n are as in (80).

Proof. We introduce two additional Gaussian sequences $X_{1,1}, X_{1,2}, \ldots, X_{1,n}$ and $X_{2,1}, X_{2,2}, \ldots, X_{2,n}$ with the following properties. Each has zero expectation and unit variance. The correlations $E(X_{i,j} X_{i,j+m}) =$

$r_{i,m}, i=1,2$, depend on m only. Finally, for $m < s(n) = n^{1/3}, r_{1,m} \leq r_m \leq r_{2,m}$, and, for $m \geq s(n), r_{1,m}$ and $r_{2,m}$ do not depend on m and the actual values are $r_{1,m} = -\rho(n)$ and $r_{2,m} = \rho(n)$, where

$$\rho(n) = \sup\{|r_m| : n^{1/3} \leq m \leq n\}.$$

By the monotonicity property of Gaussian sequences in terms of the correlations (Lemma 3.8.1),

$$P(Z_{1,n} < a_n + b_n x) \leq P(Z_n < a_n + b_n x) \leq P(Z_{2,n} < a_n + b_n x). \qquad (86)$$

On the other hand, both $\{X_{1,j}\}$ and $\{X_{2,j}\}$ differ from a Gaussian sequence with constant correlation in the first $1 \leq m \leq n^{1/3}$ of the correlation sequence $r_{i,m}, i=1,2$. Hence, by Lemma 3.8.3, the outermost terms of (86) have the same limits as if the basic random variables were equally correlated with the common values $-\rho(n)$ and $\rho(n)$, respectively. Since, from $r_m \log m \to 0$ it follows that $\rho(n) \log n \to 0$ as $n \to +\infty$, Theorem 3.8.1 yields that the two extreme terms of (86) tend to $H_{3,0}(x)$, and the statement follows. ▲

With no modification in the preceding proof, we get the following result.

Theorem 3.8.3. *With the notations of Theorem 3.8.2, let $r_m \log m \to \tau$, which is finite and positive. Then, as $n \to +\infty$, $(Z_n - a_n)/b_n$ has a limiting distribution $H(x)$ which is the convolution of $H_{3,0}(x+\tau)$ and $\Phi[x(2\tau)^{-1/2}]$.*

It is evident that none of the theorems is stated in the most general form. By the inequality of Lemma 3.8.2, one can always modify a sequence $\{r_m\}$ in such a way that a "neat" property is broken but the limiting distribution of Z_n is not affected (when normalized). But, apart from this freedom of modification of the sequence r_m, the theorems are the most general when, "for most values of m," $r_m \log m$ is bounded. Notice that Theorems 3.8.2 and 3.8.3 state, and the proofs clearly show, that if $r_m \log m$ converges, then it does not matter whether the correlations are equal or not.

The case $r_m \log m \to +\infty$ is more difficult in that the sequence r_m enters the normalizing constants needed for the extremes. It is contrary to the fact that, with constant correlation, this case was the easier part of the proof of Theorem 3.8.1. There are numerous possibilities for giving conditions which guarantee the existence of a limiting law for the maximum with suitable normalization. Out of these we give only one, which will indicate that approximation with sequences of constant correlation is again possible.

Theorem 3.8.4. *Using the notations of Theorem 3.8.2, we assume that, as $m \to +\infty$, r_m is decreasing and $r_m(\log m)^{1/3}$ tends to zero, but $r_m \log m$*

increases and tends to $+\infty$. *Then, as* $n \to +\infty$,

$$\lim P\bigl(Z_n < (1-r_n)^{1/2} a_n + x r_n^{1/2}\bigr) = \Phi(x).$$

Proof. We follow the proof of Theorem 3.8.2, in which Lemma 3.8.3 was a basic tool. We cannot use this lemma, however, since the normalizing constants are different now. In addition, it will not suffice to limit $s(n)$ by a power of n, when we approximate Z_n by $Z_{1,n}$ and $Z_{2,n}$ as in (86). The fact that $s(n)$ should be much closer to n will make the calculations a bit longer, but, at the same time, we gain somewhat by the monotonicity assumptions.

Our first step is to obtain an estimate similar to Lemma 3.8.3. Let $s = s(n) < n$ be a function which we specify later. We estimate the effect on the distribution of Z_n if we modify r_m for $m \leqslant s(n)$. We apply Lemma 3.8.2, which yields that the change of the distribution of Z_n is bounded by a constant multiple of

$$n \sum_{k=1}^{s} |r_k - r_k^*| \exp\left(-\frac{x_n^2}{1+m_k}\right), \tag{87}$$

where r_k^* is the modified sequence and $m_k = \max(r_k, r_k^*)$, and where we assumed that $m_k \leqslant M < 1$. This latter assumption is not a restriction for us, since our interest is limited to $r_k^* = r_s$ for all $k \leqslant s$ or $r_k^* = r_n$ for all $k \leqslant s$ and thus, by the monotonicity assumption, $0 < r_k^* \leqslant r_s \leqslant r_1 < 1$ (because, for a stationary Gaussian sequence, $r_1 = 1$ is not possible; it would imply that $X_1 = X_2 = X_3 = \cdots$, and thus $r_m = 1$ for all m). In the sequel we assume that r_k^* is one of the above sequences and that $x_n = (1-r_n)^{1/2} a_n + x r_n^{1/2}$. Thus $m_k = r_k$ and $|r_k - r_k^*| \leqslant r_k$. Since

$$x_n^2 = 2(1-r_n)\log n + o\bigl[(\log n)^{1/2}\bigr],$$

the expression (87) is bounded by

$$n \sum_{k=1}^{s} r_k \exp\left\{-\frac{2(1-r_n)}{1+r_k}\log n + o\bigl[(\log n)^{1/2}\bigr]\right\} = n \sum_{k=1}^{T} \cdots + n \sum_{k=T+1}^{s} \cdots$$

$$= \Sigma_1 + \Sigma_2,$$

where $T = n^t$ with some $0 < t < 1$. Just as in Lemma 3.8.2, it easily follows $\Sigma_1 = o(1)$. In Σ_2, we increase if we replace r_k by r_T for all k. Thus, if we write $n = \exp(\log n)$, the major term in the exponent of the summands of Σ_2

does not exceed

$$\left(1 - \frac{2-2r_n}{1+r_T}\right)\log n = \frac{r_T + 2r_n - 1}{1+r_T}\log n. \tag{88}$$

Since $r_T \to 0$ as $n \to +\infty$ and $0 \leq r_n \leq r_T$, we further increase (88), for large n, if we replace it by

$$(r_T + 2r_n - 1)(1 - r_T)\log n \leq (4r_T - 1)\log n.$$

Finally, since $r_m \log m$ is increasing, and since $T = n^t$,

$$r_T = \frac{r_T \log T}{\log T} \leq \frac{r_n \log n}{t \log n} = \frac{r_n}{t}.$$

The combination of the preceding estimates thus yields that, with the choice $t = \tfrac{1}{3}$, say,

$$\Sigma_2 \leq s(n)\exp\{(12r_n - 1)\log n + o[(\log n)^{1/2}]\}.$$

Let

$$s(n) = \exp\{[1 - 12r_n - (\log n)^{-1/2}]\log n\}. \tag{89}$$

It follows that the expression in (87) tends to zero as $n \to +\infty$. We therefore have proved that the limiting distribution of $[Z_n - (1 - r_n)^{1/2}a_n]r_n^{-1/2}$ is not affected if r_k is changed for all $k \leq s(n)$ either to r_s or to r_n, where $s = s(n)$ is defined in (89). Therefore, we assume $r_1 = r_2 = \cdots = r_s \geq r_{s+1} \geq \cdots \geq r_n$.

Let us now again, as in the proof of Theorem 3.8.2, introduce two new sequences, each with constant correlation. If $Z_{1,n}$ is the maximum of n Gaussian variables with zero expectation, unit variance, and constant r_s correlation, and $Z_{2,n}$ is a similar maximum except that the constant correlation is r_n, then, by Lemma 3.8.1,

$$P(Z_{2,n} < x_n) \leq P(Z_n < x_n) \leq P(Z_{1,n} < x_n), \tag{90}$$

where, as before, $x_n = (1 - r_n)^{1/2}a_n + xr_n^{1/2}$. We know from Theorem 3.8.1 that, as $n \to +\infty$, the extreme left hand side of (90) tends to $\Phi(x)$. By this same theorem, as $n \to +\infty$,

$$\lim P\left[Z_{1,n} < (1 - r_s)^{1/2}a_n + xr_s^{1/2}\right] = \Phi(x). \tag{91}$$

In view of (90), therefore, it remains to show that (91) holds if we change r_s to r_n. A criterion for such a possibility is contained in Lemma 2.2.2. It says

3.8 STATIONARY GAUSSIAN SEQUENCES

that, for our purposes, we have to show that, as $n \to +\infty$,

$$\frac{r_s}{r_n} \to 1, \quad a_n r_n^{-1/2}\left[(1-r_n)^{1/2} - (1-r_s)^{1/2}\right] \to 0. \tag{92}$$

By the monotonicity assumptions on r_m and $r_m \log m$,

$$\frac{\log s}{\log n} \leq \frac{r_n}{r_s} \leq 1,$$

from which, with $s = s(n)$ of (89), the first limit of (92) follows. For the second limit, we use $(1-r)^{1/2} = 1 - \frac{1}{2}r + O(r^2)$, as $r \to 0$. Thus, as $n \to +\infty$,

$$0 \leq a_n r_n^{-1/2}\left[(1-r_n)^{1/2} - (1-r_s)^{1/2}\right] \leq a_n r_n^{-1/2}(r_s - r_n)$$

$$\leq (2\log n)^{1/2} r_n^{-1/2}(r_s - r_n)$$

$$= \frac{(2r_n \log n)^{1/2}}{r_n \log n}(r_s \log n - r_n \log n)$$

$$\leq (2r_n \log n)^{1/2}\left(\frac{\log n}{\log s} - 1\right)$$

$$\leq (2r_n \log n)^{1/2}\left[12 r_n + (\log n)^{-1/2}\right],$$

where we have used again the explicit form of $s(n)$ from (89). This last expression tends to zero by the assumption $r_m (\log m)^{1/3} \to 0$ as $m \to +\infty$. Theorem 3.8.4 is thus established. ▲

While the proofs were challenging, and the results are interesting, from the mathematical point of view, they are far from pleasant for the applied scientist. For the mathematician, there is an easy way to decide which asymptotic law applies if, say, r_m is decreasing. But when the applied scientist estimates r_m from data, the relation of the estimated value to $\log m$ is not easily recognizable. These two viewpoints are illustrated in the following examples.

Example 3.8.1. With the basic notations of Theorems 3.8.1–3.8.4, let

$$r_m = \frac{g(m)}{\log m},$$

where, for large m, $\frac{1}{4}g(m) \sim \exp(1/\log\log m) - 1$. Then, since $e^x - 1 \sim x$ as $x \to 0$, $r_m \log m \to 0$ with the speed of $4/\log \log m$. Therefore, Theorem 3.8.2 applies, and thus the normalizing constants for Z_n are a_n and b_n of (80), and the limit law is $H_{3,0}(x)$. ▲

Example 3.8.2. Now let r_m be asymptotically $2/\log m$ as $m \to +\infty$. Then Theorem 3.8.3 applies, when the normalizing constants for Z_n remain the same as in the previous example but the limiting distribution is the convolution of $H_{3,0}(x+2)$ and $\Phi(\tfrac{1}{2}x)$. ▲

Example 3.8.3. If $r_m = g(m)/\log m$, where $g(m)$ is increasing but r_m is decreasing, and if, as $n \to +\infty$, $g(m)$ is asymptotically $\log\log m$, then all assumptions of Theorem 3.8.4 are satisfied. Consequently, we have to modify the normalizing constants to $(1 - r_n)^{1/2} a_n$ and $r_n^{1/2}$, respectively, for finding a limiting law for Z_n. The limiting distribution now is the standard normal distribution. ▲

Example 3.8.4. Assume that a sample is known to have a multivariate normal distribution, each term with zero expectation and unit variance. Furthermore, the sequence is known to be stationary and the correlations r_m to decrease with m increasing. The sample was large enough to estimate r_m for $m \leqslant 1{,}600$. From the theory we know that r_m should be compared with $\log m$, and thus, for each m, $r_m \log m$ was computed. It turned out that, for $1{,}100 \leqslant m \leqslant 1{,}600$, $1.95 \leqslant r_m \log m \leqslant 2.01$ and in fact the last 200 values of $r_m \log m$ were exactly 2.00 for two decimal digits. Having computed the previous three theoretical situations, one inclines to accept that the model of Example 3.8.2 is to be used. But then the experimenter looks at the values of $4/\log\log m$ as well as of $\log\log m$ for $1{,}400 \leqslant m \leqslant 1{,}600$, which lie in the intervals $(2.00, 2.02)$ and $(1.98, 2.00)$, respectively. He now, of course, has no choice but to recommend further investigations. He may try to make a decision on the base of values of r_m with values of m smaller than 1,100. But he may recognize no tendency there at all. Another evidence against such a decision is that the initial values of r_m have no effect on the asymptotic theory.

Finally, there is one more disturbing fact for the applied scientist here. With the above figures, r_m varied between 0.27 and 0.28 for $m \geqslant 1000$. Therefore, it was not even justified that r_m tends to zero. This difficulty can, however, be overcome, since it can be accepted quite safely that if r_m is written as $g(m)/\log m$, then $g(m)$ is "much smaller" than $\log m$, from which $r_m \to 0$ can be concluded.

The above situation rarely occurs in practice. However, the difficulties it emphasizes are very frequent. ▲

3.9. LIMITING FORMS OF THE INEQUALITIES OF SECTION 1.4

While for several applied models the mixing concept of Section 3.7 is appropriate, for others it has serious disadvantages. One disadvantage, which is only an inconvenience in some cases, is that the concept is defined in such a way that a relabeling of the random variables is not possible. In other words, if the observations came in a different order, it is not guaranteed that mixing was preserved. Another disadvantage, which makes the concept inapplicable in certain situations, is that any member of the sequence of random variables should be asymptotically independent of all others if their subscripts are sufficiently far apart. For example, if the X_j represent the random life lengths of the components of a complicated equipment, there may be no obvious way of labeling the components by which the above property would hold. This is avoided in the model that follows, which also describes a situation where asymptotic independence is stressed but with weaker assumptions than in a mixing model. It implies a wider freedom in applications. Although we describe the model in general terms, the reader may find it convenient to translate everything to "life lengths of components." The mathematical foundation of this model was laid down in Section 1.4.

For a given sequence $X_1, X_2, \ldots, X_n$ of random variables, we introduce a set E_n of so-called exceptional pairs (i,j), $i<j$, of the subscripts as follows. We place (i,j) into E_n if it is not reasonable to assume (or, in mathematical arguments, if it fails to hold) that, as x_n tends to $\max[\alpha(F_i), \alpha(F_j)]$, $P(X_i < x_n, X_j < x_n)$ is asymptotically $F_i(x_n) F_j(x_n)$. Here, as usual, $F_t(x)$ denotes the distribution function of X_t.

Example 3.9.1. Let $X_1, X_2, \ldots, X_n$ be independent. Then E_n is empty. ▲

Example 3.9.2. Let $X_1, X_2, \ldots, X_n$ be m-dependent. Then $E_n = \{(i,j) : 1 \leq i < j < i + m, j \leq n\}$. Hence, the number $N(n)$ of elements of E_n equals $(m-1)(n-m) + \binom{m}{2}$. ▲

Example 3.9.3. Let $X_1, X_2, \ldots, X_n$ be such that $X_2, X_3, \ldots, X_n$ are independent but X_1 and X_j are strongly dependent for each j. Then $E_n = \{(1,j) : 1 < j \leq n\}$. Here, $N(n) = n - 1$. ▲

Notice that the case of Example 3.9.3 is not covered by any model of the previous sections, although the asymptotic properties of the extremes can easily be reduced to the case of independence. A slight modification of it, however, will require a new argument for finding the asymptotic distribution of extremes. Let us look at an example.

Example 3.9.4. Let $X_1, X_2, \ldots, X_n$ be such that X_i and X_j cannot be considered asymptotically independent in the sense of the definition of E_n, whenever (i,j) is an element of $E_n = \{(i,j) : \text{either } i = 1 \text{ and } j > 1; \text{ or } 1 \leq i < n \text{ and } j = i + 1; \text{ or } 1 \leq i < \frac{1}{2}n \text{ and } j = 2i\}$. In this case, $N(n) \leq 2.5n$. ▲

From the definition of E_n it follows that if $1 \leq i_1 < i_2 < \cdots < i_k \leq n$ are such that no pairs from them are contained in E_n, then the events $\{X_{i_j} < x_n\}$ are pairwise asymptotically independent as x_n tends to the largest of $\alpha(F_t)$. In building our model we assume more, namely, that pairwise independence can be extended to independence. We now specify our model, in which we use the notations (34), (35), and (36). Furthermore, as in the examples above, we put $N(n)$ for the number of the elements of E_n. We call $X_1, X_2, \ldots, X_n$ an E_n-sequence, if the following three assumptions are satisfied.

Assumption 1. If the subscripts $i(k) = (i_1, i_2, \ldots, i_k)$ contain no pairs from E_n, then the difference

$$d_{i(k)}(x_n) = F^*_{i(k)}(x_n) - \prod_{t=1}^{k} F_{i_t}(x_n)$$

is negligible compared with either of the terms as $x_n \to \sup_{t \geq 1} \alpha(F_t)$.

Assumption 2. If there is exactly one pair (i_s, i_m) among the components of $i(k) = (i_1, i_2, \ldots, i_k)$ which belongs to E_n, then

$$F^*_{i(k)}(x_n) \leq \eta_k P(X_{i_s} < x_n, X_{i_m} < x_n) \prod_{\substack{t=1 \\ t \neq m, s}}^{k} F_{i_t}(x_n),$$

where η_k is a constant.

Assumption 3. $N(n) = o(n^2)$.

Notice that, for an E_n-sequence, there is no assumption on the interdependence of $(X_{i_1}, X_{i_2}, \ldots, X_{i_k})$, if $(i_1, i_2, \ldots, i_k)$ contains more than one pair from E_n. In other words, if we consider a subset of the original set of random variables, which contains at least two pairs about which we could not accept asymptotic independence, then this subset can have an arbitrary structure. Furthermore, we did not assume that the X_j are identically distributed. It should also be emphasized that Assumption 2 is far less than asymptotic independence, even in the weak sense of considering each X_j falling below a fixed number x_n, for the constant η_k can be arbitrarily large. As a final comment, let us add that Assumption 3 is a natural one. As a matter of fact, the number of all pairs of the subscripts $1, 2, \ldots, n$ is $\binom{n}{2}$, which is of the order of n^2. Hence, Assumption 3 requires that a positive percentage of all pairs cannot be exceptional.

We now state an important theorem.

Theorem 3.9.1. Let $X_1, X_2, \ldots, X_n$ be an E_n-sequence. Let x_n be such that, as $n \to +\infty$,

$$\sum_{j=1}^{n} F_j(x_n) \to a, \qquad 0 < a < +\infty. \tag{93}$$

Let us assume that there is a constant K such that, for all j and n, $nF_j(x_n) \leq K$. Let, finally,

$$\lim_{n=+\infty} \sum_{(i,j)\in E_n} P(X_i < x_n, X_j < x_n) = 0. \tag{94}$$

Then, as $n \to +\infty$,

$$\lim P(W_n < x_n) = 1 - e^{-a}. \tag{95}$$

In particular, if $F_j(x) = F(x)$ for all j, then the theory of Chapter 2 applies to W_n.

Let us define an E_n^*-sequence by changing $\{X_j < x_n\}$ to $\{X_j \geq x_n\}$, and thus $\alpha(F_j)$ to $\omega(F_j)$, in the definition of an E_n-sequence. Then $X_1, X_2, \ldots, X_n$ is an E_n^*-sequence if, and only if, $(-X_1), (-X_2), \ldots, (-X_n)$ is an E_n-sequence. Hence Theorem 3.9.1 yields the following result for Z_n.

Theorem 3.9.2. Let $X_1, X_2, \ldots, X_n$ be an E_n^*-sequence. Let x_n be such that, as $n \to +\infty$,

$$\sum_{j=1}^{n} [1 - F_j(x_n)] \to A, \qquad 0 < A < +\infty.$$

Let us assume that there is a constant K^* such that, for all n and j, $n[1 - F_j(x_n)] \leq K^*$. Furthermore, let

$$\lim_{n=+\infty} \sum_{(i,j)\in E_n^*} P(X_i \geq x_n, X_j \geq x_n) = 0.$$

Then, as $n \to +\infty$,

$$\lim P(Z_n < x_n) = e^{-A}.$$

In particular, if $F_j(x) = F(x)$ for all j, then the theory of Chapter 2 applies to Z_n.

Proof of Theorem 3.9.1. The proof is based on the inequalities of Theorem 1.4.2, in which the following notations are used. Let x_n be specified in such a way that (93) holds. Let $C_j = \{X_j < x_n\}$, $1 \leq j \leq n$. Then, of course, for $i(k) = (i_1, i_2, \ldots, i_k)$, $1 \leq i_1 < i_2 < \cdots < i_k \leq n$,

$$P(C_{i_1} C_{i_2} \cdots C_{i_k}) = F_{i(k)}^*(x_n).$$

We introduce the following sums.

$$S_{1,n}^* = S_{1,n}^{**} = S_{1,n} = \sum_{j=1}^{n} F_j(x_n)$$

and, for $k \geq 2$,

$$S^*_{k,n} = \sum_k^* F^*_{i(k)}(x_n), \qquad S^{**}_{k,n} = \sum_k^{**} F^*_{i(k)}(x_n),$$

where $\sum_k^*$ and $\sum_k^{**}$ denote summations over $i(k)=(i_1,i_2,\ldots,i_k)$, $1 \leq i_1 < i_2 < \cdots < i_k \leq n$, which contain no pairs, and at most one pair, respectively, from the exceptional set E_n. The fact that none of the C_j, $1 \leq j \leq n$, occurs is equivalent to $\{W_n \geq x_n\}$. Thus, Theorem 1.4.2 yields that, for any fixed integer $m \geq 0$,

$$1 - S^{**}_{1,n} + S^*_{2,n} - S^{**}_{3,n} + \cdots - S^{**}_{2m+1,n}$$
$$\leq P(W_n \geq x_n)$$
$$\leq 1 - S^*_{1,n} + S^{**}_{2,n} - S^*_{3,n} + \cdots + S^{**}_{2m,n}. \qquad (96)$$

We now complete the proof by showing that, for any $k \geq 1$, as $n \to +\infty$,

$$\lim(S^{**}_{k,n} - S^*_{k,n}) = 0 \qquad (97)$$

and

$$\lim S^*_{k,n} = \frac{a^k}{k!}. \qquad (98)$$

Namely, if we apply (97) and (98) in (96), we get that, for any fixed $m \geq 0$, as $n \to +\infty$,

$$\sum_{k=0}^{2m+1} (-1)^k \frac{a^k}{k!} \leq \liminf P(W_n \geq x_n)$$
$$\leq \limsup P(W_n \geq x_n) \leq \sum_{k=0}^{2m} (-1)^k \frac{a^k}{k!},$$

from which (95) is immediate by m's being arbitrary. It remains therefore to prove (97) and (98). Both hold for $k=1$, (97) by definition and (98) by condition (93). We thus assume $k \geq 2$. By Assumption 2,

$$0 \leq S^{**}_{k,n} - S^*_{k,n} \leq \eta_k S^{k-2}_{1,n} \sum_{(i,j) \in E_n} P(X_i < x_n, X_j < x_n),$$

which tends to zero on account of (93) and (94). This proves (97). Turning to (98), we first write

$$S^*_{k,n} = \sum_k^* \left[F_{i_1}(x_n) F_{i_2}(x_n) \cdots F_{i_k}(x_n) + d_{i(k)}(x_n) \right].$$

3.9 LIMITING FORMS OF THE INEQUALITIES OF SECTION 1.4

By Assumption 1 and by the condition $F_j(x_n) \leq K/n$,

$$d_{i(k)}(x_n) = o[F_{i_1}(x_n)F_{i_2}(x_n) \cdots F_{i_k}(x_n)] = o(n^{-k}).$$

Therefore,

$$\sum\nolimits_k^* d_{i(k)}(x_n) = o\left[\binom{n}{k} n^{-k}\right] = o(1).$$

Furthermore, the nonnegative difference

$$\sum_{1 \leq j_1 < j_2 < \cdots < j_k \leq n} F_{j_1}(x_n) F_{j_2}(x_n) \cdots F_{j_k}(x_n) - \sum\nolimits_k^* F_{i_1}(x_n)F_{i_2}(x_n)\cdots F_{i_k}(x_n)$$

$$\leq N(n) \left[\max_{1 \leq i < j \leq n} F_i(x_n) F_j(x_n)\right] S_{1,n}^{k-2}$$

$$\leq \frac{K^2 N(n)}{n^2} S_{1,n}^{k-2},$$

which, by Assumption 3 and by (93), tends to zero. Collecting the above estimates, we obtained

$$S_{k,n}^* = \sum_{1 \leq j_1 < j_2 < \cdots < j_k \leq n} F_{j_1}(x_n) F_{j_2}(x_n) \cdots F_{j_k}(x_n) + o(1).$$

If we appeal once more to the condition $F_j(x_n) \leq K/n$ for all j, we get

$$\left|S_{k,n}^* - \frac{1}{k!} S_{1,n}^k\right| \leq \frac{K}{n} S_{1,n}^{k-1} + o(1) \to 0$$

by $S_{1,n}$'s being bounded. Condition (93) now leads to (98), which completes the proof of (95).

The particular case $F_j(x) = F(x)$ for all j indeed reduces to the theory of Chapter 2, since (93) becomes $nF(x_n) \to a$. This is exactly the rule of Chapter 2 to determine $x_n = c_n + d_n x$ for normalizing W_n. Furthermore, the form of the limit law for W_n was exactly (95). Here, of course, $a = a(x)$. This completes the proof of the theorem. ▲

The reader is advised to go through Examples 3.9.1–3.9.4 once again and to translate the conditions of Theorems 3.9.1 and 3.9.2 to these specific models. In particular, it should be realized that, for the i.i.d. case, no new assumption is made, hence the present model extends the i.i.d. case. Furthermore, for m-dependent variables, stationarity is not needed in Theorem 3.7.1 (see also Remark 3.7.2).

Notice that E_n can be empty even though the X_j are dependent—namely, when the approximation expressed in Assumption 1 holds for all

$i(k)$. In this case the proof can be shortened by applying Theorem 1.4.1 rather than Theorem 1.4.2.

For important applications of the model, see Section 3.12.

3.10. MINIMUM AND MAXIMUM OF INDEPENDENT VARIABLES

The model of the preceding section reduced the asymptotic theory of minima and maxima of dependent variables to independent ones. However, we investigated independent variables only under the additional assumption of the variables' being identically distributed. We now supplement these results.

In view of the discussion in Section 3.3 (see particularly Example 3.3.3), we have to make some restriction on the individual terms $F_j(x)$ of the distribution functions as well as on the normalizing constants. For this section we adopt the following concepts.

Uniformity Assumption for the Minimum. We say that a sequence $F_1(x)$, $F_2(x),\ldots$ of distribution functions and sequences c_n and $d_n > 0$ of normalizing constants satisfy the uniformity assumption for the minimum if, as $n \to +\infty$,

$$\lim \max \{ F_j(c_n + d_n x) : 1 \leq j \leq n \} = 0 \qquad (99)$$

and, for any fixed number $0 < t \leq 1$,

$$\lim \sum_{j=1}^{nt} F_j(c_n + d_n x) = w(t; x) \qquad (100)$$

exists which is finite for all $0 < t \leq 1$ whenever it is finite for $t = 1$. (Recall the convention that if u is a limit in a summation, then we mean by u its integer part.)

The uniformity assumption for the maximum is similarly defined except that F_j is to be replaced by $1 - F_j$ in both limit relations.

The class of possible nondegenerate limiting distributions for the minima under the uniformity assumption is characterized in the following statement. When we use log z, we always understand $z > 0$.

Theorem 3.10.1. *Under the uniformity assumption for the minimum, a nondegenerate distribution function $L(x)$ is the limiting distribution of $(W_n - c_n)/d_n$, for some sequence $X_1, X_2, \ldots, X_n$ of independent random variables and for some sequences c_n and $d_n > 0$ of normalizing constants if, and only if, either* (i) $\log[1 - L(x)]$ *is concave or* (ii) $\omega(L)$ *is finite and* $\log \{1 - L[\omega(L) - $

3.10 MINIMUM AND MAXIMUM OF INDEPENDENT VARIABLES

$e^{-x}]\}$ is concave, or finally, (iii) $\alpha(L)$ is finite and $\log\{1 - L[\alpha(L) + e^x]\}$ is concave, where, in (ii) and (iii), $x > 0$.

By applying the above theorem to the sequence $-X_j, 1 \leq j \leq n$, we get the following result.

Theorem 3.10.2. *Under the uniformity assumption for the maximum, a nondegenerate distribution function $H(x)$ is the limiting distribution of $(Z_n - a_n)/b_n$, for some sequence $X_1, X_2, \ldots, X_n$ of independent random variables and for some sequences a_n and $b_n > 0$ of normalizing constants if, and only if, either* (i) $\log H(x)$ *is concave or* (ii) $\omega(H) < +\infty$ *and* $\log H[\omega(H) - e^{-x}]$ *is concave, where $x > 0$, or, finally,* (iii) $\alpha(H)$ *is finite and* $\log H[\alpha(H) + e^x]$, $x > 0$, *is concave.*

Proof of Theorem 3.10.1. By the basic formulas and by independence

$$P(W_n \geq y) = \prod_{j=1}^{n} [1 - F_j(y)].$$

Thus, if $P(W_n \geq c_n + d_n x) > 0$,

$$\log P(W_n \geq c_n + d_n x) = \sum_{j=1}^{n} \log[1 - F_j(c_n + d_n x)].$$

If we put

$$m_n = m_n(x) = \max\{F_j(c_n + d_n x) : 1 \leq j \leq n\},$$

the Taylor expansion

$$|\log(1 - z) + z| \leq z^2 \quad \text{for } |z| \leq \tfrac{1}{2}$$

and (99) yield that, if F_j, $1 \leq j \leq n$, c_n and d_n satisfy the uniformity assumption,

$$\log P(W_n \geq c_n + d_n x) = -[1 + O(m_n)] \sum_{j=1}^{n} F_j(c_n + d_n x). \qquad (101)$$

Furthermore, for the same reason, if $0 < t \leq 1$,

$$\log \prod_{j=1}^{nt} [1 - F_j(c_n + d_n x)] = [1 + O(m_n)] \sum_{j=1}^{nt} F_j(c_n + d_n x). \qquad (102)$$

Now, if $L(x)$ is the limiting distribution of $(W_n - c_n)/d_n$, and if the uniformity assumption holds, then, for all $x < \omega(L)$, (101) implies that

$w(1; x)$ of (100) is finite. Hence, once again by the uniformity assumption, $w(t; x)$ is finite for all $0 < t \leq 1$. This fact, in turn, on account of (102), yields that, as $n \to +\infty$,

$$P(W_{nt} \geq c_n + d_n x) \to \exp[-w(t; x)], \quad 0 < t \leq 1, \quad (103)$$

where the subscript nt is to be read as its integer part (we adopt this convention for the sequel of this proof). But, by the definition of $L(x)$,

$$P(W_{nt} \geq c_{nt} + d_{nt} x) \to 1 - L(x).$$

Comparing these last two limits, we conclude from Lemma 2.2.3 that, as $n \to +\infty$,

$$\lim \frac{d_{nt}}{d_n} = B_t, \quad \lim \frac{c_{nt} - c_n}{d_n} = A_t, \quad (104)$$

exist and $B_t > 0$. Furthermore, for $0 < t \leq 1$,

$$w(t; x) = -\log[1 - L(A_t + B_t x)]. \quad (105)$$

We now conclude the proof of one part of the theorem as follows. From (104), we specify the possible forms of B_t and A_t. Namely, we prove that, for all $0 < t \leq 1$, either

$$B_t = 1 \quad \text{and} \quad A_t = k \log t \quad (106)$$

or

$$B_t = t^m \quad \text{and} \quad A_t = k(t^m - 1), \quad (107)$$

where k and m are suitable constants. We then use the relation

$$\prod_{j=nt+1}^{n} [1 - F_j(c_n + d_n x)] \to \frac{1 - L(x)}{1 - L(A_t + B_t x)}, \quad (108)$$

which immediately follows from (103) and (105). Since the left hand side of (108) is one minus a distribution function, it is decreasing in x; consequently so is the right hand side. But if (106) applies, then the decreasing property of the right hand side of (108) is equivalent to $\log[1 - L(x)]$'s being concave. On the other hand, if (107) applies, then we write the right hand side of (108) as

$$\frac{1 - L[(x+k) - k]}{1 - L[t^m(x+k) - k]}. \quad (109)$$

Since it is decreasing both in x and in t, $(-k)$ is either $\alpha(L)$ or $\omega(L)$ according as $m<0$ or $m>0$. Therefore, the fact that (109) decreases in x implies that either $\log\{1-L[\alpha(L)+e^y]\}$ or $\log\{1-L[\omega(L)-e^{-y}]\}$ is concave, where $y>0$. We have thus obtained the three claimed classes as possibilities for $L(x)$, provided that (106) and (107) hold.

For proving (106) and (107), we appeal to (104). We write, for $0<s$, $t \leq 1$,

$$\frac{d_{nts}}{d_n} = \frac{d_{nts}}{d_{nt}} \cdot \frac{d_{nt}}{d_n}$$

and

$$\frac{c_{nts}-c_n}{d_n} = \frac{c_{nts}-c_{nt}}{d_{nt}} \cdot \frac{d_{nt}}{d_n} + \frac{c_{nt}-c_n}{d_n}.$$

These yield

$$B_{st} = B_s B_t, \qquad A_{st} = A_s B_t + A_t = A_t B_s + A_s.$$

Since $B_1 = 1$ and B_t is monotonic in t, with the substitution $s = e^{-u}$, $t = e^{-v}$, $u, v > 0$, Lemma 1.6.1 yields that either $B_t = 1$ for all t or $B_t = t^m$ with some constant $m \neq 0$. The corresponding equations for A_t are therefore

$$A_{st} = A_s + A_t \qquad (B_t = 1)$$

or

$$A_s(t^m - 1) = A_t(s^m - 1) \qquad (B_t = t^m).$$

The latter implies that $A_t/(t^m - 1)$ is a constant k, which proves (107). On the other hand, if the first case holds, we represent $A_t = \log C_t$, $C_t > 0$. Since, on account of (103) and (105), A_t is monotonic, so is C_t. Hence, the equation for A_t reduces to $C_{ts} = C_t C_s$ for C_t with C_t monotonic. Therefore, just as for B_t, $C_t = t^k$ with some constant k. Consequently, $A_t = k \log t$, as was stated in (106). This completes the proof of our claim on the possible forms of $L(x)$.

Let us now turn to the converse. Let $L(x)$ be a nondegenerate distribution function and let $\log[1 - L(x)]$ be concave. We construct a sequence $F_j(x)$, $1 \leq j \leq n$, of distribution functions and specify sequences c_n and $d_n > 0$ of real numbers with the following properties: (i) the uniformity assumption for the minimum holds and (ii) if $X_1, X_2, \ldots$ are independent random variables with distribution functions $F_1(x), F_2(x), \ldots$, then $(W_n - c_n)/d_n$ converges weakly to $L(x)$. Namely, let $F_1(x) = L(x)$ and, for $j \geq 2$, define

$$F_j(x) = 1 - \frac{1 - L(x + \log j)}{1 - L[x + \log(j-1)]}.$$

Since $L(x)$ is a distribution function and $\log[1 - L(x)]$ is concave, each $F_j(x)$ is a distribution function. We next define $c_n = -\log n$ and $d_n = 1$. Then (99) is immediate. On the other hand, under (99), the sum in (100) is asymptotically equal to the sum of $\log[1 - F_j(c_n + d_n x)]$, which in turn is the logarithm of the product of $1 - F_j(c_n + d_n x)$. This product, for our specific form of $F_j(x)$, simplifies to

$$\log\{1 - L(x - \log n + \log[nt])\},$$

where $[nt]$ signifies here the integer part of nt. This evidently converges to $\log\{1 - L(x + \log t)\}$, which is finite for all $0 < t \leq 1$ whenever it is finite for $t = 1$. Hence, the uniformity assumption for the minimum holds. As a side result, we have also obtained our second claim of $W_n + \log n$ converging weakly to $L(x)$.

If $\log\{1 - L(x)\}$ is not concave but $\omega(L) < +\infty$, and $\log\{1 - L(\omega(L) - e^{-x})\}$ is concave, or $\alpha(L)$ is finite and $\log\{1 - L[\alpha(L) + e^x]\}$ is concave for $x > 0$, the construction is similar in principle to the one in the preceding paragraph except that we now aim at the normalizing constants corresponding to (107). We therefore omit details. This concludes the proof. ▲

There are no general criteria by which one could decide if, for a given sequence $F_j(x), j \geq 1$, of distribution functions, the minimum or maximum, when suitably normalized, would have a limit law. Some special cases are represented in Exercises 21–23.

3.11. THE ASYMPTOTIC DISTRIBUTION OF THE kTH EXTREMES

In the case of i.i.d. variables, the asymptotic distribution theory of extremes did not require any addition to the theory developed for the maximum and minimum. However, for dependent random variables, several new problems may arise. The most significant new problem is that it is not at all sure that, if the maximum or minimum can be properly normalized to have an asymptotic distribution, then so can the other extremes (see Exercise 15). In the present section we give some criteria

3.11 THE ASYMPTOTIC DISTRIBUTION OF THE kTH EXTREMES

which guarantee the existence of normalizing constants with which all upper or all lower extremes have limiting distributions.

We first recall from Section 1.4 that the exact distribution of order statistics can always be reduced to the distribution of the number of occurrences in special sequences of events. As a matter of fact, if we define the events $A_j = A_j(x) = \{X_j \geq x\}$ and $B_j = B_j(x) = \{X_j < x\}$, where $X_1, X_2, \ldots$ are the basic random variables and x is a real variable, and if $\nu_n(x, A)$ and $\nu_n(x, B)$ are the numbers which occur among $A_1, A_2, \ldots, A_n$ and $B_1, B_2, \ldots, B_n$, respectively, then the distributions of $\nu_n(x, A)$ and $\nu_n(x, B)$ lead to the distribution functions of the upper and lower extremes, respectively. Thus, Theorem 1.4.1 yields the following basic limit theorem.

Theorem 3.11.1. *Let the random variables $X_1, X_2, \ldots, X_n$ and the sequences a_n and $b_n > 0$ of real numbers be such that, for any fixed integer $j \geq 0$,*

$$\lim S_j(a_n + b_n x) = u_j(x) \tag{110}$$

exist, and are finite on an interval (α, ω), where

$$S_j(x) = \sum_{1 \leq i_1 < i_2 < \cdots < i_j \leq n} P(X_{i_1} \geq x, X_{i_2} \geq x, \ldots, X_{i_j} \geq x).$$

Assume that the series

$$U_t(x) = \sum_{k=0}^{+\infty} (-1)^k \binom{k+t}{t} u_{k+t}(x), \qquad \alpha < x < \omega,$$

converges. Then, for $\alpha < x < \omega$, as $n \to +\infty$,

$$\lim P(X_{n-k+1:n} < a_n + b_n x) = \sum_{t=0}^{k-1} U_t(x).$$

Proof. Notice that $S_j(x)$ is the jth binomial moment of $\nu_n(x, A)$ (see Lemma 1.4.1). Hence, by Theorem 1.4.1, for any integer $s \geq 0$,

$$\sum_{k=0}^{2s+1} (-1)^k \binom{k+t}{t} S_{k+t} \leq P(\nu_n = t) \leq \sum_{k=0}^{2s} (-1)^k \binom{k+t}{t} S_{k+t},$$

where $S_{k+t} = S_{k+t}(a_n + b_n x)$ and $\nu_n = \nu_n(a_n + b_n x, A)$. Let us fix the integers

t and s, and let $n \to +\infty$. From the assumption (110), we get, for $\alpha < x < \omega$,

$$\sum_{k=0}^{2s+1} (-1)^k \binom{k+t}{t} u_{k+t}(x) \leq \liminf P(\nu_n = t)$$

$$\leq \limsup P(\nu_n = t)$$

$$\leq \sum_{k=0}^{2s} (-1)^k \binom{k+t}{t} u_{k+t}(x).$$

Since s is arbitrary and $U_t(x)$ is well defined for all x above, $\lim P(\nu_n = t)$ exists as $n \to +\infty$ and it equals $U_t(x)$. Formula (35) of Section 1.4 thus completes the proof. ▲

Corollary 3.11.1. *With the notations of Theorem 3.11.1, let us assume that $u_j(x) = u^j(x)/j!$ with some $u(x) \geq 0$. Then the series for $U_t(x)$ converges for all x for which $u(x)$ is finite. Furthermore, as $n \to +\infty$,*

$$\lim P(X_{n-k+1:n} < a_n + b_n x) = e^{-u(x)} \sum_{t=0}^{k-1} \frac{u^t(x)}{t!}.$$

The statement above is an easy consequence of Theorem 3.11.1 and can in fact be obtained by a substitution for $u_j(x)$. Notice that it contains, as a special case, Theorem 2.8.1. However, it can also be applied to dependent variables such as m-dependent, mixing or Gaussian sequences.

In many applications the assumption (110) turns out to be too strong. Notice that a few exceptional members of the sequence $X_1, X_2, \ldots$ can spoil a property of an "average type," while they may have no effect on the extremes. In such situations the method of the following example makes Theorem 3.11.1 still applicable.

Example 3.11.1. Let $X_1, X_2, \ldots, X_n$ be unit exponential variates. Assume that most of these variables are independent but a few among them follow a bivariate exponential distribution of the Frechet type. To be specific, let T_1 be the set of complete squares and T_2 the set of all other integers. If all subscripts belong to T_1 or if all of them to T_2, then let the corresponding X's be independent. However, if $i \in T_1$ and $j \in T_2$, we assume

$$P(X_i \geq x, X_j \geq x) = \tfrac{1}{2} e^{-x}, \qquad x \geq \log 2.$$

Let $a_n = \log n, b_n = 1$. Then $S_1(\log n + x) = e^{-x}$ but

$$S_2(\log n + x) \geq \tfrac{1}{2} n^{3/2} \left[\tfrac{1}{2} \exp(-\log n - x) \right] \to +\infty.$$

Here we calculated the contribution to S_2 only of those pairs which are dependent. Their number certainly exceeds $\frac{1}{2}n^{3/2}$ for $n \geq 5$. Therefore, Theorem 3.11.1 is not applicable even to Z_n. However, if we write

$$Z_n = \max(Z_{n,1}, Z_{n,2})$$

with

$$Z_{n,i} = \max\{X_j : 1 \leq j \leq n, j \in T_i\}, \qquad i = 1, 2,$$

then

$$P(Z_n \geq \log n + x) = P(Z_{n,2} \geq \log n + x)$$
$$+ P(Z_{n,2} < \log n + x, Z_{n,1} \geq \log n + x).$$

Estimating

$$P(Z_{n,2} < \log n + x, Z_{n,1} \geq \log n + x) \leq P(Z_{n,1} \geq \log n + x)$$
$$= 1 - P(Z_{n,1} < \log n + x)$$
$$\leq 1 - \left(1 - \frac{e^{-x}}{n}\right)^{\sqrt{n}-1}$$
$$= O(n^{-1/2}),$$

we reduce the distribution of Z_n to $Z_{n,2}$, to which the results of Chapter 2 are applicable. Should the variables be such that they are dependent, but, on T_2, Theorem 3.11.1 is applicable, we would have obtained a limit theorem for Z_n itself. The argument is similar with the other upper extremes. Evidently, the specific structure of T_1 is not essential; its role is that its number of elements is much smaller than that for T_2, when they are restricted to the first n integers. ▲

The above discussion can easily be transformed to the investigation of lower extremes by reversing the inequality in the definition of $S_j(x)$. We thus omit details.

The reader who is fimiliar with the asymptotic theory of sums of triangular arrays of (dependent) random variables will realize that the asymptotic distribution of $(X_{n-k+1:n} - a_n)/b_n$ can also be obtained as a part of that theory. Indeed, if $Y_{j,n}$ denotes the indicator variable of the event $\{X_j \geq a_n + b_n x\}$, then

$$\nu_n(a_n + b_n x) = \sum_{j=1}^{n} Y_{j,n}.$$

However, such an approach already assumes that we know a_n and $b_n > 0$. Therefore, an essential part of the present theory does not follow from the theory of sums (the so-called central limit problem). In addition, the analytic tools applied in the theory of sums usually require different types of assumptions than the present direct and elementary approach.

3.12. SOME APPLIED MODELS

The applied scientist faces a more difficult problem than the mathematician, since he must choose an appropriate model for the specific situation he is investigating. The mathematician can make assumptions under which neat solutions can be found, but the applied scientist must confront situations in which a slight error may lead to a huge loss. The problem investigated more or less specifies the random variables $X_1, X_2, \ldots, X_n$ (Sections 1.1 and 1.2) but seldom specifies their structure. It is usually not hard to decide if they are independent and/or identically distributed. The more difficult decision is the next step: what type of dependence to use and what the common distribution is if the variables are identically distributed. In the present section we describe some general rules and discuss some specific models as well.

3.12.1. Exact Models Via Characterization Theorems

The most pleasant case in applications arises when simple nonmathematical properties completely specify the model to be applied. We have described a number of situations of this nature in Section 1.6. We add one more example here, in the solution of which we use asymptotic theory. The emphasis in the example is that we make only logical, but nonmathematical, assumptions, from which we deduce that there is a unique model for the problem.

An equipment composed of similar components breaks down when the first component does. The components function independently of each other. We accept that the random time X to the first breakdown has a distribution similar to those of the components X_j, $1 \leq j \leq n$. Experience shows that $nE(X) = E(X_j)$. We deduce that the distribution both of the components and of the equipment is exponential.

Notice that the assumptions can be translated as X_j, $1 \leq j \leq n$, are i.i.d. and $X = W_n$, the minimum of the X_j. Furthermore, the distribution of W_n is of the same type as the distribution of X_1. That is, if $P(X_1 < x) = F(x)$, then $P(W_n < x) = F(C_n + D_n x)$ with some constants C_n and $D_n > 0$. Finally, $nE(W_n) = E(X_1)$. The first part of the assumptions is logical; that is, n is arbitrary. The equation $nE(W_n) = E(X_1)$, however, is based on

3.12 SOME APPLIED MODELS

experience; therefore n is either a given value or one of a few possible values. Hence, in the solution, we proceed as follows. By assumption, for all n,

$$F(C_n + D_n x) = P(W_n < x) = 1 - [1 - F(x)]^n,$$

or

$$[1 - F(c_n + d_n x)]^n = 1 - F(x),$$

where $c_n = -C_n/D_n$ and $d_n = 1/D_n$. Thus, by Theorem 2.4.2, $F(x)$ is one of the three types $L_{1,\gamma}(x)$, $L_{2,\gamma}(x)$, or $L_{3,0}(x)$. In fact, since $X_j \geq 0$, $\alpha(F)$ is finite, therefore $F(x)$ is of the type $L_{2,\gamma}(x)$. Evaluating $E(W_n)$ and $E(X_1)$ for $L_{2,\gamma}(x)$, the assumed equation for these expectations yields $\gamma = 1$, which is the exponential distribution (see also Theorem 1.6.2).

Unfortunately, the assumptions are valid for a very special equipment only. Very rarely can one arrive at a unique model to describe failures or other practical situations. In the next subsection, however, we give a method which leads to a unique type of distribution for some important random quantities.

3.12.2. Exact Distributions Via Asymptotic Theory (Strength of Materials)

Let a piece of metal have random strength X. Let its distribution be $F(x)$. We assume that X is proportional to the size of a particular sheet of metal. In other words, if we theoretically cut the sheet into n equal parts, then each piece will have the same distribution and this distribution is of the same type as $F(x)$. Let $X_1, X_2, \ldots, X_n$ be the strengths of the n equal pieces of the original sheet. We also assume that the sheet breaks at its weakest point, that is, $X = W_n$, the minimum of X_j, $1 \leq j \leq n$. Finally, we assume that the pieces close to each other may be strongly dependent, but the dependence weakens with distance. We shall show that X has a Weibull distribution.

Again, the assumptions are logical ones, free of mathematical niceties. Some of these assumptions easily translate into mathematical formulas, but there is no unique way of reformulating the dependence structure which weakens with distance. We have two models to choose from: the mixing model of Section 3.7, and the so-called E_n-sequences of Section 3.9. We should reject the mixing model for two reasons. The more serious is that, if the sheet is divided into $n = m^2$ equal parts in such a way that—assuming the sheet is rectangular—both edges are divided into m equal parts, then the neighboring pieces on the sheet will not be successive terms in the sequence $X_1, X_2, \ldots, X_n$. Therefore, indices' being far apart does not mean distant neighbors. The second objection is the choice of $\tau(s,u)$, which does not have any practical equivalent. On the other hand, the model of Section 3.9 is very appropriate. Its assumptions are made in general terms, and one does not have to accept mathematical conditions which cannot be checked.

Let us go through the model of Section 3.9, as it applies to the present problem.

Let the sheet be rectangular, and let us divide each edge into m equal parts; this divides the sheet into $n=m^2$ pieces. Let us denote by X_{ij} the strength of the piece obtained by the ith horizontal and the jth vertical division. We shall refer to this peice as the (i,j)th piece. We say that the (i_1,j_1)th and (i_2,j_2)th pieces are s-close neighbors if both $|i_1-i_2|$ and $|j_1-j_2|$ are smaller than s. By assumption, if two pieces are close, their strengths are dependent. Thus, for a given s, let $E_{n,s}$ be the set of all pairs $\{(i_1,j_1),(i_2,j_2)\}$ that are s-close. Now, our assumption is that, if s is large, then elements which are not in $E_{n,s}$ are almost independent. This is exactly the Assumptions 1 and 2 of Section 3.9, expressing almost independence in the weakest mathematical terms.

Let us look at Assumption 3. This is related to the choice of s in $E_{n,s}$. Since the number $N(n)$ of terms of $E_{n,s}$ is of the order of ns^2, the assumption $N(n)=o(n^2)$ means that $s=o(m)$. But this is reasonable in terms of the practical model, if we look at a middle piece of the original sheet. If s were as large as m, then this middle piece would be dependent on all other pieces, since its s-close neighbors would exhaust the whole sheet. In other words, if, with increasing m, the dependent neighbors do not cover the whole, or a positive percentage of the original sheet for a fixed piece, then $s=o(m)$, which is exactly Assumption 3.

Therefore, we can conclude that the sequence X_{ij} is an E_n-sequence and they are identically distributed. This makes Theorem 3.9.1 applicable, which states that the minimum of an identically distributed E_n-sequence has the same asymptotic properties as in the case of i.i.d. variables. But, for each n, the minimum of X_{ij} is X itself. Consequently, the asymptotic distribution of the minimum, on the one hand, is one of the three possible types $L_{1,\gamma}(x)$, $L_{2,\gamma}(x)$, and $L_{3,0}(x)$; on the other hand, it is $F(x)$. It is evident that $\alpha(F)$ is finite; thus $F(x)$ is of the same type as $L_{2,\gamma}(x)$—that is, $F(x)$ is a Weibull distribution.

While we made the deduction above for the special problem of strength of metals, the method can be formulated as a general principle. If a random measurement X is the minimum of X_j, $1 \leq j \leq n$, where the X_j are the X-values of members of a division of an item measured by X, then X has a Weibull distribution under the following conditions. The distributions of X and X_j belong to the same type, and, if the sizes of two members of a division are equal, then the corresponding X-values are identically distributed. Furthermore, distant neighbors in the division are asymptotically independent.

Among the assumptions in the previous general Weibull model was that X is the minimum of the X-values of the members of the division. This fact is frequently referred to as the weakest-link principle. Whenever X can be related to the strength of a chain, where the division corresponds to the links, then the above property expresses the fact that a chain is as strong as its weakest link. It should be emphasized, however, that this principle is only a part of the model, which alone would not permit us to obtain the exact distribution of X.

Here, we have obtained that the type of the distribution function $F(x)$ of X is Weibull. Hence,

$$F(x) = L_{2,\gamma}(C + Dx) = 1 - \exp\left[-(C + Dx)^{\gamma}\right]$$

for $C + Dx > 0$, and $F(x) = 0$ otherwise. Here, C and $D > 0$ are unknown parameters. We thus see that $F(x)$ contains three unknown parameters C, D, and $\gamma > 0$. Their values should be determined from observations, using one of the several available statistical methods.

3.12.3. Strength of Bundles of Threads

While the model of the preceding subsection is applicable to a large variety of industrial problems related to strength, a new model is needed for the strength of bundles of threads.

Consider a bundle of n parallel threads of equal length. Let $X_1, X_2, \ldots, X_n$ denote the strength of the individual threads. We assume that the X_j are i.i.d. Furthermore, let us assume that a free load on the bundle is distributed equally on the individual threads. Evidently, the bundle will not break under a load S if there are at least k threads in the bundle each of which can withstand a load S/k. In other words, if $X_{1:n} \leqslant X_{2:n} \leqslant \cdots \leqslant X_{n:n}$ are the ordered strengths of the individual threads, then the strength S_n of the bundle is represented by

$$S_n = \max\{(n - k + 1)X_{k:n} : 1 \leqslant k \leqslant n\}.$$

The variables $X_{k:n}$ are strongly dependent and none of our models describes the asymptotic behavior of S_n. Without specifying the actual distribution $F(x)$ of the X_j, we can show that S_n, properly normalized, is asymptotically normal under very mild conditions on $F(x)$. In fact, the following result is true.

Let $F(x)$ be absolutely continuous with finite second moment. Assume that $x[1 - F(x)]$ has a unique maximum at $x = x_0 > 0$ and let $\Theta = x_0[1 - F(x_0)]$. If, in a neighborhood of x_0, $F(x)$ has a positive, continuous second derivative, then, as $n \to +\infty$,

$$\lim P(S_n < n\Theta + x\sqrt{n}) = (2\pi)^{-1/2} \int_{-\infty}^{x} e^{-t^2/2} dt.$$

The differentiability properties can be replaced by mathematically weaker assumptions, but those would not make the conclusion more readily accessible to the applied scientist. However, relaxing the independence of the X_j to m-dependence has the advantage that if the bundle is made of threads, m of which were cut from the same longer thread, then one does not have to assume the independence of these m pieces. The conclusion is known to hold under the condition of m-dependence.

We do not prove the above statement in full. However, we wish to point out that the asymptotic properties of S_n can be reduced to the so-called central limit problem, which is known to lead to the asymptotic normal distribution. To see the method of the reduction, let us first introduce the empirical distribution function $K_n(x)$. It is defined by $K_n(x) = \nu_n(x)/n$, where $\nu_n(x)$ is the number of $j \leq n$ for which $X_j < x$. Then

$$\frac{S_n}{n} = \max\left\{\frac{(n-k+1)X_{k:n}}{n} : 1 \leq k \leq n\right\}$$

$$= \max_{1 \leq j \leq n} \{X_j[1 - K_n(X_j)]\}$$

$$= \sup\{x[1 - K_n(x)] : -\infty < x < +\infty\}.$$

Under our conditions, one can now estimate

$$P\left\{\sqrt{n} \left|\left(\frac{S_n}{n} - \Theta\right) - (K_n(x_0) - F(x_0))\right| \geq \epsilon\right\}$$

and conclude that it tends to zero for any fixed $\epsilon > 0$. On account of Lemma 2.2.1, the asymptotic distributions of $\sqrt{n}(S_n/n - \Theta)$ and $\sqrt{n}(K_n(x_0) - F(x_0))$ thus coincide. But the latter, by the simplest form of the central limit theorem, is asymptotically normal. This gives our claim.

3.12.4. Approximation by i.i.d. Variables

The early applications of the asymptotic theory of extremes to practical problems were based on the assumption that the basic random variables $X_1, X_2, \ldots, X_n$ are i.i.d. Although this assumption is rarely correct, we have

3.12 SOME APPLIED MODELS

described several dependent models when the asymptotic results remain the same as for i.i.d. variables (Theorem 3.5.1 with degenerate $U(y;x)$; Theorem 3.6.2 again with degenerate $U(y;x)$; Corollary 3.6.3; Section 3.7; Theorem 3.8.1 with $\tau=0$; and Section 3.9). It should be noted that the model of Section 3.5 does not require that the basic random variables are identically distributed. The approximation may still be valid when the extendibility requirement is satisfied. Although this condition, and the assumptions of several of the other models, are difficult, if not impossible, to check in a given practical situation, the value of these results cannot be overemphasized. They provide conditions under which an approximation by i.i.d. variables is valid. At the same time, the additional results on each of the mentioned models clearly state that, when these conditions are violated, the asymptotic properties of the extremes may significantly deviate from those known in the i.i.d. case. Here, conditions that cannot be checked are accepted by scientists as reasonable; at the same time, an alternative is given that leads to different results. On the contrary, when one assumes $X_1, X_2, \ldots, X_n$ to be i.i.d., in most cases it is clear that the starting point is already wrong, which spoils the credibility of any conclusion.

For the sequel of the present subsection, let us assume that we deal with a problem in which one of the mentioned models was found reasonable, and we thus accept that the asymptotic distributions are one of the limited possibilities of Chapter 2.

The actual application of an asymptotic model is as follows. One may use a number of observations to find a statistical estimate of the common distribution function $F(x)$ and then compute the normalizing constants by which the extremes are transformed. The distribution of this transformed variable is then assumed to be the asymptotic distribution corresponding to $F(x)$. Or, one may use directly one of the extreme value distributions to replace the exact distribution of the extremes in question, and then estimate its parameters from a set of observations.

Before going on, we should reemphasize that all limit theorems determine only a type of limiting distribution. Hence, if $H(x)$ is a limit law, then all functions $H(Ax+B), A>0$, are actual candidates in a particular problem. Therefore a limiting distribution will contain two or more unknown parameters. Both methods are widely applied and neither is superior to the other. Both, by the nature of statistical decisions, have errors in the final conclusion. These errors, which can be very significant (see, for example, Examples 2.6.3 and 3.8.4), are due to lack of more accurate statistical methods rather than to the theory discussed here. We therefore do not analyze this problem further.

We illustrate with actual solutions to some practical problems.

Air pollution. Let X_j be the concentration of a pollutant in the jth time interval of a predetermined length. It is reasonable to assume that the X_j are identically distributed but successive X_j values are dependent. However, the dependence weakens as the time passes. As a first approximation, m-dependence is reasonable. More cautious people would incline toward mixing or toward the model of Section 3.9. In any case, the approximation by i.i.d. variables is reasonable. As soon as this has been agreed upon, it does not matter which model was accepted, if only individual distributions and asymptotic extreme value distributions are of interest. At the time of this writing, the most widely accepted conclusion is that the common distribution $F(x)$ of the X_j is lognormal. (There is no theoretical justification for this. Conclusions of such nature are arrived at by making observations, assuming different distributions for the observed quantity, and accepting that distribution the graph of which is closest to the observations. Such comparisons are done with a few families of distributions, and so the distribution accepted should be understood as being the most likely one out of those that have been tried. It may be of interest to remark that, before 1969, $F(x)$ was believed to be different from lognormal.)

The general lognormal distribution has several parameters which are estimated from random observations. With these estimated parameters, the X_j are transformed into the standard form that we adopted in Section 2.3.3. We now compute a_n and b_n by the formulas of Section 2.3.3 and, using $H_{3,0}(x)$ as the distribution function of $(Z_n - a_n)/b_n$, we compute the probability that this particular pollutant concentration would remain below a given level during a period of n intervals of the length specified at the definition of X_j. If this probability is not close to one, then society is justified in requesting preventive measures from those responsible for this particular pollution.

Floods. Let $X_1, X_2, \ldots$ be the daily maximum discharges of river R at city C. Then Z_{365} is the annual maximum flood. Notice that X_j itself can be decomposed into smaller units of time, and by this the annual maximum flood will be the maximum of a much larger number of quantities. Without specifying the actual distribution of floods in these time intervals, we can safely assume the annual maximum flood to follow an asymptotic extreme value distribution. It is again reasonable to accept that discharges are less and less dependent as times of records are far apart. Hence, the model of Section 3.9 is acceptable to support the assumption of approximating the model by i.i.d. variables (m-dependence or mixing are not completely acceptable, since relations of successive days are not uniform all year round). Therefore, the annual maximum flood $Z_{365} = Z$ is distributed as one of the types of $H_{1,\gamma}(x), H_{2,\gamma}(x)$, or $H_{3,0}(x)$. The concept of types of distributions brings in two parameters, as x is to be transformed to $Ax+B$. There is, however, no theoretical reason to favor any of the three types, and in fact all three have been used at different times and at different locations. When one has decided the actual form to use, then its parameters are to be estimated from observations on Z. With these estimates, the distribution of Z is completely specified (remember that it applies to R at C). This distribution is then used to predict future water discharges.

Droughts. If again $X_1, X_2, \ldots$ are the daily discharges of river R at city C, their annual minima $W = W_{365}$ are called droughts. With the argument adopted in connection with floods one can conclude that the type of the distribution of W is one of the three types for i.i.d. variables. Furthermore, since $W \geq 0$, we obtain as a single possibility,

$$P(W<x) = L_{2,\gamma}(Ax+B) = 1 - \exp\left[-(Ax+B)^\gamma\right],$$

where $Ax+B>0$ and $\gamma>0$. From observations, $A, B>0$ and $\gamma>0$ are to be estimated to make the distribution $P(W<x)$ specific. It can then be used for the prediction of droughts in the future, which is very important for irrigation projects.

3.12.5. Time to First Failure of Equipments with a Large Number of Components

Several of the examples in the development of the asymptotic theory in the present book are given in terms of failures of equipment. In particular, we should stress the model of Section 3.1, which was later shown to be equivalent to an arbitrary failure model whenever failure can be expressed as the maximum or minimum of certain random variables (see Corollary 3.2.1). Although the direct application of this equivalence is not yet possible without further developing the theory, it shows one significant fact: approximation by i.i.d. variables is not justified in describing the random time to first failure of an equipment in the general case. There are, of course, cases when such approximation is justified; we shall comment on these after showing that, in fact, the time to first failure of any equipment can be expressed as the maximum or minimum of well-defined random variables.

By an equipment we mean a structure of components which was assembled in such a way that each component serves some purpose and the repair of nonfunctioning components means an improvement (these terms have accurate mathematical meanings). We call a subset of components a minimal path set if it is a minimal set of components such that when each functions, so does the equipment. Let the equipment contain n minimal path sets R_j. Let X_j be the life length of R_j. Then the life length Z_n of the equipment is evidently the maximum of the X_j. From the definition it is also clear that X_j is the minimum of the life length Y_s of the components belonging to R_j. In summary,

$$Z_n = \max\{X_j : 1 \leq j \leq n\}, \qquad X_j = \min\{Y_s : s \in R_j\}.$$

If n is large, then the distribution of Z_n can be approximated by an extreme value distribution. The X_j are, however, usually dependent, since R_j and R_m, $j \neq m$, may have components in common. In most cases, even an approximation by i.i.d. variables is unjustified. In the case of series systems ($n = 1$) or of parallel systems (the R_j are disjoint), one can of course assume independence. In these cases, Section 3.10 is applicable. By Theorem 3.10.1, if the system is series and the absolutely continuous distribution $L(x)$ of Z_n satisfies $L(x) < 1$ for all x, then, for independent components, $L(x)$ has a monotonic hazard rate. The hazard rate is defined as the ratio $L'(x)/[1 - L(x)]$.

In the general case, the conditions of Theorem 3.9.1 may be satisfied. If not, Theorem 1.4.2 can provide good estimates. The construction of the sets E_n is evident: if R_i and R_j contain common elements, then let (i,j) belong to E_n. Some examples of this nature have been worked out in the text without actual reference to minimal path sets.

Another potential application of the asymptotic theory of extremes is to the theory of competing risks. This theory is developed for providing a probabilistic model for life distributions when an individual can die from one of n causes. Let X_j be the individual's hypothetical length of life that would apply if only the ith cause of death were present. The actual life length is therefore W_n, the minimum of the X_j, $1 \leq j \leq n$. If n is large, an asymptotic model for W_n can give a good approximation. Because a forthcoming book deals with this subject (see the survey), we do not discuss it further here.

The fields of application of extreme value theory are not limited to those that we mentioned in the present section. However, each model typifies a class of solutions in spite of the very specific terms used in it. By the methods described, one should see that asymptotic theory may provide exact solutions to some problems (3.12.1 and 3.12.2). On the other hand, when approximations are used, the statistician faces a more difficult problem than in the case of averages (estimation of location or scale). The distinction between two distributions that are uniformly close to each other may be irrelevant when the mean is estimated, but they can lead to substantially different decisions in extreme value theory (see Example 2.6.3).

3.13. SURVEY OF THE LITERATURE

The need for deviating from the assumption of the basic random variables' being i.i.d has been stressed ever since the theory was applied to specific problems. Even if the approximation by i.i.d. variables is good, the credibility of a solution should be questioned if it starts by assuming that the variables are i.i.d. when they are not. Therefore, the results of this section are very valuable in providing conditions for a good approximation by i.i.d variables as well as providing alternate models when such an approximation is unjustified. It should be noted that there do exist limiting distributions for the extremes in very simple models which are unrelated to the basic three possibilities of the model of i.i.d variables.

Two classes of results should be very pleasing to scientists. One is the class of distributions (51), which, when written in detail for the maximum, can be reformulated as a classical distribution but with an arbitrary distribution assigned to its parameter. This is the Bayesian thinking, and thus (51) is in fact a natural class of distributions to be used. The other result, Theorem 3.10.1, yields as a limiting distribution of the minimum any distribution with a monotonic hazard rate. Scientists in reliability theory find this appealing, in that the choice of such functions as the population distribution is theoretically justified. The two quoted results are for special models, but when we add Sections 3.5 and 3.9, where the conclusions are the same, a very wide justification is given for using these distributions. The other sections, on the other hand, provide justifications for approximations by i.i.d. variables.

In its construction the present chapter is new, unifying and bringing together for the first time the scattered literature. Some of the proofs are new as well. In particular, the material has not previously been related to applied fields. The possibility of bringing these results together, however, is due to very active research in this field. These contributions are analyzed below.

The pioneer in deviating from independence was S. M. Berman. In a series of papers, Berman (1962b, c) and (1964) established the basic results for extremes of segments of infinite sequences of exchangeable variables and for stationary sequences. His investigation for exchangeable variables is along the line of Corollary 3.6.1, his result being more general than that. In particular, he obtained that, under (i), (50) is also necessary for the existence of the limiting distribution of $(Z_n - a_n)/b_n$. Because we established the more general result Theorem 3.6.2, which is due to J. Galambos (1975e), we did not state Berman's result in its generality. Theorem 3.6.1 is due to B. deFinetti (1930). Useful and interesting inequalities for the distribution of extremes of exchangeable sequences are obtained in B. B. Bhattacharyya (1970). An application of results for exchangeable extremes is given in W. Dziubdziela and B. Kopocinski (1976a).

For stationary sequences, Berman (1964) established a general result, which we did not reproduce here. He then applied his method to special cases and founded a technique for Gaussian sequences. His inequality of Lemma 3.8.2 became basic for later investigations. Its proof is based on Lemma 3.8.1, which is due to R.L. Plackett (1954) (it is also reobtained in M. Sibuya (1960) and D. Slepian (1962)). One of the major results of Berman is Theorem 3.8.2. The method of its proof is also applicable to obtain the result stated in Exercise 16. M. Nisio (1967) gave some extensions. J. Pickands, III (1967b) gave an example of the type of Theorem 3.8.4, which was later proved under some assumptions by Y. Mittal and D. Ylvisaker (1975). Theorem 3.8.3 is also due to them. In its full generality, Theorem 3.8.4 was obtained by W. McCormick and Y. Mittal (1976). The fact that the normal distribution is a possible limit for Z_n was observed by Berman (1962c). Exact results on the distribution of Z_n were given by A. Kudo (1958).

While the mentioned results above enlarged the set of possible limit laws, the results of Section 3.7 concentrate on dependent models, the limits for which are the classical ones. This approach started with the work of G. S. Watson (1954), who gave the sufficiency part of Corollary 3.7.1. Its necessity was established by G. O'Brien (1974a). The first extensive study of mixing sequences is due to R.M. Loynes (1965), who used nice ideas to arrive at (63) under somewhat different conditions than those in Theorem 3.7.1. His approach was further analyzed by O'Brien (1974a), who, in addition to obtaining Corollary 3.7.1, gave examples which show that the conditions in Loynes's approach cannot be dropped. The form of Theorem 3.7.1 is due to M. R. Leadbetter (1974), whose concept of mixing in tails only is adopted here. Not only does his approach extend the class of possible structures to be used, but his method of proof also adds to the theory. In particular, he established that the theory of Gaussian sequences is an integral part of his method, if one uses Berman's inequality. In addition, within his structure, kth extremes, lower extremes, and joint distribution of extremes can be handled by the same method (see Exercises 13 and 14). He shows, in fact, that the number of X_j which exceed $a_n + b_n x$ is close to a binomial distribution, from which the kth extremes are easy to handle. Earlier, under the stronger assumption of strong mixing (not only in tails), I. A. Ibragimor and Yu. A. Rozanov (1970, p. 250) have shown that Gaussian sequences with zero expectation and unit variance satisfy the condition of Exercise 16 and thus Z_n for such sequences behaves as in the case of i.i.d. normal variates.

It should be remarked that, if one considers mixing sequences without further assumptions, then it is possible that $(Z_n - a_n)/b_n$ converges weakly but $X_{n-1:n}$ cannot be normalized to obtain a limit law. Such an example is given by T. Mori (1976a) (see Exercise 15). Therefore, the theory of kth extremes is not always automatic from Z_n itself. Another paper of G. O'Brien (1974b) and the Poisson limit theorem of W. Philipp (1971) are also significant contributions to this area. For joint distributions of Z_n and $X_{n-1:n}$, see the papers of R. E. Welsch (1971, 1972, and 1973), C. M. Deo (1973a), G. F. Newell (1964), and Mori (1976a).

The limit theorem of Section 3.9 is due to J. Galambos (1972). In a special case, it was first obtained in Galambos (1970), in which Watson's result was covered but stationarity was a restriction. Section 3.9 reduces the case of dependent variables to independent ones, which were treated by M. L. Juncosa (1949) and by D. Mejzler (1950, 1953, and 1956). The main results of Section 3.10 are essentially due to Mejzler. Some special classes, supplementing the result of Juncosa, are investigated by J. Tiago de Oliveira (1976).

A completely new approach was proposed by J. Galambos (1975e). It was made possible by the recognition in Galambos (1973a) that the number of occurrences in a given sequence of events can always be reduced to exchangeable ones. For exchangeable events, on the other hand, we have D. G. Kendall's (1967) representation, which leads to problems of limit laws for special mixtures. This approach is used in Sections 3.1–3.6, and the results could have been applied to other structures as well. It would not have simplified the approach there, however, and we would not have had diversified ideas, had we selected the approach of using the theoretical model as the main theorem in each case. That model has a great potential to predict what to expect for a given model. For actually proving a prediction, one can then choose one of several possibilities, including those presented in the different sections. The basic tool, Lemma 3.2.1, is due to D. G. Kendall. So is Theorem 3.2.2, which is mathematically more specific in the form of (21). Kendall used these results to obtain a Poisson limit theorem, but, for our special interest, it reduces to Berman's result quoted earlier. (Kendall's result is more general in that it can be applied to arbitrary exchangeable events.) Corollary 3.4.1 is essentially due to C. J. Ridler-Rowe (1967), and Theorem 3.4.2 was obtained in a somewhat more special form by A. Benczur (1968). In the case when the binomial moments converge, special cases of Sections 3.5 and 3.6 were obtained by Galambos (1974). In the same paper, the inversion formula of L. Takács (1965) or (1967b) is extended, making it applicable to limiting forms.

Discussion of a class of dependent sequences, the so-called chain dependent sequences (defined on a Markov chain), would require the introduction of several concepts outside the present book's main focus; we therefore limit ourselves to a list of papers dealing with this class: S. I. Resnick and M. F. Neuts (1970), A. J. Fabens and M. F. Neuts (1970), S. I. Resnick (1971b and 1972b), and G. E. Denzel and G. O'Brien (1975). For the same reason, maxima of sums of random variables are not treated here. The reader is referred to writings by L. Takács—a book (1967a), a review article (1975), and a manuscript (1977), which is to appear as a book.

In the field of applications, B. Epstein, E. J. Gumbel, W. Weibull, and H. E. Daniels were the pioneers. Their work is properly acknowledged by Gumbel (1958). Gumbel's book takes a technical approach; much is told in easier terms in papers by Epstein (1948) and Daniels (1945). A newer review is by Epstein (1960). Out of the special topics, the book by Barlow and Proschan (1975) provides an account of works in reliability, stressing dependence. This book presents several useful inequalities under minimal assumptions. For the theory of competing risks, see the new monograph by H. A. David and M. L. Moeschberger (1978). The papers dealing with air pollution problems are steadily increasing in number. Out of these, let us mention R. I. Larsen (1969), who influenced development for many years by choosing the lognormal distribution as the distribution of pollutant concentration. The wide acceptance of this observation is due to the same fact shared by the Weibull distribution: the three parameters make it general enough to fit data even from widely different distributions. A set of papers by R. Barlow (1971), N. D. Singpurwalla (1972), and Barlow and Singpurwalla (1974) added much mathematical background to the theory.

The fact that the same set of data can be fitted to different distributions is clearly demonstrated by T. Kunio et al. (1974), who try their test with several distributions. (I am indebted to Professor T. Kunio of Keio University, Yokohama, for translating for me the essential part of their paper.) On the other hand, Margaret Greig (1967) obtains that the assumption of (even multivariate) normality is inadequate in her problem (this remark is intended for that small group of readers who believe in the superiority of the normal distribution).

The fact that the choice of the population distribution has a great effect on decisions when the extremes govern the laws is a serious problem in medical research and for the government agencies which regulate food or drug safety standards. If two groups use different distributions that are accepted by the same set of data, they may come to different conclusions in regard to safety. The major difficulty is that the range of values which can be measured in experiments is too small to permit one to distinguish on a statistical basis between several models. This point is clearly demonstrated in the report of the Food and Drug Administration Advisory Committee on Protocols for Safety Evaluation (1971) and in the paper by H. A. Guess and K. S. Crump (1977).

The strength of bundles is a major statistical problem. Our description is based on recent extensions by M. W. Suh et al. (1970) and P. K. Sen (1973a, b) of H. E. Daniels' (1945) work. See also S. L. Phoenix and H. M. Taylor (1973).

Queueing and reliability applications are presented in D. J. Clough and S. Kotz (1965), M. Morrison and F. Tobias (1965), E. C. Posner (1965), D. J. Clough (1969), R. Harris (1970), F. Downton (1971), D. L. Iglehart (1972), A. J. Stam (1973), and A. G. Pakes (1975) (see also Chapter 6).

An interesting technique for studying the asymptotic theory of extremes in occupancy problems is developed in the papers by V. F. Kolchin (1969) and G. I. Ivcenko (1971).

As basic references for statistical questions, see Gumbel (1958), A. E. Sarhan and B. G. Greenberg (1962), and H. A. David (1970). These are, however, for the i.i.d. samples only. For the theory of distributions, see the four volumes of N. L. Johnson and S. Kotz (1968/1972). Special questions of a statistical nature are treated in W. L. Stevens (1939), B. Afonja (1972), and P. K. Sen (1970).

The surveys by E. Marszal and J. Sojka (1974), J. Tiago de Oliveira (1975), M. R. Leadbetter (1975), and J. Galambos (1977a) stress different points of the theory. A survey on applications of extremes in hydrology is given by A. Jensen (1969). The new mathematical approaches to dependent samples by B. Gyires (1975) and by I. Berkes and W. Philipp (1977) may lead to simplified proofs and new results.

3.14. EXERCISES

1. In the failure model of Section 3.1, let us purchase five components out of a lot of 100 items in the store. Let $Y_1, Y_2, \ldots, Y_{100}$ be the life lengths of the items in the store, out of which $Y_1, Y_2, \ldots, Y_{60}$ are i.i.d. exponential variates with expectation of 50 units. The rest, $Y_{61}, \ldots, Y_{100}$, are independent of the others, and their joint distribution is given by

$$P(Y_j < x_j, 61 \leq j \leq 100) = \int_1^2 \prod_{j=61}^{100} (1 - e^{-u(x_j - 40)}) du,$$

where $x_j \geq 40$. Find $P(W_5 \geq x)$.

2. Write formula (3) in detail for $n = N - 5 = 6$. Choose different functions $H_k(x)$ and compare the results with the case of $H_k(x) = F^k(x)$.

3. Let $A_1, A_2, \ldots, A_n$ be exchangeable events. Show that if $P^2(A_1) > P(A_1 A_2)$, then

$$n \leq \frac{P(A_1) - P(A_1 A_2)}{P^2(A_1) - P(A_1 A_2)}.$$

Hence conclude that, for an infinite sequence of exchangeable events, $P^2(A_1) \leq P(A_1 A_2)$. (Use the method preceding Definition 3.2.2.)

4. Prove formula (14).

5. Out of an urn of M white and $n-M$ red balls, select R without replacement. Let C_j be the event that the jth choice results in a white ball. Show that the events $C_1, C_2, \ldots, C_t, t = \min(M, R)$, are exchangeable and

$$P(C_{i_1} C_{i_2} \cdots C_{i_k}) = \frac{M(M-1) \cdots (M-k+1)}{n(n-1) \cdots (n-k+1)}.$$

6. In Lemma 3.4.2, let $M = Ny/n$. Show that the convergence stated in the lemma is uniform on any finite interval $0 \leq y \leq B$.

7. Prove the following version for the lower extremes. Let $X_1, X_2, \ldots$ be an infinite sequence of exchangeable random variables. Let $Y(x)$, x real, be the set of random variables occurring in the deFinetti representation (47). Then there are normalizing constants c_n and $d_n > 0$, which are characteristic to the lower extremes, if, and only if,

$$\lim_{n = +\infty} P[nY(c_n + d_n x) < y] = U(y; x)$$

exists for all continuity point y of $U(y; x)$. Here, $U(y; x)$ is a distribution function in y. Give the formula for the limiting distribution of the normalized lower extremes.

8. Let $D(x)$ be a distribution function such that, with some constants c_n and $d_n > 0$, $[1 - D(c_n + d_n x)]^n$ converges to $1 - L(x)$, where $L(x)$ is a nondegenerate distribution function. Using the notations of Exercise 7, let us assume that, as $x \to \alpha(D)$,

$$\lim P\left[\frac{Y(x)}{D(x)} < y\right] = U^*(y)$$

exists for all continuity points y of $U^*(y)$, where $U^*(y)$ is continuous at $y = 0$. Show that the normalized lower extremes $(X_{k:n} - c_n)/d_n$ converge weakly to $E_k(x)$, where

$$1 - E_k(x) = \sum_{t=0}^{k-1} \frac{1}{t!} \{-\log[1 - L(x)]\} \int_0^{+\infty} z^t [1 - L(x)]^z dU^*(z).$$

9. Restate Theorem 3.7.1 for stationary sequences which are mixing in the lower tail and conclude that $(W_n - c_n)/d_n$ converges to the same distribution as if the basic variables were independent.

10. Let $X_1, X_2, \ldots, X_n$ be i.i.d. variables with common exponential distribution. Let $Y_1, Y_2, \ldots, Y_n$ be another sequence of i.i.d. unit exponential variates, where

$$P(X_j<x, Y_j<x) = 1 - 2e^{-x} + (2e^x - 1)^{-1}$$

and X_j is independent of all other Y's and Y_j is independent of all other X's. Thus the sequence $X_1, Y_1, \ldots, X_n, Y_n$ is a two-dependent sequence. Show that the maximum Z_{2n} of this combined sequence can be normalized to have a nondegenerate limiting distribution but the result is different than if the X's and Y's were completely independent.

11. Let $X_1, X_2, \ldots, X_n$ be an m-dependent stationary sequence with distribution $F(x)$. Let c_n and $d_n > 0$ be such that, as $n \to +\infty$,

$$nF(c_n + d_n x) \to e^x, \quad nP(X_1 < c_n + d_n x, X_i < c_n + d_n x) \to 0,$$

where $1 \leq i \leq m$. Show that, as $n \to +\infty$,

$$\lim P(W_n < c_n + d_n x) = 1 - \exp(e^{-x}).$$

12. Show that, for the stationary m-dependent sequence $X_1, X_2, \ldots, X_n$, $(Z_n - a_n)/b_n$ and $(W_n - c_n)/d_n$ are asymptotically independent, whenever the conditions of both Corollary 3.7.1 and Exercise 11 are satisfied.

13. Show the validity of the conclusion of the preceding exercise for stationary mixing sequences if the conditions of Theorem 3.7.1 and of Exercise 9 are satisfied.

14. Show that, under the conditions of Theorem 3.7.1, $(X_{n-k:n} - a_n)/b_n$ converges weakly. Also obtain that the limiting distribution is the same as if the basic variables were independent.

15. Let $Y_1, Y_2, \ldots$ be i.i.d. unit exponential variates. Let $f(x) = 1$ on each of the intervals $[2^{2j}, 2^{2j+1}), j \geq 0$, and zero otherwise. Define $X_j = \max[Y_{j-1}, Y_j - f(Y_j)]$. Show that the sequence $X_1, X_2, \ldots$ is a stationary, two-dependent sequence for which $Z_n - \log n$ converges weakly, but Z_n and $X_{n-1:n}$ cannot be normalized to have a joint limiting distribution.

[T. Mori (1976a)]

16. Let $X_1, X_2, \ldots$ be a stationary Gaussian sequence with $E(X_j) = 0, V(X_j) = 1$, and $E(X_1 X_j) = r_j$. Let

$$\sum_{j=1}^{+\infty} r_j^2 < +\infty.$$

Show that, with the usual normalization a_n and $b_n > 0$ for i.i.d. standard normal variates, $(Z_n - a_n)/b_n$ converges weakly to $H_{3,0}(x)$.

[S. M. Berman (1964)]

17. Let $X_1, X_2, \ldots$ be random variables for which

$$\sum_{1 \leq i_1 < i_2 < \cdots < i_k \leq n} P(X_{i_j} \geq a_n + b_n x, 1 \leq j \leq k)$$

converges to e^{-kx}, where a_n and $b_n > 0$ are suitable constants. Show that $(Z_n - a_n)/b_n$ converges weakly to the logistic distribution $1/(1 + e^{-x})$.

[J. Galambos (1974)]

18. Let $Y \geq 0$ be a random variable with distribution function $U^*(z)$, which is continuous at zero. Let S_1, S_2, and S_3 be random variables with distribution functions $H_{1,\gamma}(x)$, $H_{2,\gamma}(x)$, and $H_{3,0}(x)$, respectively. Assume that S_j and Y are independent. Show that (51), for the case of $Z_n(k=1)$, can be interpreted as $(Z_n - a_n)/b_n$ converges weakly to the distribution of one of $S_1 Y^{1/\gamma}$, $S_3 + \log Y$, or $S_2 Y^{-1/\gamma}$.

[S. M. Berman (1962b)]

19. Show that none of the distributions in (51) with $k = 1$ is the normal distribution.

[S. M. Berman (1962b)]

20. Evaluate explicitly (51) for $U^*(z) = 1 - e^{-z}, z > 0$.

21. Let $F(x)$ be a distribution function with $\alpha(F) = 0$ and $\omega(F) = +\infty$. Let $\lambda_k > 0$ be a given sequence of numbers. Let $X_1, X_2, \ldots$ be independent random variables with $P(X_k < x) = F(\lambda_k x)$. Assume that, as $t \to 0^+$, $F(tx)/F(t) \to x^r$ with some $r > 0$. Let $\Sigma \lambda_k^r = +\infty$ and $\lambda_n = o(\Sigma_{k=1}^n \lambda_k^r)$. Show that there is a sequence $d_n > 0$ such that W_n/d_n converges weakly to $1 - \exp(-x^r)$.

[M. L. Juncosa (1949)]

22. Let $X_j, E(X_j) = 1/\lambda_j, j \geq 1$, be independent exponential variates. Show that, with suitable $d_n > 0$, W_n/d_n is unit exponential.

23. With the notations of Exercise 21, let us assume that $F(x)$ and the sequence $\lambda_k > 0$ satisfy the following properties:

(i) $\alpha(F) = -\infty$ and $F(x+y)/F(y) \to e^x$ as $y \to -\infty$,
(ii) $0 < \lambda_k \leq M < +\infty$ and $\nu_n(y) \to g(y)$, where $n\nu_n$ is the number $k \leq n$ with $0 < \lambda_k \leq yM$.

Show that there is a sequence c_n such that $W_n - c_n$ converges weakly to $1 - \exp[-u(x)]$, where

$$u(x) = \int_0^1 e^{xy} dg(y).$$

[M. L. Juncosa (1949)]

CHAPTER 4

Degenerate Limit Laws; Almost Sure Results

For a sequence of random variables, a degenerate asymptotic distribution expresses that the sequence in question is asymptotically close to a constant. As a matter of fact, if, as $n \to +\infty$, $P(Y_n < x)$ converges to the degenerate distribution $F(x) = 0$ if $x \leq c$ and $F(x) = 1$ for $x > c$, then, for any $\varepsilon > 0$,

$$\lim P(|Y_n - c| \geq \varepsilon) = 0 \qquad (n \to +\infty).$$

In the first section of the present chapter we shall investigate this property for normalized extremes. We shall consider two types of normalizations: purely additive or purely multiplicative. That is, we seek conditions under which there are sequences of constants a_n and $b_n > 0$ such that, if E_k is one of the extremes, then either $E_k - a_n$ or $(1/b_n)E_k$ converges weakly to a degenerate distribution function. In the second part of the chapter we strengthen the results to $E_k - a_n$ or $(1/b_n)E_k$ converging almost surely to zero or one, respectively. This will be obtained through a more general set of problems: we shall determine constants a_n and $b_n > 0$ such that, as $n \to +\infty$, the lim sup or lim inf of $(E_k - a_n)/b_n$ is finite and different from zero almost surely. Most of the results will be stated for i.i.d. variables, but several general results will also be formulated and some special dependent systems will be mentioned.

4.1. DEGENERATE LIMIT LAWS

In the present section we deduce limit theorems with degenerate limiting distributions. We start with a simple general rule.

Lemma 4.1.1. *Let Y_n be a sequence of random variables and u_n and $v_n > 0$ be two sequences of numbers such that, as $n \to +\infty$, $(Y_n - u_n)/v_n$*

converges weakly to a nondegenerate distribution $T(x)$. Then

(i) *if $u_n/v_n \to +\infty$ with n,*

$$\lim_{n=+\infty} P(Y_n < u_n z) = \begin{cases} 1 & \text{if } z > 1, \\ 0 & \text{if } z \leq 1; \end{cases} \qquad (1)$$

(ii) *if $v_n \to 0$ as $n \to +\infty$,*

$$\lim_{n=+\infty} P(Y_n < u_n + z) = \begin{cases} 1 & \text{if } z > 0, \\ 0 & \text{otherwise}. \end{cases} \qquad (2)$$

Proof. By assumption, as $n \to +\infty$,

$$\lim P(Y_n < u_n + v_n x) = T(x) \qquad (3)$$

for all continuity points x of $T(x)$. Let $\varepsilon > 0$ be an arbitrary number and let x_1 and x_2 be such that $T(x_1) \geq 1 - \varepsilon$ and $T(x_2) < \varepsilon$. Then, for part (i), we proceed as follows. Let $z > 0$ be fixed. Since $u_n/v_n \to +\infty$ with n, $(u_n/v_n)z > x_1$ for all large n. Hence, in view of (3), as $n \to +\infty$,

$$\liminf P(Y_n < u_n(1+z)) \geq \liminf P(Y_n < u_n + v_n x_1) \geq 1 - \varepsilon.$$

Since $\varepsilon > 0$ is arbitrary, the limit itself exists and equals one, which is the first claim in (1). On the other hand, for $z < 0$, $(u_n/v_n)z \to -\infty$, and thus $(u_n/v_n)z < x_2$ for all large n. We thus get, for $n \to +\infty$,

$$\limsup P(Y_n < u_n(1+z)) \leq \limsup P(Y_n < u_n + v_n x_2) < \varepsilon,$$

which yields the second limit of (1).

The proof of part (ii) is similar. We write

$$u_n + z = u_n + v_n\left(\frac{z}{v_n}\right),$$

which will be greater than $u_n + x_1 v_n$ for $z > 0$ and smaller than $u_n + x_2 v_n$ for $z < 0$ by the assumption $v_n \to 0$. This completes the proof. ▲

Notice that both (1) and (2) express that, in probability, Y_n is "close" to the constant u_n. Borrowing the term of laws of large numbers from the theory of sums of random variables, we shall refer to a limit theorem with a degenerate limit distribution as a weak law of large numbers. For distinguishing the situations of (1) and (2), we shall speak of a multiplicative or of an additive weak law of large numbers.

Definition 4.1.1. If, for a sequence Y_n of random variables and for a sequence u_n of numbers, (1) holds, then we say that (Y_n, u_n) satisfies the multiplicative weak law of large numbers, to be abbreviated to MWL. On the other hand, if (2) is valid, then (Y_n, u_n) is said to satisfy the additive weak law of large numbers, or AWL.

Lemma 4.1.1 gives conditions under which a weak convergence to a nondegenerate distribution implies weak laws of large numbers. We can therefore apply several theorems and results of examples from Chapters 2 and 3 and combine them with Lemma 4.1.1. Before formulating this possibility into general theorems, let us list some weak laws for extremes.

Example 4.1.1. Let $X_1, X_2, \ldots$ be a stationary sequence of Gaussian variates with $E(X_j) = 0$ and $V(X_j) = 1$. Let $r_m = E(X_1 X_{m+1})$, and assume that $r_m = o(1/\log m)$ as $m \to +\infty$. Then, by Theorem 3.8.2 and Lemma 4.1.1, (Z_n, a_n) satisfies both the MWL and the AWL, where a_n is given in (80) of Section 3.8. Notice that here the AWL is a stronger statement than the MWL. ▲

Example 4.1.2. Let $X_1, X_2, \ldots, X_n$ be i.i.d. unit exponential variates. Then, by Example 1.3.1 and Lemma 4.1.1, $(W_n, 0)$ satisfies the AWL and $(W_n, 1/n)$ satisfies the MWL. The normalizing constants are, of course, not unique. Since Lemma 4.1.1 makes reference to a weak convergence to a nondegenerate limiting distribution, the normalizing constants 0 and $1/n$ above can be modified to the same extent as in the corresponding weak convergence theorems. For example, in view of Lemma 2.2.2, (W_n, u_n) also satisfies the MWL whenever $nu_n \to 1$ as $n \to +\infty$. Let us also remark that, in this example, the MWL is the stronger statement. ▲

Example 4.1.3. Let $X_1, X_2, \ldots, X_n$ be i.i.d. with the common Weibull distribution

$$F(x) = 1 - \exp(-\sqrt{x}).$$

Then, by Corollary 1.3.1 and by Lemma 4.1.1, $(Z_n, (\log n)^2)$ satisfies the MWL, while no AWL follows. ▲

The examples above well illustrate the fact that no general rule can be given for the interrelation of the two types of weak laws of large numbers. However, if the normalizing constants u_n in the definitions are limited to sequences which tend to infinity, then a criterion for one type of weak laws can be transformed into a criterion for the other one. One such possibility will be stated after the proof of Theorem 4.1.2.

We now state a result for maxima.

Theorem 4.1.1. *Let $X_1, X_2, \ldots, X_n$ be identically distributed random variables with common distribution function $F(x)$ such that $\omega(F) = +\infty$. Assume* that there are constants a_n and $b_n > 0$ such that, as $n \to +\infty$,

$$\lim F^n(a_n + b_n x) = H_{3,0}(x). \tag{4}$$

If $(Z_n - a_n)/b_n$ converges weakly to a nondegenerate distribution function, then (Z_n, a_n) satisfies the MWL.

Remark 4.1.1. We did not assume that the X_j are independent. The theorem, therefore, can be applied to several models of Chapter 3. All assumptions are satisfied if the X_j can be approximated by i.i.d. variables for which the maximum, when normalized by a_n and b_n, is attracted to $H_{3,0}(x)$.

Proof of Theorem 4.1.1. By Theorem 2.4.3(iii), assumption (4) implies that all conditions of Theorem 2.1.3 hold. In particular, a_n and $b_n > 0$ can be evaluated by its formulas, implying that $a_n \to +\infty$ with n. Furthermore, as $n \to +\infty$,

$$\lim \frac{1 - F(a_n + b_n x)}{1 - F(a_n)} = e^{-x},$$

where x is an arbitrary real number. But then, by $F(a_n) \to 1$, the numerator $1 - F(a_n + b_n x)$ should tend to zero. This in turn implies that $a_n + b_n x \to +\infty$. It thus follows that, for any fixed x, $a_n + b_n x > 0$ for all sufficiently large n. That is, for any negative x, as $n \to +\infty$, $a_n/b_n > -x$, which means that $a_n/b_n \to +\infty$ with n. This, together with the assumption that $(Z_n - a_n)/b_n$ converges weakly to a nondegenerate distribution, makes part (i) of Lemma 4.1.1 applicable. The proof is complete. ▲

We now prove a theorem which leads to an AWL. Although we state it for i.i.d. variables, it can easily be transformed into a statement for several dependent systems.

Theorem 4.1.2. *Let $X_1, X_2, \ldots, X_n$ be i.i.d. random variables with common distribution function $F(x)$. Let $\omega(F) = +\infty$. Then there is a sequence u_n of real numbers for which (Z_n, u_n) satisfies the AWL if, and only if, for all positive x, as $y \to +\infty$,*

$$\lim \frac{1 - F(y + x)}{1 - F(y)} = 0. \tag{5}$$

Proof. The fact that there is a sequence u_n such that (Z_n, u_n) satisfies the AWL means that, as $n \to +\infty$,

$$\lim F^n(u_n + x) = 1, \qquad x > 0, \tag{6a}$$

and
$$\lim F^n(u_n + x) = 0, \qquad x < 0. \tag{6b}$$

Therefore, we have to prove the equivalence of (5) and (6) for some sequence u_n. We shall show that we can always take $u_n = \sup\{z : 1 - F(z) \geq 1/n\}$.

Let us first assume that (6) holds. Then, for $x > 0$, $F(u_n + x) \to 1$. But, by the definition of u_n and by the assumption $\omega(F) = +\infty$, $u_n \to +\infty$. Hence, for $x < 0$ as well, $F(u_n + x) \to 1$. We can thus take logarithm in (6), which, by using the Taylor expansion

$$\log v = \log[1 - (1 - v)] \sim v - 1 \qquad \text{as} \quad v \to 1,$$

yields

$$\lim_{n = +\infty} n[1 - F(u_n + x)] = \begin{cases} 0 & \text{if } x > 0, \\ +\infty & \text{if } x < 0. \end{cases} \tag{7}$$

From the definition of u_n, $n[1 - F(u_n)] \geq 1$. Substituting this relation into (7) leads to (5) for the special sequence $y = u_n$. Let now $y \to +\infty$. Let n be such that $u_n < y < u_{n+1}$. Thus, for any $x > 0$,

$$\frac{1 - F(y + x)}{1 - F(y)} \leq \frac{1 - F(u_n + x)}{1 - F(u_{n+1})} \leq (n+1)[1 - F(u_n + x)] \to 0,$$

which establishes (5).

Conversely, if (5) holds, then it holds for $y = u_n$ as defined earlier. But then, for $x > 0$,

$$(n-1)[1 - F(u_n + x)] \leq \frac{1 - F(u_n + x)}{1 - F(u_n)} \to 0,$$

which is the first limit of (7). Working backward from (7), we immediately get (6a). For proving (6b), we again argue with its equivalent form in (7). If, for a given $x < 0$, we choose $y = u_n + x$ in (5), then we obtain

$$\frac{1}{n[1 - F(u_n + x)]} \leq \frac{1 - F(u_n)}{1 - F(u_n + x)} \to 0.$$

This is the second limit of (7), and now (6b) follows. The proof is completed. ▲

Let us remark that, for i.i.d. variables, when the common distribution function $F(x)$ is such that $\omega(F) = +\infty$, an AWL for Z_n can always be transformed into a MWL by the following transformation. Define

$$F^*(x) = \begin{cases} F(\log x) & \text{if } x > 0, \\ 0 & \text{otherwise} \end{cases}$$

Now, if, with some sequence u_n, (Z_n, u_n) satisfies the AWL, when the common distribution function is $F(x)$, then (Z_n^*, u_n^*) satisfies the MWL, where the star signifies the common distribution to be $F^*(x)$. This remark, together with Theorem 4.1.2, gives additional cases when the conclusion of Theorem 4.1.1 holds with some sequence a_n.

4.2. BOREL-CANTELLI LEMMAS

The lemmas to be developed in the present section will help us in establishing almost sure results for normalized extremes. These lemmas are known in the literature as different forms of a Borel-Cantelli lemma.

Lemma 4.2.1. *Let $A_1, A_2, \ldots$ be an infinite sequence of events. Let $\nu = \nu(A)$ be the number of those events A_j which occur. If*

$$\sum_{j=1}^{+\infty} P(A_j) < +\infty, \tag{8}$$

then $P(\nu < +\infty) = 1$.

Proof. Let $B_N = \cup_{j=N}^{+\infty} A_j$. Then

$$\{\nu = +\infty\} \subset B_N, \qquad N = 1, 2, \ldots.$$

Thus, by Exercise 6 of Chapter 1, for all $N \geq 1$,

$$P(\nu = +\infty) \leq P(B_N) \leq \sum_{j=N}^{+\infty} P(A_j).$$

Letting $N \to +\infty$, we get from (8) that $P(\nu = +\infty) = 0$, which was to be proved. ▲

Lemma 4.2.2. *We use the notations of the preceding lemma. Let us put*

$$S_{1,n} = \sum_{j=1}^{n} P(A_j), \qquad S_{2,n} = \sum_{1 \leq i < j \leq n} P(A_i A_j).$$

If

$$\sum_{j=1}^{+\infty} P(A_j) = +\infty \tag{9}$$

4.2 BOREL-CANTELLI LEMMAS

and if, as $n \to +\infty$,

$$\lim \frac{S_{2,n}}{S_{1,n}^2} = \frac{1}{2}, \tag{10}$$

then $P(\nu = +\infty) = 1$.

Proof. Again introducing $B_N = \cup_{j=N}^{+\infty} A_j$, we have

$$\{\nu = +\infty\} = \bigcap_{N=1}^{+\infty} B_N.$$

We shall show that, for each N, $P(B_N) = 1$, from which the relation above yields the conclusion of the lemma.

For a fixed N, let us define

$$S_{1,n}(N) = \sum_{j=N}^{n} P(A_j), \qquad S_{2,n}(N) = \sum_{N \leq i < j \leq n} P(A_i A_j).$$

Then, (9) and (10) imply that, for any fixed N, as $n \to +\infty$,

$$\lim S_{1,n}(N) = +\infty, \qquad \lim \frac{S_{2,n}(N)}{S_{1,n}^2(N)} = \frac{1}{2}. \tag{11}$$

Now, by the following special form of Theorem 1.4.3 (see Exercise 15 of Chapter 1),

$$P(B_N) \geq P\left(\bigcup_{j=N}^{n} A_j\right) \geq \frac{S_{1,n}^2(N)}{2S_{2,n}(N) + S_{1,n}(N)}, \tag{12}$$

we get from (11), by letting $n \to +\infty$, that $P(B_N) = 1$. This completes the proof. ▲

The following statement is a special case of Lemma 4.2.2.

Lemma 4.2.3. *Let $A_1, A_2, \ldots$ be an infinite sequence of events such that*

$$P(A_i A_j) \leq P(A_i) P(A_j), \qquad i \neq j.$$

Then (9) implies $P(\nu = +\infty) = 1$. In particular, if the events A_j are independent, then the conclusion holds.

Proof. With the notations of Lemma 4.2.2, our condition implies

$$2S_{2,n} \leq S_{1,n}^2. \tag{13}$$

On the other hand, as a corollary to Theorem 1.4.3 we got in Exercise 15 of Chapter 1 (see (12) above) that, for any sequence of events,

$$2S_{2,n} + S_{1,n} \geq S_{1,n}^2.$$

This inequality, combined with (13), yields (10). Consequently, Lemma 4.2.2 is applicable and the proof is complete. ▲

A direct consequence of Lemma 4.2.1 to extremes is formulated in the following theorems. As usual, $Z_n = \max(X_1, X_2, \ldots, X_n)$ and $W_n = \min(X_1, X_2, \ldots, X_n)$. We use the abbreviation i.o. for "infinitely often."

Theorem 4.2.1. *Let $X_1, X_2, \ldots$ be an infinite sequence of random variables and let $F_j(x)$ be the distribution function of X_j. Let u_n be a nondecreasing sequence of real numbers such that either, for any fixed m,*

$$P(X_j < u_m, j \geq m) = 0, \tag{14}$$

or $u_n \to +\infty$ with n. Then

$$P(Z_n \geq u_n \text{ i.o.}) = P(X_n \geq u_n \text{ i.o.}). \tag{15}$$

In particular, if

$$\sum_{j=1}^{+\infty} [1 - F_j(u_n)] < +\infty, \tag{16}$$

then

$$P(Z_n \geq u_n \text{ i.o.}) = 0. \tag{17}$$

Proof. Consider the sequence $A_n = \{Z_n \geq u_n\}$ of events. We shall show that, by our assumptions, for almost all points of the probability space, infinitely many of the A_n, $n \geq 1$, occur if, and only if, infinitely many of the events $\{X_j \geq u_j\}$ occur. This evidently implies (15). Hence, the combination of (16) and Lemma 4.2.1 yields (17).

Since $Z_n \geq X_n$, if $X_n \geq u_n$ infinitely often, then Z_n also exceeds u_n for these same infinitely many values of n. Therefore, only the converse implication needs proof. Let $Z_n \geq u_n$. Then, for a $j \leq n$, $X_j \geq u_n$. Since u_n is nondecreasing, $X_j \geq u_j$. Let us denote by $j(n)$ the largest of such values of $j \leq n$. Clearly, $j(n)$ is a nondecreasing sequence of n. Since $X_{j(n)} \geq u_n \geq u_{j(n)}$, there corresponds an infinite sequence of $j(n)$ to an infinite sequence of n, whenever u_n is unbounded. However, if u_n is bounded, then, in principle, the same value $j = j(n)$ can be taken for infinitely many values of n. But such a case means that, for a fixed m, $X_k < u_m$ for all $k \geq m$. By (14), its probability is zero. Hence, (15) follows and the proof is complete. ▲

We also state the corresponding result for the minimum. As usual, it does not require a separate proof.

Theorem 4.2.2. *With the notations of the preceding theorem, let*

$$P(X_j \geq t_m, j \geq m) = 0, \quad m \geq 1, \tag{18}$$

where t_n is a nonincreasing sequence of real numbers. Then

$$P(W_n < t_n \text{ i.o.}) = P(X_n < t_n \text{ i.o.}). \tag{19}$$

Consequently,

$$\sum_{j=1}^{+\infty} F_j(t_n) < +\infty$$

implies

$$P(W_n < t_n \text{ i.o.}) = 0.$$

The assumption (14) in Theorem 4.2.1 is necessary. For example, if X_1 is a unit exponential variate which is independent of X_j, $j \geq 2$, where these latter variables are i.i.d. with $P(X_2 < 1) = 1$, then, with $u_n = 2$ for each n, (16) holds but

$$P(Z_n \geq u_n \text{ i.o.}) = P(Z_n \geq 2, n \geq 1) = P(X_1 \geq 2) > 0.$$

4.3. ALMOST SURE ASYMPTOTIC PROPERTIES OF EXTREMES OF I.I.D. VARIABLES

Throughout this section, $X_1, X_2, \ldots$ are i.i.d. variables with common nondegenerate distribution function $F(x)$. The sequence u_n is always assumed to be nondecreasing and t_n to be nonincreasing. Now, (14) and (18) are not necessary in Theorems 4.2.1 and 4.2.2, respectively. Indeed, if $\omega(F) < +\infty$ and $u_n \geq \omega(F)$, then (15) is evident, namely, both sides equal zero. On the other hand, for any fixed value $u_m < \omega(F)$, (14) is immediate. With a similar argument, (18) is not needed either in Theorem 4.2.2. Now, therefore, not only Lemma 4.2.1 but Lemma 4.2.3 also is applicable. We thus get, as a corollary to (15), the following result.

Corollary 4.3.1. *With the standard assumptions of the present section,*

$$P(Z_n \geq u_n \text{ i.o.}) = 0 \quad \text{or} \quad 1$$

according as

$$\sum_{n=1}^{+\infty} [1 - F(u_n)] < +\infty \quad \text{or} \quad = +\infty.$$

The investigation of the reversed inequalities $\{Z_n \leq u_n\}$ is far more difficult. A somewhat lengthy proof will lead to the statement that follows.

Theorem 4.3.1. *In addition to the standard assumptions of the present section, let us assume that $n[1 - F(u_n)]$ is also nondecreasing. Furthermore, let $F(x)$ be continuous. Then, for $u_n < \omega(F)$,*

$$P(Z_n \leq u_n \text{ i.o.}) = 0 \quad \text{or} \quad 1 \tag{20}$$

according as

$$\sum_{j=1}^{+\infty} [1 - F(u_j)] \exp\{-j[1 - F(u_j)]\} < +\infty \quad \text{or} \quad = +\infty. \tag{21}$$

The proof will be split up into a number of steps, stated as lemmas, several of which are of independent interest.

Remark 4.3.1. Statements like $P(Z_n \leq u_n \text{ i.o.}) = 0$ or 1 are evidently not affected by changing the value of u_n for a finite number of n. Consequently, any assumption on the sequence u_n is to be valid for all but a finite number of n. This remark applies to all statements of the present section.

Lemma 4.3.1. *For any nondecreasing sequence u_n, the probability occurring in (20) is either 0 or 1.*

Proof. Let $u_n \to u < \omega(F)$. Then

$$P(Z_n \leq u_n \text{ i.o.}) \leq P(Z_n \leq u \text{ i.o.}) \leq F^N(u)$$

for any $N \geq 1$. Since $F(u) < 1$, our claim follows. We thus assume $u_n \to \omega(F)$ for the remainder of the proof.

Let M be any fixed positive integer. Define $Z_{n,M} = \max(X_M, X_{M+1}, \ldots, X_n)$. We shall show that the complement D of the event $\{Z_n \leq u_n \text{ i.o.}\}$ can be expressed by the sequence $Z_{n,M}$ for any M. Therefore, D is independent of $X_1, X_2, \ldots, X_{M-1}$ for any M. Let C_M be an event which depends on X_j, $1 \leq j < M$, only. We have got

$$P(C_M D) = P(D) P(C_M).$$

Since M is arbitrary, we can let $M \to +\infty$, yielding

$$P(CD) = P(D) P(C), \tag{22}$$

where C is any event that can be approximated by events of the type $C_M, M \geq 1$. But any event related to the sequence $X_1, X_2, \ldots$ is an event of the type of C, and thus D can be taken as C in (22). That is, $P(D) =$

$P^2(D)$, yielding $P(D) = 0$ or 1. By the definition of D, this result is equivalent to the lemma.

It remains now to show that D is indeed expressible in terms of $Z_{n,M}$, $n \geq M$, whatever be $M > 0$. As a matter of fact, we shall show that, for any fixed $M > 0$, with probability one,

$$D = \{Z_n > u_n \text{ for all large } n\} = \{Z_{n,M} > u_n \text{ for all large } n\}.$$

Since $Z_n \geq Z_{n,M}$, the inequality $Z_{n,M} > u_n$ implies $Z_n > u_n$. Consequently, only the reversed implication is to be proved.

Assume that $Z_n > u_n$ for all $n \geq N$, but there is an infinite sequence $n(k)$ of positive integers such that $Z_{n(k),M} \leq u_{n(k)}$. Then, for at least one $j = j(k) < M, X_j > u_{n(k)}$, $k \geq 1$. Since M is bounded, the same j occurs in the preceding argument for infinitely many k. Thus, by u_n's being nondecreasing and $u_n \to \omega(F)$, we get

$$P(Z_n > u_n \text{ for all large } n, \text{ but } Z_{n,M} \leq u_n \text{ i.o.}) = 0,$$

what was to be proved. ▲

Lemma 4.3.2. *For proving Theorem 4.3.1, it can be assumed that, for $n \geq 3$,*

$$\frac{1}{2} \leq \frac{n[1 - F(u_n)]}{\log \log n} \leq 2. \tag{23}$$

Proof. Let u_n be an arbitrary sequence satisfying the conditions of Theorem 4.3.1. Introduce an additional sequence v_n, which equals u_n if (23) holds. On the other hand, if (23) fails, then v_n is defined by

$$v_n = \begin{cases} \sup\left\{v : F(v) \leq 1 - \dfrac{2 \log \log n}{n}\right\} \\ \inf\left\{v : F(v) \geq 1 - \dfrac{\frac{1}{2} \log \log n}{n}\right\} \end{cases}$$

according as the upper or the lower inequality of (23) is violated. By this definition, both v_n and $n[1 - F(v_n)]$ are nondecreasing, and (23) holds with v_n replacing u_n (recall that $F(x)$ is continuous). Therefore, if Theorem 4.3.1 had been proved under the additional condition (23), then

$$P(Z_n \leq v_n \text{ i.o.}) = 0 \quad \text{or} \quad 1 \tag{24}$$

according as

$$\sum_{j=1}^{+\infty} h(v_j) < +\infty \quad \text{or} \quad = +\infty, \tag{25}$$

where, with a general variable x_j,

$$h(x_j) = [1 - F(x_j)] \exp\{-j[1 - F(x_j)]\}. \tag{26}$$

We next show that the series (21) and (25) converge or diverge simultaneously. We distinguish two cases. First, let $u_n > v_n$ for infinitely many values of n. Let this sequence be denoted by $n(t)$, $t \geq 1$. We then evaluate

$$\sum_{j=1}^{+\infty} h(u_j) = \lim_{t = +\infty} \sum_{j=1}^{n(t)} h(u_j),$$

and we show that this always diverges. First note that, by the monotonicity of $n[1 - F(u_n)]$,

$$\sum_{j=1}^{n} h(u_j) \geq \exp\{-n[1 - F(u_n)]\} \sum_{j=1}^{n} [1 - F(u_j)].$$

Once again, by the monotonicity of $n[1 - F(u_n)]$ and by $u_m < \omega(F)$, $j[1 - F(u_j)] \geq 2[1 - F(u_2)] = c > 0$, we get

$$\sum_{j=1}^{n} h(u_j) \geq c \exp\{-n[1 - F(u_n)]\} \sum_{j=1}^{n} \frac{1}{j}.$$

Finally, for $n = n(t)$, $u_n > v_n$, which means $n[1 - F(u_n)] < \frac{1}{2} \log \log n$, and thus

$$\sum_{j=1}^{n(t)} h(u_j) \geq c[\log n(t)]^{-1/2} \log n(t) \to +\infty \quad \text{with } t.$$

Since, by definition, $j[1 - F(v_j)] \geq \frac{1}{2} \log \log j$ and, for $n = n(t)$, $n[1 - F(v_n)] = \frac{1}{2} \log \log n$, an even simpler argument yields that the series in (25) also diverges. For showing the equivalence of (21) and (25) in the remaining cases, that is, when ultimately $u_n \leq v_n$, we have only to observe that only equal values $u_n = v_n$ can make (21) and (25) diverge. Namely, by $h(x_j)$ of (26) being increasing, whenever $1 - F(x_j) > 1/j$,

$$\sum_{1} h(u_j) \leq \sum_{1} h(v_j) \leq \sum_{j=1}^{+\infty} \frac{2 \log \log j}{j} (\log j)^{-2} < +\infty,$$

where Σ_1 is summation over those j for which $v_j > u_j$.

We now complete the proof of the lemma as follows. Assume that Theorem 4.3.1 has been proved under the additional assumption (23). Then, for an arbitrary sequence u_n, satisfying the conditions of Theorem 4.3.1 but not necessarily (23), we first introduce the sequence v_n. If (21) converges, then, as was shown, so does (25). This implies $P(Z_n \leqslant v_n \text{ i.o.}) = 0$. In addition, in the course of the preceding arguments we have seen that the convergence of (25) yields $v_n \geqslant u_n$ ultimately. Hence

$$P(Z_n \leqslant u_n \text{ i.o.}) \leqslant P(Z_n \leqslant v_n \text{ i.o.}) = 0.$$

Finally, let (21) diverge. Then so does (25), and thus $P(Z_n \leqslant v_n \text{ i.o.}) = 1$. Let $T = \{n : n \geqslant 1, v_n > u_n\}$. Recall that if $n \in T$, then $1 - F(v_n) = (2 \log \log n)/n$, and if for all n this were satisfied, then (25) would converge and thus

$$P\left(Z_n \leqslant F^{-1}\left[1 - \frac{2\log\log n}{n}\right] \text{ i.o.}\right) = 0. \tag{27}$$

Combining all these with evident estimates, we get

$$P(Z_n \leqslant u_n \text{ i.o.}) \geqslant P(Z_n \leqslant u_n \text{ i.o. and } Z_n \leqslant v_n \text{ i.o.})$$

$$= P(Z_n \leqslant v_n \text{ i.o.}) - P(Z_n \leqslant v_n \text{ i.o. and } Z_n > u_n \text{ ultimately})$$

$$= 1 - P(Z_n > u_n \text{ ultimately and } Z_n \leqslant v_n \text{ i.o. for } n \in T)$$

$$\geqslant 1 - P(Z_n \leqslant v_n \text{ i.o. for } n \in T) = 1,$$

where the last estimate follows from (27). This completes the proof. ▲

Lemma 4.3.3. *Let u_n be a nondecreasing sequence for which $\Sigma_{n=1}^{+\infty}[1 - F(u_n)] = +\infty$ and (21) converges. Then $P(Z_n \leqslant u_n \text{ i.o.}) = 0$.*

Note that if (23) holds, then the series above always diverges. Hence, in this case, the only assumption is the convergence of (21).

Proof. In view of Lemma 4.2.3, the independence of the X_j and the condition $\Sigma_{n=1}^{+\infty}[1 - F(u_n)] = +\infty$ imply $P(X_n > u_n \text{ i.o.}) = 1$. This, in turn, yields $P(Z_n > u_n \text{ i.o.}) = 1$. Therefore

$$P(Z_n \leqslant u_n \text{ i.o.}) = P(Z_n \leqslant u_n, Z_{n+1} > u_{n+1} \text{ i.o.}). \tag{28}$$

Since u_n is nondecreasing and the X_j are independent, we get

$$\sum_{n=1}^{+\infty} P(Z_n \leq u_n, Z_{n+1} > u_{n+1})$$

$$= \sum_{n=1}^{+\infty} P(Z_n \leq u_n, X_{n+1} > u_{n+1})$$

$$= \sum_{n=1}^{+\infty} F^n(u_n)[1 - F(u_{n+1})]$$

$$\leq \sum_{n=1}^{+\infty} F^n(u_n)[1 - F(u_n)].$$

On account of the upper inequality of Lemma 1.3.1, the last series above is smaller than the series of (21), which is assumed to converge. Hence, Lemma 4.2.1 and (28) conclude the proof. ▲

In the proof of the next theorem, we shall argue on the following subsequence $m(n)$, $n \geq 2$, of the positive integers. Let $\tau > 0$ be a constant to be determined later. We then define $m(n)$ as the integer part of $\exp(\tau n / \log n)$. An important property of $m(n)$ is that, for moderate values of $t > 0$, as $n \to +\infty$,

$$\frac{m(n+t) - m(n)}{m(n+t)} \log \log m(n+t) \geq \tau \lambda, \qquad \lambda > 0. \tag{29}$$

Furthermore, if i_n is an arbitrary sequence of integers such that $i_n = o(\log n)$, then, as $n \to +\infty$,

$$\frac{m(n+i_n) - m(n)}{m(n+i_n)} \sim \frac{\tau i_n}{\log(n+i_n)}. \tag{30}$$

We now conclude the proof of Theorem 4.3.1 by proving the following result.

Theorem 4.3.2. *Let u_n be a nondecreasing sequence satisfying (23). Let (21) diverge. Then $P(Z_n \leq u_n \text{ i.o.}) = 1$.*

Proof. We shall in fact prove that, under the above conditions, $P(Z_{m(n)} \leq u_{m(n)} \text{ i.o.}) = 1$. By Lemma 4.3.1, it suffices to prove that

$$P(Z_{m(n)} \leq u_{m(n)} \text{ i.o.}) > 0. \tag{31}$$

Since, for any sequence A_n of events, $\{A_n \text{ i.o.}\} = \cap_{M=1}^{+\infty} \cup_{n=M}^{+\infty} A_n$, (31) follows if we show the existence of an absolute constant $c>0$, for which

$$P\left(\bigcup_{n=M}^{+\infty} \{Z_{m(n)} \leq u_{m(n)}\}\right) \geq c, \quad M \geq M_0.$$

(See Appendix I for probabilities of monotonic sequences.) We shall actually show that, for large $M>0$, there is an integer $M'>M$ such that

$$P\left(\bigcup_{n=M}^{M'} \{Z_{m(n)} \leq u_{m(n)}\}\right) \geq c > 0, \quad (32)$$

where c does not depend on M or M'.

By the identity, which we state for arbitrary events A_j,

$$P\left(\bigcup_{j=M}^{M'} A_j\right) = \sum_{j=M}^{M'} P\left(A_j \cap \bigcap_{t=j+1}^{M'} A_t'\right)$$

$$= \sum_{j=M}^{M'} P(A_j) - \sum_{j=M}^{M'} P\left(A_j \cap \bigcup_{t=j+1}^{M'} A_t\right),$$

(32) follows, if we prove that there is a $\Delta>0$, not depending on M and M', such that

$$\sum_{n=M}^{M'} P(Z_{m(n)} \leq u_{m(n)}) \geq \Delta > 0 \quad (33)$$

and, with another absolute constant $0<\delta<1$, for all $M_0 \leq M \leq n < M'$,

$$P\left(Z_{m(n)} \leq u_{m(n)} \text{ and } \bigcup_{t=n+1}^{M'} \{Z_{m(t)} \leq u_{m(t)}\}\right) \leq \delta P(Z_{m(n)} \leq u_{m(n)}). \quad (34)$$

We first prove that the divergence of (21) implies

$$\sum_{n=2}^{+\infty} P(Z_{m(n)} \leq u_{m(n)}) = +\infty. \quad (35)$$

Evidently, (35) is a stronger statement than (33). We start with the observation that, since the basic random variables X_j are i.i.d. and u_j satisfies (23), by Lemma 1.3.1,

$$P(Z_{m(n)} \leq u_{m(n)}) = F^{m(n)}(u_{m(n)}) \sim \exp\{-m(n)[1-F(u_{m(n)})]\}. \quad (36)$$

Hence, (35) is equivalent to

$$\sum_{n=2}^{+\infty} \exp\{-m(n)[1-F(u_{m(n)})]\} = +\infty. \tag{37}$$

Now, from (21) and because the sequence u_j is nondecreasing (we use the notation (26)),

$$+\infty = \sum_{n=2}^{+\infty} h(u_n) = \sum_{n=2}^{+\infty} \sum_{t=m(n)}^{m(n+1)-1} h(u_t)$$

$$\leq \sum_{n=2}^{+\infty} [1-F(u_{m(n)})][m(n+1)-m(n)]\exp\{-m(n)[1-F(u_{m(n+1)})]\}.$$

By (23) and (30), and by $m(n+1)/m(n) \to 1$,

$$[1-F(u_{m(n)})][m(n+1)-m(n)] \to \tau \qquad \text{as } n \to +\infty.$$

Hence, the above estimate immediately yields (37). In (36), when compared with (23), we also get that, as $n \to +\infty$,

$$\lim P(Z_{m(n)} \leq u_{m(n)}) = 0.$$

Therefore, (35) yields that, for any sufficiently large $M > 0$, we can find an $M' > M$ such that

$$0 < \Delta \leq \sum_{n=M}^{M'} P(Z_{m(n)} \leq u_{m(n)}) \leq 2\Delta, \tag{38}$$

where $\Delta > 0$ is an arbitrary prescribed number.

Turning to (34), we first use the elementary estimate

$$P\left(Z_{m(n)} \leq u_{m(n)} \text{ and } \bigcup_{t=n+1}^{M'} \{Z_{m(t)} \leq u_{m(t)}\}\right)$$

$$\leq \sum_{t=n+1}^{M'} P(Z_{m(n)} \leq u_{m(n)}, Z_{m(t)} \leq u_{m(t)}). \tag{39}$$

Once again, by the nondecreasing property of u_j and by the X_j's being i.i.d., for $t > n$,

$$P(Z_{m(n)} \leq u_{m(n)}, Z_{m(t)} \leq u_{m(t)}) = P(Z_{m(n)} \leq u_{m(n)}) F^{m(t)-m(n)}(u_{m(t)}). \tag{40}$$

4.3 EXTREMES OF I.I.D. VARIABLES

Furthermore, by Lemma 1.3.1 and by (23),

$$F^{m(t)-m(n)}(u_{m(t)}) \leq \exp\left\{-\frac{1}{2}\frac{m(t)-m(n)}{m(t)}\log\log m(t)\right\}, \quad (41)$$

or, by using the opposite inequalities in Lemma 1.3.1 and in (23), for any $t \geq T > n$,

$$F^{m(t)-m(n)}(u_{m(t)}) \leq F^{m(t)}(u_{m(t)})\exp\left\{\frac{2m(n)}{m(T)}\log\log m(T)\right\}. \quad (42)$$

Now, for completing the proof of (34), and thus the theorem itself, we apply the estimates (39)–(42). We split the region $n < t \leq M'$ into three subsets; namely, $n < t \leq k$, where $k - n$ is moderate and thus (29) is applicable. Furthermore, we choose a $T > k$, for which the exponent on the right hand side of (42) remains bounded for $T < t \leq M'$. Finally, for the middle terms $k < t \leq T$, we apply (30). We thus get by (40), (41), and (29), for $k = k(n) > n$,

$$\sum_{t=n+1}^{k} P(Z_{m(n)} \leq u_{m(n)}, Z_{m(t)} \leq u_{m(t)}) \leq P(Z_{m(n)} \leq u_{m(n)}) \sum_{t=n+1}^{+\infty} e^{-(1/2)(t-n)\tau\lambda}$$

$$\leq P(Z_{m(n)} \leq u_{m(n)})\frac{e^{-(1/2)\tau\lambda}}{1-e^{-(1/2)\tau\lambda}};$$

on the other hand, (40), (41), and (30) yield

$$\sum_{t=k+1}^{T} P(Z_{m(n)} \leq u_{m(n)}, Z_{m(t)} \leq u_{m(t)}) \leq P(Z_{m(n)} \leq u_{m(n)}) \sum_{t=k+1}^{T} e^{-(1/4)(t-n)\tau}$$

$$\leq P(Z_{m(n)} \leq u_{m(n)})(T-k)\exp\left[-\tfrac{1}{4}(k-n)\tau\right];$$

and, finally, by (40), (42), and (38), for our choice of T,

$$\sum_{t=T+1}^{M'} P(Z_{m(n)} \leq u_{m(n)}, Z_{m(t)} \leq u_{m(t)}) \leq P(Z_{m(n)} \leq u_{m(n)})\rho \sum_{t=T+1}^{M'} F^{m(t)}(u_{m(t)})$$

$$\leq 2\Delta\rho P(Z_{m(n)} \leq u_{m(n)}),$$

where $\rho > 0$ is the bound of the exponential function on the right hand side of (42). Note that, by increasing the value of T, ρ can be brought arbitrarily close to one. If we combine the three estimates above, (39) yields (34), where

$$\delta = \frac{e^{-(1/2)\tau\lambda}}{1-e^{-(1/2)\tau\lambda}} + (T-k)\exp\left[-\frac{1}{4}(k-n)\tau\right] + 2\Delta\rho.$$

We are free to choose τ and λ, where $0<\lambda<1$. Thus, the first term can be made smaller than $\frac{1}{4}$, say, by a proper choice. For the second term, we observe that the conditions on the choice of k and T are satisfied if both $T-k$ and $k-n$ tend to infinity as fast as positive powers of $\log n$. Hence, the second term actually tends to zero. Consequently, if M_0 is large enough, then, for all $n \geqslant M \geqslant M_0$, the second term is also smaller than $\frac{1}{4}$. As remarked earlier, $\rho < \frac{3}{2}$, say, for large T. Therefore, since (38) holds with arbitrary $\Delta > 0$ for all $M \geqslant M_0 = M_0(\Delta)$, there is no restriction on Δ. Thus, with $\Delta = \frac{1}{12}$, the third term above does not exceed $\frac{1}{4}$. That is, (34) holds with $0 < \delta < \frac{3}{4}$, which completes the proof. ▲

Lemmas 4.3.2 and 4.3.3 and Theorem 4.3.2 evidently imply Theorem 4.3.1.

Let us also state the corresponding result for the minimum of i.i.d. We combine the results into one theorem.

Theorem 4.3.3. *Let v_n be a nonincreasing sequence. Then $P(W_n < v_n$ i.o.$) = 0$ or 1 according as $\sum_{n=1}^{+\infty} F(v_n)$ converges or diverges.*

In addition, if $F(x)$ is continuous and if v_n is nonincreasing and either $nF(v_n)$ is nondevreasing or, for all large n,

$$\frac{1}{2} \leqslant \frac{nF(v_n)}{\log\log n} \leqslant 2,$$

then $P(W_n \geqslant v_n) = 0$ or 1 according as

$$\sum_{n=1}^{+\infty} F(v_n)\exp[-nF(v_n)] < +\infty \quad \text{or} \quad = +\infty.$$

This statement, of course, follows from Corollary 4.3.1 and Theorems 4.3.1–4.3.2 by turning to the negative of the original sequence of random variables.

Example 4.3.1. Let $X_1, X_2, \ldots$ be independent random variables with uniform distribution on $(0,1)$. That is, their common distribution function $F(x) = x$ for $0 \leqslant x \leqslant 1$. Then, for any $y > 0$,

$$P\left(Z_n \geqslant 1 - \frac{y\log\log n}{n} \quad \text{i.o.}\right) = P\left(W_n \leqslant \frac{y\log\log n}{n} \quad \text{i.o.}\right) = 1.$$

We first note that, for any increasing u_n,

$$P(Z_n \geq u_n \text{ i.o.}) = P(W_n \leq 1 - u_n \text{ i.o.}). \tag{43}$$

Let us put $u_n = 1 - (y \log \log n)/n$. Then u_n increases and

$$\sum_{n=2}^{+\infty} [1 - F(u_n)] = \sum_{n=2}^{+\infty} (1 - u_n) = y \sum_{n=2}^{+\infty} \frac{\log \log n}{n} = +\infty$$

for any $y > 0$. An appeal to Corollary 4.3.1, combined with (43), results in the statement of the example. ▲

Example 4.3.2. Let $X_1, X_2, \ldots$ be i.i.d. uniform variates on the interval $(0, 1)$. Then, for all y,

$$P\left(Z_n \leq 1 - \frac{(1+y)\log\log n}{n} \text{ i.o.}\right)$$

$$= P\left(Z_n \leq 1 - \frac{\log\log n}{n} - \frac{(2+y)\log\log\log n}{n} \text{ i.o.}\right),$$

and this common value is 0 or 1 according as $y > 0$ or ≤ 0.

Indeed, since $F(x) = x$ for $0 \leq x \leq 1$,

$$u_n = 1 - \frac{(1+y)\log\log n}{n} \quad \text{and} \quad n[1 - F(u_n)] = (1+y)\log\log n$$

are increasing in n. Furthermore,

$$\sum_{n=2}^{+\infty} [1 - F(u_n)] \exp\{-n[1 - F(u_n)]\} = (1+y) \sum_{n=2}^{+\infty} \frac{\log\log n}{n(\log n)^{1+y}}$$

converges or diverges according as $y > 0$ or $y \leq 0$. Hence, by Theorem 4.3.1, the claim about the value of

$$P\left(Z_n \leq 1 - \frac{(1+y)\log\log n}{n} \text{ i.o.}\right)$$

follows. If we now choose in Theorem 4.3.1

$$u_n = 1 - \frac{\log\log n}{n} - \frac{(2+y)\log\log\log n}{n},$$

then again both u_n and $n[1-F(u_n)]$ increase in n. The series (21) becomes

$$\sum_{n=2}^{+\infty} \frac{\log\log n + (2+y)\log\log\log n}{(n\log n)(\log\log n)^{2+y}},$$

which again converges or diverges, depending on whether $y>0$ or $y\leq 0$. We thus get what was stated. ▲

Example 4.3.3. Let $X_1, X_2, \ldots$ be unit exponential variates. Then

$$P\left(\lim_{n=+\infty} \frac{Z_n}{\log n} = 1\right) = 1.$$

As a matter of fact, by Corollary 4.3.1,

$$\sum_{n=2}^{+\infty} \left[1 - F(\log n + 2\log\log n)\right] = \sum_{n=2}^{+\infty} \frac{1}{n(\log n)^2} < +\infty$$

implies

$$P(Z_n < \log n + 2\log\log n \quad \text{ultimately}) = 1.$$

On the other hand, for $u_n = \log n - \log\log n$, the series (21) becomes

$$\sum_{n=2}^{+\infty} \frac{\log n}{n^2} < +\infty.$$

We can apply Theorem 4.3.1, since u_n and $n[1-F(u_n)] = \log n$ both increase. Thus

$$P(Z_n \geq \log n - \log\log n \quad \text{ultimately}) = 1.$$

It now follows that, with probability one, $Z_n/\log n \to 1$. ▲

Example 4.3.4. Let $X_1, X_2, \ldots$ be i.i.d. with common distribution function $F(x) = 1 - 1/x$, $x \geq 1$. Then, with arbitrary increasing sequence $d_n > 0$, either

$$P\left(\limsup \frac{Z_n}{d_n} = +\infty\right) = 1$$

or

$$P\left(\limsup \frac{Z_n}{d_n} = 0\right) = 1.$$

In both statements, lim sup is taken as $n \to +\infty$. Since

$$\limsup\left\{\frac{Z_n}{d_n} = +\infty\right\} = \bigcap_{t=1}^{+\infty} \{Z_n \geq t d_n \quad \text{i.o.}\},$$

we have from Corollary 4.3.1,

$$P\left(\limsup \frac{Z_n}{d_n} = +\infty\right) = 1 \quad \text{if, and only if,} \quad \sum_{n=1}^{+\infty} \frac{1}{t d_n} = +\infty$$

for all $t \geq 1$. But the latter condition is evidently true, whatever be the value of t, if the sum of $1/d_n$ diverges. On the other hand, if $\sum_{n=1}^{+\infty} 1/d_n < +\infty$, then, for any $t \geq 1$, $P(Z_n \leq d_n/t \text{ ultimately}) = 1$. This is, however, equivalent to $P(\limsup Z_n/d_n = 0) = 1$. ▲

Examples 4.3.3 and 4.3.4 express very distinct properties of the two distributions occurring in them. Since, in mathematical investigations, the actual order of magnitude of Z_n and W_n plays an important role, we analyze this problem further in the next section.

4.4. LIM SUP AND LIM INF OF NORMALIZED EXTREMES

In the present section we again assume that the basic random variables $X_1, X_2, \ldots$ are independent, but in some statements we will not need that they are identically distributed. We put $F_j(x) = P(X_j < x)$. Our aim is to find conditions under which the lim sup or lim inf of the extremes, when normalized, is almost surely a finite, nonzero constant.

We first establish a general result.

Theorem 4.4.1. *Let $X_1, X_2, \ldots$ be independent random variables. Let $B_n > 0$ be a nondecreasing sequence of real numbers which tends to infinity with n. Let S be the set of numbers $k > 0$ for which*

$$\sum_{j=1}^{+\infty} \left[1 - F_j(kB_j)\right] \tag{44}$$

diverges. Then

$$P\left(\limsup \frac{Z_n}{B_n} = t\right) = 1, \tag{45}$$

where $t = 0$ if S is empty and $t = \sup S$ otherwise.

Proof. Let us first observe that, for any $n \geq 1$, $X_1 \leq Z_n$. Thus, in view of $B_n \to +\infty$ with n,

$$P\left(\limsup \frac{Z_n}{B_n} \geq 0\right) = P\left(\liminf \frac{Z_n}{B_n} \geq 0\right) = 1. \tag{46}$$

Next we show that, for any $k > 0$,

$$P(Z_n \geq kB_n \text{ i.o.}) = 1 \quad \text{or} \quad 0 \tag{47}$$

according as (44) diverges or converges. Indeed, since $B_n \to +\infty$ with n, we can apply (15) with $u_n = kB_n$. Thus, by independence and by Lemmas 4.2.1 and 4.2.3, (47) follows. Thus, if, for arbitrary $k > 0$, (44) diverges, then (45) holds with $t = +\infty$, which also equals $\sup S$. Now let S be empty. Then, by the criterion for (47), for arbitrary $k > 0$,

$$P\left(\limsup \frac{Z_n}{B_n} < k \quad \text{ultimately}\right) = 1.$$

This is evidently equivalent to (45) with $t = 0$.

It remains to investigate the case when S is nonempty and $t = \sup S < +\infty$. The definition of S and the equivalence of (47) to the divergence or convergence of (44) yield that, for any $\varepsilon > 0$,

$$P(Z_n \geq (t+\varepsilon)B_n \text{ i.o.}) = 0 \quad \text{and} \quad P(Z_n \geq (t-\varepsilon)B_n \text{ i.o.}) = 1,$$

and thus (45) is valid. The theorem is established. ▲

For the liminf, we prove a weaker statement.

Theorem 4.4.2. *Let $X_1, X_2, \ldots$ and $B_n > 0$ be as in Theorem* 4.4.1. *Then*

$$P\left(\liminf \frac{Z_n}{B_n} = t^*\right) = 1, \tag{48}$$

where $t^* \geq 0$ and t^* is a constant, possibly infinite.

Proof. The fact that $t^* \geq 0$ has been established in (46).

Again let $k > 0$ be an arbitrary number. Since, in the proof of Lemma 4.3.1, only the independence of $X_1, X_2, \ldots$ was used, but nothing about their distributions, we have

$$P(Z_n \leq kB_n \text{ i.o.}) = 1 \quad \text{or} \quad 0. \tag{49}$$

Let $S^* = \{k : k > 0 \text{ and the value in (49) is one}\}$. Set $t^* = \inf S^*$ if S^* is nonempty and $t^* = +\infty$ if S^* is empty. It is now an easy repetition of the last steps in the preceding proof to conclude that (48) holds with t^* just defined. The proof is completed. ▲

We now turn to i.i.d. variables. In this case, we have Theorem 4.3.1 to decide the actual case that applies in (49). Therefore, t^*, defined in terms of S^*, can be determined which, as was shown, is the value occurring in (48). This fact will be used in the forthcoming arguments. The sequence a_n will always signify

$$a_n = \inf\left\{y : F(y) \geq 1 - \frac{1}{n}\right\}. \tag{50}$$

Theorem 4.4.3. *Let $X_1, X_2, \ldots$ be i.i.d. random variables with common continuous distribution function $F(x)$. Let $\omega(F) = +\infty$. Then, for any $B_n \sim a_n$,*

$$P\left(\liminf \frac{Z_n}{B_n} \leq 1\right) = P\left(\limsup \frac{Z_n}{B_n} \geq 1\right) = 1.$$

Proof. The assumption $\omega(F) = +\infty$ implies that $a_n \to +\infty$ with n. Therefore, it suffices to prove the theorem with $B_n = a_n$. We shall apply that, by the continuity of $F(x)$, $F(a_n) = 1 - 1/n$.

We first apply Theorem 4.4.1. Since, for any $k \leq 1$,

$$\sum_{j=1}^{+\infty}[1 - F(ka_j)] \geq \sum_{j=1}^{+\infty}[1 - F(a_j)] = \sum_{j=1}^{+\infty}\frac{1}{j} = +\infty,$$

we immediately get the claim about $\limsup Z_n/B_n$.

Turning to $\liminf Z_n/a_n$, let us observe that $n[1 - F(a_n)] = 1$ and thus both a_n and $n[1 - F(a_n)]$ are nondecreasing. Theorem 4.3.1 is thus applicable, which yields that

$$\sum_{j=1}^{+\infty}[1 - F(a_j)]\exp\{-j[1 - F(a_j)]\} = \frac{1}{e}\sum_{j=1}^{+\infty}\frac{1}{j} = +\infty$$

implies $P(Z_n \leq a_n \text{ i.o.}) = 1$. This is equivalent to our statement on $\liminf Z_n/B_n$. The proof is completed. ▲

Corollary 4.4.1. *Let $X_1, X_2, \ldots$ be i.i.d. with continuous distribution function $F(x)$. Let $\omega(F) = +\infty$. If $P(\lim Z_n/B_n = 1) = 1$, then $B_n \sim a_n$.*

Example 4.4.1. Let $X_1, X_2, \ldots$ be i.i.d. standard normal variates. Put, as in Section 2.3.2,

$$a_n^* = (2\log n)^{1/2} - \frac{\frac{1}{2}(\log\log n + \log 4\pi)}{(2\log n)^{1/2}},$$

$$b_n = (2\log n)^{-1/2}.$$

Then, with probability one, there is an integer n_0 such that, for all $n \geq n_0$, and for arbitrary $\delta > 0$,

$$a_n^* - b_n[\log\log\log n + \log(1+\delta)] \leq Z_n \leq a_n^* + (1+\delta)b_n \log\log n. \quad (51)$$

In particular,

$$P\left(\lim_{n=+\infty}(Z_n - \sqrt{2\log n}) = 0\right) = P\left(\lim_{n=+\infty}\frac{Z_n}{\sqrt{2\log n}} = 1\right) = 1.$$

The method of arriving at (51) is generally applicable to any continuous distribution function $F(x)$. Therefore, we state each step in general terms, and then the calculations are carried out when $F(x)$ is the standard normal distribution function. We shall make several references to Section 2.3.2.

First, let us determine an asymptotic expression for a_n defined in (50). For the normal distribution, this has been done in Section 2.3.2, where a_n was found to be asymptotically equal to a_n^* above. Next, guided by Lemma 4.3.2, we determine $u_n = a_n^* - s_n$ such that (23) holds and (21) converges. Since, by definition, $n[1 - F(a_n^*)] \sim 1$, $s_n > 0$. If we consider (23) alone, then formula (62) of Section 2.3.2 yields

$$1 - F(a_n^* - s_n) \sim \frac{C\left\{\exp\left[-\frac{1}{2}(a_n^* - s_n)^2\right]\right\}}{a_n^* - s_n}$$

$$\sim \frac{a_n^*}{a_n^* - s_n}[1 - F(a_n^*)]\exp\left(a_n^* s_n - \frac{1}{2}s_n^2\right).$$

Now, if $s_n \to 0$ (which holds in our case, and which we can always achieve by finding an accurate enough expression for a_n, namely, by stopping in its approximation at a term which itself tends to zero), then the above relation reduces to

$$1 - F(a_n^* - s_n) \sim \frac{1}{n}e^{a_n^* s_n}.$$

Thus, writing $s_n = s_n^*/\sqrt{2\log n}$, then

$$1 - F(a_n^* - s_n) \sim \frac{1}{n} e^{s_n^*}. \tag{52}$$

An appeal to (23) thus yields

$$\log\tfrac{1}{2} \leqslant s_n^* - \log\log\log n \leqslant \log 2.$$

We thus choose this difference by which (21) converges. If $t = s_n^* - \log\log\log n$, then with $u_n = a_n^* - s_n$,

$$[1 - F(u_n)]\exp\{-n[1 - F(u_n)]\} \sim \frac{e^t \log\log n}{n}(\log n)^{-e^t}.$$

Hence, with $t = \log(1+\delta)$, $0 < \delta < 1$, the nth term of (21) is of the magnitude

$$\frac{\log\log n}{n(\log n)^{1+\delta}},$$

the sum of which is well known to converge. The lower inequality of (51) is thus implied by Lemma 4.3.3. The upper inequality is much simpler, and the preceding calculations can be used. We now apply Corollary 4.3.1 with $u_n = a_n^* - s_n$, where $s_n = s_n^*/\sqrt{2\log n} < 0$. By (52), if $s_n^* = -(1+\delta)\log\log n$, $\delta > 0$,

$$1 - F(u_n) \sim \frac{1}{n(\log n)^{1+\delta}},$$

the sum of which converges, which is sufficient for the validity of the upper inequality of (51). ▲

The stated limits are special cases of (51). They show that both the AWL and MWL of large numbers can be strengthened to almost sure convergence. Such situations will be referred to as strong laws of large numbers. We speak of additive strong laws (ASL) and multiplicative strong laws (MSL) according as $Z_n - a_n \to 0$ or $Z_n/a_n \to 1$, respectively, both limits being valid almost surely. In the MSL, it is, of course, always assumed that $a_n \neq 0$. The ASL and MSL are similarly defined for arbitrary extremes.

The reader can realize from the calculations leading to (51) that the orders of magnitude of both bounds are accurate. Therefore, the different orders of magnitude in the two bounds are not just matters of convenience.

Remark 4.4.1. The result (51) is pleasing to the mathematician but disappointing to the applied scientist. For the mathematician it is neat that we developed a method by which accurate bounds can be set on Z_n to be valid with probability one. But let us consider the applied scientist's point of view. The mathematician instructed him in Section 2.3.2 that, for a sample of n independent observations on a standard normal variate, he should first calculate a_n^* and b_n, as given in Example 4.4.1. Then, if n is large, the distribution of $(Z_n - a_n^*)/b_n$ is well approximated by $H_{3,0}(x)$. The scientist by now has gathered $n = 10{,}000$ observations and he computes a_n^* and b_n. His actual Z_n is just slightly smaller than a_n^*; in fact, $(Z_n - a_n^*)/b_n = -0.81$, say. The approximation by $H_{3,0}(x)$ yields

$$P(Z_n < a_n^* - 0.8 b_n) \sim .108.$$

As a double check, he also computes the lower bound of (51) with a small δ. Since $\log\log\log 10{,}000 = 0.7977$, (51) implies that $(Z_n - a_n^*)/b_n$ practically never would fall below -0.8.

These seemingly contradictory conclusions are not contradictions at all. The approximation by $H_{3,0}(-0.8)$ above is accurate (see Theorem 2.10.1). On the other hand, the inequalities of (51) are valid, with probability one, for all $n \geq n_0$, but n_0 itself is a random variable depending both on the actual sample and on the choice of $\delta > 0$. In other words, n_0 varies from sample to sample, and a fixed n can be large for one set of observations and thus exceed n_0, while in other cases the same n remains smaller than n_0. Consequently, practical conclusions cannot be drawn from (51). Its theoretical implications, however, are interesting.

In Example 4.4.1 we described a method of obtaining the exact order of magnitude of the bounds on Z_n, which are valid with probability one. If one is interested in the major terms of these bounds only, then the method becomes simple. In particular, strong laws of large numbers are immediate.

Theorem 4.4.4. *Let $X_1, X_2, \ldots$ be i.i.d. random variables with common distribution function $F(x)$. Let $\omega(F) = +\infty$. Let a_n be the sequence defined at (50). Then*

$$P\left(\lim_{n = +\infty} \frac{Z_n}{a_n} = 1\right) = 1 \qquad (53)$$

if, and only if, for arbitrary $k > 1$,

$$\sum_{n=1}^{+\infty} \left[1 - F(ka_n)\right] < +\infty. \qquad (54)$$

Proof. By Theorem 4.4.1, (53) implies (54). Therefore, we have to prove the converse implication. Since the series of (54) diverges for $k=1$ by the definition of a_n, one more appeal to Theorem 4.4.1 yields that the validity of (54) implies

$$P\left(\limsup_{n=+\infty} \frac{Z_n}{a_n} = 1\right) = 1.$$

Hence, it remains to prove that the validity of (54) is also sufficient for

$$P\left(\liminf_{n=+\infty} \frac{Z_n}{a_n} = 1\right) = 1.$$

We apply Lemma 4.3.3 with $u_n = ta_n$, where $0 < t < 1$. For

$$\sum_{n=1}^{+\infty} [1 - F(u_n)] \geq \sum_{n=1}^{+\infty} [1 - F(a_n)] \geq \sum_{n=1}^{+\infty} \frac{1}{n} = +\infty,$$

it suffices to show that (21) converges for each fixed $t < 1$.

We first show that (54) holds if, and only if, for arbitrary $0 < t < 1$,

$$\int_1^{+\infty} \frac{dF(y)}{1 - F(ty)} < +\infty. \tag{55}$$

Let $G(y) = \inf\{x : F(x) \geq 1 - 1/y\}$. Then $a_n = G(n)$ and $G(y)$ is a nondecreasing function. Thus, by

$$\int_n^{n+1} \{1 - F[kG(y)]\} \, dy \leq 1 - F(ka_n) \leq \int_{n-1}^n \{1 - F[kG(y)]\} \, dy,$$

(54) is equivalent to

$$\int_1^{+\infty} \{1 - F[kG(y)]\} \, dy < +\infty.$$

But

$$\int_1^{+\infty} \{1 - F[kG(y)]\} \, dy = \int_1^{+\infty} \left[\int_{kG(y)}^{+\infty} dF(x)\right] dy$$

$$= \int_{kG(1)}^{+\infty} \left\{\int_1^{1/[1-F(x/k)]} dy\right\} dF(x),$$

the convergence of which is indeed equivalent to (55). A somewhat more unpleasant but not difficult calculation now shows that, as a consequence of (55), (21) converges. Out of this calculation, we mention only that one should start with the observation that the convergence of (21) is equivalent to

$$\sum_{n=1}^{+\infty}\left[1-F(u_{n+1})\right]\exp\left\{-(n+1)\left[1-F(u_n)\right]\right\}<+\infty,$$

where $u_n = ta_n$, $0 < t < 1$. By the monotonicity of $G(y)$, the series above is dominated by

$$\sum_{n=1}^{+\infty}\int_n^{n+1}\left\{1-F\left[tG(y)\right]\right\}\exp\left\{-y\left[1-F(tG(y))\right]\right\}dy.$$

Writing this as a single integral and substituting $x = G(y)$, one arrives at an integral which converges whenever (55) holds. This completes the proof. ▲

The results of this section can be extended in several directions. Since they do not require essential new ideas, they are collected among the problems for solution. In particular, results will be found for specific distributions, for extremes other than the maxima as well as for dependent cases.

4.5. THE ROLE OF EXTREMES IN THE THEORY OF SUMS OF RANDOM VARIABLES

The classical theory of probability is associated with the theory of sums of i.i.d. random variables. In particular, the laws of large numbers (for sums) and the asymptotic normality were the centers of investigation, in both of which it is assumed that at least the variance of the population is finite. In such cases, the contribution of the extreme terms to the sum of i.i.d. random variables is negligible. This explains why theory of sums was not related to the theory of extremes for a long period of time. However, with the recognition of the importance of distributions for which the expectation itself is not finite, the interrelations of the two theories emerged. The present section is devoted to this relation. We give all results for i.i.d. random variables, but those familiar with the theory of sums of dependent random variables (stationary, mixing, and exchangeable sequences are the best developed) will recognize the corresponding relations with the appropriate reference to Chapter 3.

Throughout this section $X_1, X_2, \ldots$ are i.i.d. random variables with common distribution function $F(x)$. In addition to our standard relations, we put $T_n = X_1 + X_2 + \cdots + X_n$.

Let us start with an example. Let

$$F(x) = \frac{1}{2} + \frac{1}{\pi}\arctan x,$$

the standard Cauchy distribution. Then routine calculation shows that the distribution of $(1/n)T_n$ is also $F(x)$ (one can use the standard convolution formula or characteristic functions or Laplace transforms). We emphasize from this result only that the normalizing constant by which T_n is divided in order to get a nondegenerate (limiting) distribution is of magnitude n. Next, we observe that the order of magnitude of b_n in the relation

$$\lim_{n=+\infty} P\left(\frac{Z_n}{b_n} < x\right) = \exp\left(-\frac{1}{x}\right), \qquad x > 0, \tag{56}$$

is also n. Indeed, in Section 2.3.4 we found (56) together with $b_n = \tan(\pi/2 - \pi/n)$. But, in view of $\cos(\pi/2 - x) = \sin x$ and, for $x \to 0$, $\sin(\pi/2 - x) \to 1$ and $(\sin x)/x \to 1$,

$$b_n = \tan\left(\frac{\pi}{2} - \frac{\pi}{n}\right) \sim \frac{1}{\sin(\pi/n)} = \frac{n}{\pi} \frac{\pi/n}{\sin(\pi/n)} \sim \frac{n}{\pi},$$

as stated. The fact stressed here suggests a close relation between T_n and Z_n for the Cauchy distribution. This relation actually is due partly to $E(|X_1|) = +\infty$ and partly to the behavior of $1 - F(x)$ as $x \to +\infty$. It is, however, not essential that X_1 and $(1/n)T_n$ are identically distributed. In fact, the following result is true.

Theorem 4.5.1. *Let $X_1, X_2, \ldots$ be i.i.d. random variables with common distribution function $F(x)$. Let $F(0) = 0$ and $\omega(F) = +\infty$. Assume that there are constants a_n, $b_n > 0$, A_n and $B_n > 0$ such that $(Z_n - a_n)/b_n$ and $(T_n - A_n)/B_n$ converge weakly to some nondegenerate distribution functions $H(x)$ and $U(x)$, respectively. Then $b_n/B_n \sim \tau$, $0 < \tau < +\infty$, if, and only if, $F(x)$ belongs to the domain of attraction of $H_{1,\gamma}(x)$ with some $0 < \gamma < 2$.*

Remark 4.5.1. The assumption $F(0) = 0$ can be dropped. The conclusion then becomes that $F(x)$ should belong to the domain of attraction of both $H_{1,\gamma}(x)$ and $L_{1,\delta}(x)$, where $0 < \gamma < 2$ and $\gamma \leq \delta$. This exact statement follows from the theory of weak convergence of $(T_n - A_n)/B_n$. It is, however, quite plausible that it should be so. Namely, if $F(x)$ were such that $|W_n|$ would become large compared with Z_n, then T_n would be comparable with W_n rather than Z_n.

We also wish to remark that no general rule can be expected about the relation of a_n and A_n. As a matter of fact, if we add a fixed number to each X_j, then T_n increases by a multiple of n, but Z_n increases only by the fixed constant.

Finally, notice that the conclusion of the present theorem and Theorems 2.4.1, 2.4.3, 2.1.1, and 2.7.3 imply that if b_n and B_n are proportional to each other, then $a_n = 0$ and $E(X_1^a) = +\infty$ or $< +\infty$ according as $a \geq \gamma$ or $0 < a < \gamma < 2$.

Proof. By $\omega(F) = +\infty$ and by $(Z_n - a_n)/b_n$ converging weakly to a nondegenerate distribution function $H(x)$, the theorems of Section 2.4 yield that $F(x)$ is in the domain of attraction of either $H_{1,\gamma}(x)$ for some $\gamma > 0$ or $H_{3,0}(x)$. Should $F(x)$ belong to the domain of attraction of $H_{3,0}(x)$, then Theorem 2.7.3 would yield $E(X_1^a) < +\infty$ for all $a > 0$. In particular, $E = E(X_1)$ and $V = V(X_1)$ were finite. Then the classical central limit theorem would imply $A_n = nE$, $B_n = (nV)^{1/2}$, and $U(x) = \Phi(x)$, the standard normal distribution function. On the other hand, if $E(X^a) < +\infty$, then by (129) of Chapter 2,

$$+\infty > \int_1^{+\infty} x^{a-1}[1 - F(x)]\,dx$$

$$= \sum_n \int_{a_n + b_n}^{a_{n+1} + b_{n+1}} \ldots$$

$$\geq \sum_n (a_n + b_n)^{a-1}[1 - F(a_{n+1} + b_{n+1})],$$

where summation is over $n \geq k$ with $a_k + b_k \geq 1$. Now, since $F(x)$ is assumed to belong to the domain of attraction of $H_{3,0}(x)$, Step 1 of Section 2.5 and the convergence above imply, for $n \to +\infty$,

$$\frac{(a_n + b_n)^{a-1}}{n} \to 0 \quad \text{or} \quad a_n + b_n = o(n^{1/(a-1)}).$$

We have shown in the proof of Theorem 4.1.1 that $b_n = o(a_n)$. Therefore $b_n = o(n^{1/(a-1)})$, for all $a > 1$. This now gives that b_n and $B_n = (nV)^{1/2}$ are not proportional. Consequently, $F(x)$ can belong only to the domain of attraction of $H_{1,\gamma}(x)$ with some $\gamma > 0$. Thus, by Theorem 2.4.3, for all $x > 0$, as $t \to +\infty$,

$$\lim \frac{1 - F(tx)}{1 - F(t)} = x^{-\gamma}.$$

We rewrite this relation as

$$1 - F(x) = \frac{L(x)}{x^\gamma}, \quad \frac{L(tx)}{L(t)} \to 1 \quad (t \to +\infty). \tag{57}$$

Let us now compare b_n and B_n. If $\gamma > 2$, then, in view of Theorem 2.7.3 again, E and V are finite and thus $B_n = (nV)^{1/2}$. But, by (22) of Chapter 2, for any $\gamma > 0$,

$$n[1 - F(b_n)] = \frac{nL(b_n)}{b_n^\gamma} \to 1 \quad \text{as} \quad n \to +\infty. \tag{58}$$

Since, for the slowly varying function $L(x)$, $L(x) < x^\varepsilon$ for any $\varepsilon > 0$ (see Appendix III), $b_n \leq n^{1/(\gamma - \varepsilon)}$. Hence, for $\gamma > 2$, we can choose $\varepsilon > 0$ such that $1/(\gamma - \varepsilon) < \frac{1}{2}$. Consequently, $b_n / B_n \to 0$ as $n \to +\infty$.

Now let $0 < \gamma < 2$. Then it is known from the theory of sums of i.i.d. variables (see Ibragimov and Linnik (1971)) that $U(x)$ exists and B_n is such that

$$\frac{nL(B_n)}{B_n^\gamma} \to s, \quad 0 < s < +\infty, \quad n \to +\infty.$$

This, combined with (58), yields

$$\left(\frac{b_n}{B_n}\right)^\gamma \sim s \frac{L(b_n)}{L(B_n)} = s \frac{L[B_n(b_n/B_n)]}{L(B_n)} = s \frac{L(b_n)}{L[b_n(B_n/b_n)]}.$$

The right hand side always tends to s because either b_n / B_n or B_n / b_n is bounded (see Appendix III). Therefore, so does the left hand side, which was to be proved for $0 < \gamma < 2$.

There remains only the case $\gamma = 2$. Now, the variance of X_1 is not finite, but $U(x)$ is again the standard normal distribution. In this case, a somewhat more accurate result than in the case of $\gamma < 2$ is needed from the theory of sums to conclude that Z_n / B_n has a degenerate distribution, which is degenerate at zero. Therefore, b_n and B_n cannot be of the same order of magnitude. For the details on the theory of sums, we again refer to Ibragimov and Linnik (1971). This concludes the proof. ▲

It follows from Theorem 2.8.1 that if $F(x)$ belongs to the domain of attraction of $H_{1,\gamma}(x)$ with some $0 < \gamma < 2$, and if $F(0) = 0$, then all upper extremes $X_{n-k:n}$, when normalized by B_n of Theorem 4.5.1, have a nondegenerate limit law. It has an interesting implication. Namely, however large n is, a constant plus a finite number of upper extremes will have the same magnitude as T_n. Even more surprising is the case when A_n can be taken as zero. For example, if $1 < \gamma < 2$, then $E(X_1) = E < +\infty$, and translating each X_j by $-E$ gives $A_n = 0$. We get from Theorem 4.5.1 that, to any $x > 0$, there is a unique $y > 0$, such that $U(x) = H(y)$ (both are strictly increasing continuous functions) and thus, as $n \to +\infty$,

$$\lim P(T_n < b_n x) = \lim P(Z_n < b_n y),$$

where $b_n = \inf\{z : 1 - F(z) \leq 1/n\}$. If one takes several, but a finite number of, upper extremes, x and y will be closer and closer to each other.

Let us continue our investigation for i.i.d. random variables. We assume $X_1 \geq 0$ as before. Let now $E(X_1^a) = +\infty$ for all $a > 0$. Then, by Theorem 2.7.3, there are no sequences a_n and $b_n > 0$ for which $(Z_n - a_n)/b_n$ would have a nondegenerate limit distribution. A similar observation is known in the theory of sums. However, in this case, the relation of T_n and Z_n is even closer.

Theorem 4.5.2. *Let $X_1, X_2, \ldots$ be i.i.d. random variables with common absolutely continuous distribution function $F(x)$. Let $F(0) = 0$ and $\omega(F) = +\infty$. Assume that, for all $x > 0$, as $t \to +\infty$,*

$$\lim \frac{1 - F(tx)}{1 - F(t)} = 1. \tag{59}$$

Then

$$\lim_{n = +\infty} E\left(\frac{T_n}{Z_n}\right) = 1. \tag{60}$$

Proof. Put $R_n = T_n/Z_n$. Let the joint density function of $X_1, X_2, \ldots, X_n$ and Z_n be $g(x_1, x_2, \ldots, x_n, z)$. Since Z_n is one of the X_j, and, by the assumption of continuity, ties can be neglected, $z = x_j$ for some $1 \leq j \leq n$. Because of symmetry,

$$g(x_1, x_2, \ldots, x_n, x_j) = g(x_1, x_2, \ldots, x_n, x_1) \quad \text{for all } j$$

and

$$g(x_1, x_2, \ldots, x_n, x_1) = \begin{cases} f(x_1) f(x_2) \cdots f(x_n) & \text{if } x_1 = \max\{x_j\}, \\ 0 & \text{otherwise,} \end{cases}$$

where $f(x) = F'(x)$, the density of $F(x)$. From basic formulas of expectation we thus get

$$E(e^{itR_n}) = n \int_0^{+\infty} \int_0^{x_1} \cdots \int_0^{x_1} \left\{\exp\left[\frac{it}{x_1} \sum_{j=1}^n x_j\right]\right\} \prod_{j=1}^n f(x_j) \, dx_2 \cdots dx_n \, dx_1$$

$$= n e^{it} \int_0^{+\infty} \left[\int_0^{x_1} e^{itu/x_1} f(u) \, du\right]^{n-1} f(x_1) \, dx_1.$$

If we differentiate with respect to t and set $t=0$, the left hand side yields $iE(R_n)$. Therefore

$$E(R_n) = n\int_0^{+\infty}\left(\int_0^{x_1} f(u)\,du\right)^{n-1} f(x_1)\,dx_1$$

$$+ n(n-1)\int_0^{+\infty}\left(\int_0^{x_1} f(u)\,du\right)^{n-2}\left(\int_0^{x_1}\frac{u}{x_1} f(u)\,du\right) f(x_1)\,dx_1$$

$$= 1 + n(n-1)\int_0^{+\infty} F^{n-2}(x)\left[\int_0^x \frac{u}{x} f(u)\,du\right] f(x)\,dx. \tag{61}$$

For estimating the second term above, we first rewrite the inner integral by substituting $v = u/x$ and then integrating by parts. We get

$$\int_0^x \frac{u}{x} f(u)\,du = \int_0^1 vx f(vx)\,dv$$

$$= \int_0^1 [1 - F(vx)]\,dv - [1 - F(x)]$$

$$= [1 - F(x)]\left[\int_0^1 \frac{1 - F(vx)}{1 - F(x)}\,dv - 1\right]$$

$$= [1 - F(x)]\int_0^1 \left[\frac{1 - F(vx)}{1 - F(x)} - 1\right]dv.$$

In view of (59), the results of Appendix III are applicable. In particular, as $x \to +\infty$,

$$\lim \int_0^1 \left[\frac{1 - F(vx)}{1 - F(x)} - 1\right]dv = 0.$$

Let us put $I(x)$ for this last integral. Let now A be such that, for all $x \geq A$, $I(x)$ does not exceed a given $\varepsilon > 0$. For $x \leq A$, this same integral is, of course, bounded. In addition, if $x \leq A$, then $n(n-1)F^{n-2}(x) \leq n(n-1) \times F^{n-2}(A) \to 0$ as $n \to +\infty$, because $\omega(F) = +\infty$. Therefore, if we write the second term of the extreme right hand side of (61) as

$$n(n-1)\int_0^{+\infty} F^{n-2}(x)[1 - F(x)]I(x)f(x)\,dx,$$

we can easily estimate it and conclude that it tends to zero as $n \to +\infty$. Indeed, if we cut the integration at A, $n(n-1)F^{n-2}(A) \to 0$ guarantees that as x varies from zero to A, the integral will be small. On the other hand, by the choice of A, the integral from A to $+\infty$ will be smaller than

$$\varepsilon n(n-1)\int_{A}^{+\infty} F^{n-2}(x)[1-F(x)]f(x)\,dx$$

$$= \varepsilon\{1 - nF^{n-1}(A)[1-F(A)]\},$$

which tends to ε as $n \to +\infty$. Since $\varepsilon > 0$ is arbitrary, we obtained (60), which was to be proved. ▲

Let us conclude this section by remarking that results similar to those in the preceding section have been developed for sums of i.i.d. variables as well. The reader who is familiar with the theory of sums will immediately recognize the strong relation of the criteria for $P(T_n \leq u_n \text{ i.o.}) = 1$ or 0 to those for $P(Z_n \leq u_n \text{ i.o.}) = 1$ or 0; however, we do not discuss it here.

4.6. SURVEY OF THE LITERATURE

The degenerate limit laws, under names different from ours, were first investigated by B. V. Gnedenko (1943), although the case of normal populations was recorded earlier. Theorem 4.1.2 and its equivalent form by transforming F to F^* (see the last paragraph of Section 4.1) are due to him. J. Geffroy (1958/1959) continued his work and obtained extensions in two directions—to the kth extremes and to almost sure results. Geffroy's results include the necessity part of Theorem 4.4.4, the relations between Z_n and $X_{n-k:n}$, and the case $c = 0$ of Exercise 17. This line of work was continued, and substantially improved results were obtained by O. Barndorff-Nielsen (1961 and 1963). The major idea of Section 4.3 is due to him. So is Theorem 4.4.4 and its extension to $X_{n-k:n}$ as stated in Exercise 9. Actually, Barndorff-Nielsen shows that the assumption of continuity in Section 4.3 is not necessary. On the other hand, in Theorem 4.3.2 he needs that $n[1 - F(u_n)]$ is nondecreasing. This restriction was dropped by H. Robbins and D. Siegmund (1972), who reobtained Barndorff-Nielsen's result. Both papers use the same method of proof, which is also adopted in this book and is basically due to P. Erdös (1942), who considered the iterated logarithm theorem for sums of i.i.d. variables. It should be remarked that the monotonicity of $n[1 - F(u_n)]$ cannot be disposed of in Theorem 4.3.1, as shown by the example of Exercise 21.

De Haan (1970) gave equivalent forms of the results of Gnedenko for weak laws. He has also shown that Geffroy's results on absolutely continuous populations in connection with weak laws are essentially necessary conditions, not only sufficient. For smooth enough functions, de Haan and Hordijk (1972) obtained easily applicable results for the almost sure lim sup and lim inf of normalized extremes. They partly extend results of Pickands III (1967a). Pickands (1968) related weak laws of large numbers to convergence of moments. S. Resnick and R. J. Tomkins (1973) analyze the relation of B_n to $a_n = \inf\{x : 1 - F(x) \leq 1/n\}$, where B_n is such that either lim inf or lim sup of Z_n/B_n is positive and finite. Some of these results have already been present in the work of Barndorff-Nielsen (1963). C. W. Anderson (1970) obtains an interesting extension of weak laws to discrete populations. This result is quoted in Exercise 19. R. F. Green (1976a) observes the strong relation of weak laws to the statistical problem of outliers. A representative of his results is given in Exercise 11.

Out of dependent systems, systematic studies are available only for Gaussian sequences. Some general cases are mentioned in the present chapter which do not assume normality, and other possibilities are indicated in Exercises 13 and 14. In addition, the technique of the present chapter can be translated to several dependent models of Chapter 3 without much effort [for a case of strong mixing, see W. Philipp (1976)]. For the normal sequences, the first general result is obtained by S. M. Berman (1962a). His work is extended in several subsequent papers: Pickands III (1967b and 1969), M. Nisio (1967), C. M. Deo (1971), P. I. Judickaja (1974), Y. Mittal (1974), and Y. Mittal and D. Ylvisaker (1976). W. Philipp's result (1971) on sums of mixing variables can be transformed to results on extremes, although further work is needed on these to be useful in extreme value theory.

It is clear from Sections 4.3 and 4.4 that a dependent structure can be handled without additional difficulty if Borel-Cantelli lemmas are available for it. Apart from very abstract forms, Lemma 4.2.2 appears to be the most general available Borel-Cantelli lemma, which is due to P. Erdös and A. Rényi (1959). Another form, which we did not quote, is due to W. Philipp (1967) and can be used to estimate the rank of an order statistic for which the normalization is essentially the same as for maxima. General asymptotic bounds (for arbitrary dependent systems) are established by T. L. Lai and H. Robbins (1977) in their investigations on maximal dependent systems.

The relation of extremes and sums is somewhat surprising. Theorem 4.5.1 presents a new viewpoint of a result known in the theory of sums (see Ibragimov and Linnik (1971) and J. L. Mijnheer (1975)). What is actually surprising is Theorem 4.5.2, which is due to D. A. Darling (1952). For more extensive studies on this line, see A. A. Bobrov (1954) and D. Z. Arov and A. A. Bobrov (1960). For the effect of the extremes in strong laws for sums see T. Mori (1976b and 1977). A. V. Nagaev (1970 and 1971) studies the effect on Z_n if the sum of the observations is assumed to be large. In agreement with Tiago de Oliveira's result (see the survey of Chapter 2), the condition has no effect on Z_n.

The question of the behavior of $\sum_{k=1}^{n} Z_k$ or $\sum_{k=1}^{n} W_k$ was raised first by U. Grenander (1965). He obtained a normalization of $\log n$ for the latter sum. Grenander's work induced much further research, and extensions are found in the papers O. Frank (1966), T. Hoglund (1972), P. Deheuvels (1974), and M. Ghosh et al. (1975).

Interestingly, the theory becomes very difficult if stationarity is dropped. R. Mucci (1977) makes extensive study of almost sure results for Z_n and W_n for independent but not identically distributed variables. Because this permits the inclusion of a few variables which may control Z_n and W_n, very varying results can be obtained without restrictive assumptions. Theorems 4.4.1 and 4.4.2 are essentially due to him.

In a much more analytical manner, several theorems can be unified by studying different representations of functions and their inverses. Although in Chapter 6 we present such mathematical approaches, the reader may be interested at this stage in the works by W. Vervaat (1972) and A. A. Balkema (1973). See also the surveys of W. Vervaat (1973b and 1977).

4.7. EXERCISES

1. Show that if (Z_n, u_n) satisfies the AWL, then so does (Z_n, u_n^*), where u_n^* satisfies $u_n - u_n^* \to 0$ as $n \to +\infty$.

2. Show that if (Z_n, u_n), $u_n > 0$, satisfies the multiplicative law of large numbers, then so does (Z_n, u_n^*), whenever $u_n^*/u_n \to 1$ as $n \to +\infty$.

3. Let $X_1, X_2, \ldots, X_n$ be i.i.d. nonnegative random variables with common distribution function $F(x)$. Let $\omega(F) = +\infty$. Show that if (Z_n, u_n) satisfies the MWL, then u_n can be taken as $\inf\{x : 1 - F(x) \leq 1/n\}$.

4. Show that if, in Exercise 3, Z_n satisfies the MWL or the AWL, then $E(X_1)$ is finite.

5. Transform Theorem 4.1.2 by introducing the function $F^*(x)$ of the last paragraph of Section 4.1 and obtain a necessary and sufficient condition for Z_n satisfying an MWL.

6. Show that if, in Exercise 3, (Z_n, u_n) satisfies an MWL, then so do $(X_{n-k:n}, u_n)$ for all fixed k.

7. Obtain conditions for (W_n, v_n) satisfying a weak law from those known for (Z_n, u_n).

8. Extend Exercise 6 to additive weak laws and conclude that if (Z_n, u_n) satisfies the AWL, then, for any $\varepsilon > 0$, $P(Z_n - X_{n-k:n} \geq \varepsilon) \to 0$, where $k \geq 1$ is a fixed integer.

9. Let $X_1, X_2, \ldots, X_n$ be i.i.d random variables whose common distribution function $F(x)$ satisfies $\omega(F) = +\infty$. Let $k \geq 1$ be an integer and assume that, for every $\varepsilon > 0$,

$$\int_{-\infty}^{+\infty} \frac{[1-F(x)]^{k-1}}{[1-F(x-\varepsilon)]^k} dF(x) < +\infty.$$

Show that $X_{n-k+1:n}$ satisfies the additive strong law of large numbers.

[O. Barndorff-Nielsen (1963)]

10. Let $F(x)$ be a distribution function with $\omega(F) = +\infty$. Let $0 < t < 1$ be such that, for every $\varepsilon > 0$,

$$\int_1^{+\infty} \frac{dF(y)}{1-F[(t-\varepsilon)y]} < +\infty, \qquad \int_1^{+\infty} \frac{dF(y)}{1-F[(t+\varepsilon)y]} = +\infty.$$

Show that if $X_1, X_2, \ldots, X_n$ are i.i.d. with common distribution function $F(x)$, then, as $n \to +\infty$,

$$\limsup \frac{Z_n}{a_n} = \frac{1}{t}$$

almost surely, where $a_n = \inf\{x : 1 - F(x) \leq 1/n\}$. Hence, conclude that Z_n/B_n cannot converge to a constant with probability one, whatever be the value of $B_n > 0$.

[S. I. Resnick and R. J. Tomkins (1973)]

11. Call a distribution $F(x)$ outlier prone if there exist $\varepsilon > 0$, $\delta > 0$, and $n_0 \geq 1$ such that for all integers $n \geq n_0$,

$$P(Z_n - X_{n-1:n} \geq \varepsilon) \geq \delta,$$

where X_j, $j \geq 1$ are i.i.d. with common distribution $F(x)$. Prove that if $\omega(F) = +\infty$, then $F(x)$ is outlier prone if, and only if, there exist constants $c > 0$ and $d > 0$ such that

$$\frac{1-F(x+c)}{1-F(x)} \geq d \qquad \text{for all } x.$$

[R. F. Green (1976a)]

12. Compare the behavior of the difference $Z_n - X_{n-1:n}$ when the population distribution is normal with the case of exponential population, assuming that the basic random variables $X_j, j \geq 1$, are i.i.d.

13. Let $X_1, X_2, \ldots$ be identically distributed random variables. Assume that their multivariate distribution results in the bivariate marginals

$$P(X_i < x, X_j < y) = 1 - e^{-x} - e^{-y} + e^{-x-y}\left[1 + \tfrac{1}{2}(1-e^{-x})(1-e^{-y})\right]$$

(the so-called Morgenstern bivariate exponential distribution). Show that, for any nondecreasing sequence u_n, $P(Z_n \geq u_n \text{ i.o.})$ is either 0 or 1. (Apply Lemma 4.2.2 and Theorem 4.2.1.)

14. For the sequence of random variables of Exercise 13 prove that, with probability one, as $n \to +\infty$,

$$\limsup \frac{Z_n}{\log n} = 1.$$

15. Let $X_1, X_2, \ldots$ be standard normal variates. Let $b_n = (2\log n)^{1/2}$. Show that, whatever be the interdependence of the X's, it has probability one that, as $n \to +\infty$,

$$\limsup \frac{b_n(Z_n - b_n)}{\log \log n} \leq \frac{1}{2}.$$

[J. Pickands, III (1969)]

16. In Exercise 15, assume that $E(X_i X_{i+k}) = r_k$, such that

$$\sum_{k=1}^{+\infty} r_k^2 < +\infty.$$

Show that the conclusion of Exercise 15 can be improved to equality.

[J. Pickands, III (1969)]

17. Let $X_1, X_2, \ldots$ be i.i.d. random variables with common distribution function $F(x)$. Put

$$g(x) = \frac{[1-F(x)]\log\log[1/(1-F(x))]}{F'(x)},$$

where $F'(x)$ is the derivative of $F(x)$, which is assumed to be positive for all large x. Assume that, as $x \to +\infty$,

$$\lim \frac{g(x)}{x} = c, \qquad 0 \leq c < +\infty.$$

Prove that, with probability one,

$$\liminf \frac{Z_n}{b_n} = 1, \qquad \limsup \frac{Z_n}{b_n} = e^c,$$

where b_n is defined by $F(b_n) = 1 - 1/n$.

[L. de Haan and A. Hordijk (1972)]

18. Apply the above criterion to the distribution function

$$F(x) = 1 - \exp(-\log x \log \log x), \qquad x \geq e.$$

19. Let $X_1, X_2, \ldots$ be i.i.d. discrete random variables which take nonnegative integers. Assume that if n is large, then $P(X_1 = n) > 0$. Let $F(x)$ be the common distribution function and assume that, as $n \to +\infty$,

$$\frac{1 - F(n+1)}{1 - F(n)} \to 0.$$

Show that there is a sequence a_n of integers such that, as $n \to +\infty$,

$$\lim P(Z_n = a_n \text{ or } a_n + 1) = 1.$$

[C. W. Anderson (1970)]

20. Compare Z_n when the basic random variables X_j, $j \geq 1$, are i.i.d. Poisson or geometric variates.

21. Let X_j, $j \geq 1$, be i.i.d. uniform variates on the interval $(0, 1)$. Define $n_k = 2^{2^k}$, and $\lambda_n = \exp(-2 \log k / n_k)$ for $n_k \leq n < n_{k+1}$. Show that $P(Z_n \leq \lambda_n \text{ i.o.}) = 0$ but

$$\sum_{n=3}^{+\infty} F^n(\lambda_n) \frac{\log \log n}{n} = +\infty.$$

[O. Barndorff-Nielsen (1961)]

CHAPTER 5

Multivariate Extreme Value Distributions

If measurements of several characteristics are taken on the same members of the population, then the observed random quantities follow some type of multivariate distribution. Let the number of characteristics be m and the corresponding random quantities be $(X^{(1)}, X^{(2)}, \ldots, X^{(m)})$. We shall abbreviate vectors of this kind by a boldfaced letter $\mathbf{X}$ and the dimension m will always be specified in advance. Observations on $\mathbf{X}$ will be denoted by $\mathbf{X}_1, \mathbf{X}_2, \ldots$ and the components of $\mathbf{X}_j$ by $X_{t,j}$. That is, $X_{t,j}$ is the tth component of $\mathbf{X}_j$, or, $\mathbf{X}_j = (X_{1,j}, X_{2,j}, \ldots, X_{m,j})$. Let $\mathbf{X}_j, 1 \leq j \leq n$, be n observations. The order statistics of the tth component are $X_{t,1:n} \leq X_{t,2:n} \leq \cdots \leq X_{t,n:n}$. As in the previous chapters, we also use the notation $W_{t,n} = X_{t,1:n}$ and $Z_{t,n} = X_{t,n:n}$. Our main interest in this chapter is to investigate the existence of the asymptotic distribution of $(W_{1,n}, W_{2,n}, \ldots, W_{m,n})$ and $(Z_{1,n}, Z_{2,n}, \ldots, Z_{m,n})$, which we also refer to by the vector notations $\mathbf{W}_n$ and $\mathbf{Z}_n$, respectively. Some results will also be obtained on other extremes of multivariate observations and on so-called concomitants of order statistics. These will be accurately defined later.

For numerical vectors $\mathbf{x} = (x_1, x_2, \ldots, x_m)$, the components are signified by a subscript. Basic arithmetical operations are always meant component-wise. Thus

$$\mathbf{x} < \mathbf{y} \quad \text{means} \quad x_t < y_t, \quad 1 \leq t \leq m,$$

$\mathbf{x} + \mathbf{y} = (x_1 + y_1, x_2 + y_2, \ldots, x_m + y_m)$, $\mathbf{xy} = (x_1 y_1, x_2 y_2, \ldots, x_m y_m)$, and $\mathbf{x}/\mathbf{y} = (x_1/y_1, x_2/y_2, \ldots, x_m/y_m)$. The special vectors $\mathbf{0} = (0, 0, \ldots, 0)$ and $\mathbf{1} = (1, 1, \ldots, 1)$ will frequently be used.

The distribution function $F(\mathbf{x}) = F(x_1, x_2, \ldots, x_m)$ of the vector $\mathbf{X}$ is defined as

$$F(\mathbf{x}) = P(\mathbf{X} < \mathbf{x}) = P(X^{(1)} < x_1, X^{(2)} < x_2, \ldots, X^{(m)} < x_m).$$

We can now formulate our problem. We seek conditions on $F(\mathbf{x})$, under which there are sequences $\mathbf{a}_n$ and $\mathbf{b}_n$ of vectors such that each component of

$\mathbf{b}_n$ is positive and

$$P(\mathbf{Z}_n < \mathbf{a}_n + \mathbf{b}_n \mathbf{z}) = H_n(\mathbf{a}_n + \mathbf{b}_n \mathbf{z})$$

converges weakly to a nondegenerate m-dimensional distribution function $H(\mathbf{z})$ (see the next section for definitions). The problem can also be stated in terms of $\mathbf{W}_n$, which can again be reduced to $\mathbf{Z}_n$ by turning to $(-\mathbf{X}_j)$.

5.1. BASIC PROPERTIES OF MULTIVARIATE DISTRIBUTIONS

Let $\mathbf{X} = (X^{(1)}, X^{(2)}, \ldots, X^{(m)})$ be a random vector and let $\mathbf{x} = (x_1, x_2, \ldots, x_m)$ be an arbitrary point of the m-dimensional (Euclidean) space. Then the distribution function $F(\mathbf{x}) = F(x_1, x_2, \ldots, x_m)$ is defined as

$$F(\mathbf{x}) = P\big(X^{(1)} < x_1, X^{(2)} < x_2, \ldots, X^{(m)} < x_m\big).$$

Then elementary properties of probability immediately yield that $F(\mathbf{x})$ is nondecreasing in each of its variables x_j, $1 \leq j \leq m$. Furthermore, if $x_j \to -\infty$ for one j, then $F(\mathbf{x}) \to 0$. On the other hand, if $x_j \to +\infty$, then $F(\mathbf{x})$ tends to an $(m-1)$-dimensional distribution, which is the distribution function of the vector obtained from $\mathbf{X}$ by removing its jth component. We can, of course, repeat this limit procedure a finite number of times, by which we arrive at the marginal distribution $F_j(x)$ of $X^{(j)}$. Namely, if we let each x_t, $t \neq j$, tend to $+\infty$, the limit of $F(\mathbf{x})$ is $F_j(x)$. As a particular consequence of this observation, we obtained that $F(\mathbf{x})$ uniquely determines all marginals. The converse is not true, however. There are several possibilities for $F(\mathbf{x})$ for given marginals $F_j(x)$, $1 \leq j \leq m$, although one is not completely free in choosing $F(\mathbf{x})$. The following simple theorem shows that the distributions $F_j(x)$, $1 \leq j \leq m$, do impose restrictions on $F(\mathbf{x})$.

Theorem 5.1.1 (The Fréchet Bounds). *Let $F(\mathbf{x})$ be an m-dimensional distribution function with marginals $F_j(x), 1 \leq j \leq m$. Then, for all $x_1, x_2, \ldots, x_m$,*

$$\max\left(0, \sum_{j=1}^{m} F_j(x_j) - m + 1\right) \leq F(x_1, x_2, \ldots, x_m) \leq \min(F_1(x_1), \ldots, F_m(x_m)).$$

Proof. The proof is a simple observation on probabilities of events. Put $A_j = \{X^{(j)} < x_j\}$ and $B_j = \{X^{(j)} \geq x_j\}$. Then

$$F(x_1, x_2, \ldots, x_m) = P(A_1 A_2 \cdots A_m) \leq P(A_j), \qquad 1 \leq j \leq m,$$

which yields the upper inequality. The lower inequality is a special case of Theorem 1.4.1. If we write ν_m for the number of $j \leq m$ for which B_j occurs, then

$$P(\nu_m = 0) = F(x_1, x_2, \ldots, x_m). \tag{1}$$

Applying Theorem 1.4.1, we get

$$P(\nu_m = 0) \geq 1 - S_{1,m} = 1 - \sum_{j=1}^{m} P(B_j) = 1 - \sum_{j=1}^{m} [1 - F_j(x_j)],$$

which leads to the lower inequality upon our observing that, if this expression is negative, then it can be replaced by zero. The proof is complete. ▲

Although, for large m, the lower bound tends to be trivial, in the bivariate case ($m = 2$) it proved to be a useful guide in actually constructing bivariate distributions with given marginals.

We can set up further inequalities if we use higher-dimensional distributions, not just univariate marginals. In fact, all the inequalities of Chapter 1 can be restated in view of (1). We shall restate only Theorem 1.4.1, since we shall make several references to these inequalities. We first introduce two notations. Let

$$G_{\mathbf{j}(k)}(x_{j_1}, x_{j_2}, \ldots, x_{j_k}) = P(B_{j_1} B_{j_2} \cdots B_{j_k}),$$

where $B_j = \{X^{(j)} \geq x_j\}$ and $\mathbf{j}(k)$ signifies the vector $(j_1, j_2, \ldots, j_k)$. Furthermore, we put $S_0(\mathbf{x}) = 1$ and, for $k \geq 1$,

$$S_k(\mathbf{x}) = \sum_{1 \leq j_1 < j_2 < \cdots < j_k \leq m} G_{\mathbf{j}(k)}(x_{j_1}, x_{j_2}, \ldots, x_{j_k}).$$

Theorem 1.4.1 and (1) now yield the following relations.

Theorem 5.1.2. *Let $m \geq 2$. Then*

$$F(x_1, x_2, \ldots, x_m) = \sum_{k=0}^{m} (-1)^k S_k(x_1, x_2, \ldots, x_m). \tag{2}$$

In addition, for any integer $0 \leq s \leq (m-1)/2$,

$$\sum_{k=0}^{2s+1} (-1)^k S_k(\mathbf{x}) \leq F(\mathbf{x}) \leq \sum_{k=0}^{2s} (-1)^k S_k(\mathbf{x}). \tag{3}$$

Example 5.1.1 (Bivariate Exponential Distributions). Let (X, Y) be a two-dimensional vector, where both X and Y are unit exponential variates.

5.1 BASIC PROPERTIES OF MULTIVARIATE DISTRIBUTIONS

Let $F(x,y)$ be their bivariate distribution function. We put

$$G(x,y) = P(X \geq x, Y \geq y). \tag{4}$$

Then, by (2),

$$F(x,y) = 1 - e^{-x} - e^{-y} + G(x,y).$$

One has much freedom in the choice of $G(x,y)$ to arrive at different bivariate exponential distributions. What one has to consider are (i) the probabilistic meaning of $G(x,y)$ in (4), (ii) the Fréchet bounds, and (iii) the fact that both $g_y(x) = G(x,y) - e^{-x}$ and $g_x(y) = G(x,y) - e^{-y}$ should be nondecreasing functions.

We now list the most commonly used bivariate exponential distributions.

The Morgenstern distribution:

$$G(x,y) = e^{-x-y}\left[1 + \alpha(1 - e^{-x})(1 - e^{-y})\right];$$

Gumbel's type I distribution:

$$G(x,y) = \exp(-x - y + \Theta xy);$$

Gumbel's type II distribution:

$$G(x,y) = \exp\left[-(x^m + y^m)^{1/m}\right];$$

The Marshall-Olkin distribution:

$$G(x,y) = \exp\left[-x - y - \lambda \max(x,y)\right] \qquad \lambda > 0;$$

Mardia's distribution:

$$G(x,y) = (e^x + e^y - 1)^{-1}.$$

It should be noted that any continuous distribution $T(x)$ can be transformed to an exponential distribution by turning to the random variable $[-\log T(X)]$ from X. Hence, other bivariate exponential distributions are immediate. ▲

We conclude the section with two definitions. The first is a direct extension of the univariate concept of weak convergence. We say that a sequence $F_n(\mathbf{x})$ of m-dimensional distributions converges weakly to $F(\mathbf{x})$, if, for all continuity points $\mathbf{x}$ of $F(\mathbf{x})$, $F_n(\mathbf{x}) \to F(\mathbf{x})$ as $n \to +\infty$. Second, we call an m-dimensional distribution function $F(\mathbf{x})$ nondegenerate if all of its univariate marginals are nondegenerate.

5.2. WEAK CONVERGENCE OF EXTREMES FOR I.I.D. RANDOM VECTORS: BASIC RESULTS

All boldfaced letters signify vectors of the same dimension m. We use as standard notations those of the introduction to the present chapter. Throughout this section, $\mathbf{X}_1, \mathbf{X}_2, \ldots, \mathbf{X}_n$ are independent random vectors with common distribution function $F(\mathbf{x})$. Thus

$$H_n(\mathbf{z}) = P(Z_{1,n} < z_1, Z_{2,n} < z_2, \ldots, Z_{m,n} < z_m) = F^n(\mathbf{z}) \tag{5}$$

and

$$L_n^*(\mathbf{y}) = P(W_{1,n} \geq y_1, W_{2,n} \geq y_2, \ldots, W_{m,n} \geq y_m) = G^n(\mathbf{y}),$$

where

$$G(\mathbf{y}) = P(X^{(1)} \geq y_1, X^{(2)} \geq y_2, \ldots, X^{(m)} \geq y_m).$$

Again, any problem on $\mathbf{W}_n$ is equivalent to one on $\mathbf{Z}_n$ by changing the basic vector $\mathbf{X}$ to $(-\mathbf{X})$. We therefore concentrate on $\mathbf{Z}_n$.

Our aim is to give conditions on $F(\mathbf{x})$ under which there are sequences $\mathbf{a}_n$ and $\mathbf{b}_n > \mathbf{0}$ of vectors (an inequality with vectors is meant componentwise) such that

$$H_n(\mathbf{a}_n + \mathbf{b}_n \mathbf{x}) \to H(\mathbf{x}), \tag{6}$$

where the limit is in the sense of weak convergence and $H(\mathbf{x})$ is a nondegenerate distribution function. The following lemma shows that the choice of $\mathbf{a}_n$ and $\mathbf{b}_n > \mathbf{0}$ has actually been settled in Chapter 2.

Lemma 5.2.1. *Let $F_n(\mathbf{x})$ be a sequence of m-dimensional distribution functions. Let the tth univariate marginal of $F_n(\mathbf{x})$ be $F_{tn}(x_t)$. If $F_n(\mathbf{x})$ converges weakly to a nondegenerate continuous distribution function $F(\mathbf{x})$, then, for each t with $1 \leq t \leq m$, $F_{tn}(x_t)$ converges weakly to the tth marginal $F_t(x_t)$ of $F(\mathbf{x})$.*

Proof. Let x_t be an arbitrary fixed number. Let $\mathbf{x}$ be an m-dimensional vector whose tth component is x_t. Now, by assumption, for any $\varepsilon > 0$ there is an integer n_0 such that, for all $n \geq n_0$,

$$|F_n(\mathbf{x}) - F(\mathbf{x})| < \varepsilon. \tag{7}$$

In principle, n_0 depends on both ε and $\mathbf{x}$. But just as in Lemma 2.10.1, we can prove that for continuous limit $F(\mathbf{x})$, weak convergence is uniform in $\mathbf{x}$. Hence, n_0 is a function of ε alone. We can, therefore let the components of $\mathbf{x}$ vary without affecting (7). We next use the elementary considerations

5.2 WEAK CONVERGENCE OF EXTREMES: BASIC RESULTS

of Section 5.1, by which, for sufficiently large $x_j, j \neq t$,

$$|F(\mathbf{x}) - F_t(x_t)| < \varepsilon. \tag{8}$$

Let now $n \geq n_0$ be fixed. Let us choose each $x_j, j \neq t$, large enough so that, in addition to (8), the inequality

$$|F_n(\mathbf{x}) - F_{tn}(x_t)| < \varepsilon$$

also holds. Thus, for all $n \geq n_0$,

$$|F_{tn}(x_t) - F_t(x_t)|$$
$$\leq |F_{tn}(x_t) - F_n(\mathbf{x})| + |F_n(\mathbf{x}) - F(\mathbf{x})| + |F(\mathbf{x}) - F_t(x_t)|$$
$$< 3\varepsilon.$$

Since $\varepsilon > 0$ is arbitrary, $F_{tn}(x_t) \to F_t(x_t)$ as claimed. This completes the proof. ▲

We shall prove later that all limiting distribution functions of multivariate extremes are continuous. Hence, Lemma 5.2.1 tells us that we can appeal to Sections 2.1, 2.2, and 2.4 for determining the components of $\mathbf{a}_n$ and $\mathbf{b}_n$ whenever (6) holds.

Example 5.2.1. Let (X, Y) have a bivariate exponential distribution in the sense of Example 5.1.1. If $(\mathbf{Z}_n - \mathbf{a}_n)/\mathbf{b}_n$ converges weakly to a nondegenerate distribution $H(z_1, z_2)$, then we can choose $\mathbf{a}_n = (\log n, \log n)$ and $\mathbf{b}_n = (1, 1)$.

Indeed, by assumption, each component is a unit exponential variate. Therefore (Example 1.3.1), each component of $\mathbf{Z}_n$ can be normalized by $a_n = \log n$ and $b_n = 1$. We now apply Lemma 5.2.1, which gives the statement of the example. ▲

Example 5.2.2. For the Marshall-Olkin distribution (Example 5.1.1), the limit distribution of $\mathbf{Z}_n - \mathbf{a}_n$ is $H(z_1, z_2) = H_{3,0}(z_1)H_{3,0}(z_2)$. On the other hand, for Mardia's distribution,

$$H(z_1, z_2) = H_{3,0}(z_1)H_{3,0}(z_2)\exp\left[(e^{z_1} + e^{z_2})^{-1}\right].$$

We saw in the preceding example that $\mathbf{a}_n = (\log n, \log n)$. Thus

$$F(\log n + z_1, \log n + z_2) = 1 - \frac{e^{-z_1} + e^{-z_2}}{n} + G(\log n + z_1, \log n + z_2).$$

Now, for the Marshall-Olkin distribution, the last term is $O(n^{-3})$ and, for

Mardia's distribution,

$$G(\log n + z_1, \log n + z_2) = (ne^{z_1} + ne^{z_2} - 1)^{-1}.$$

The elementary relation

$$\lim_{n=+\infty} \left(1 + \frac{u}{n} + o\left(\frac{1}{n}\right)\right)^n = e^u,$$

combined with (5) and (6), thus yields the claimed forms for $H(z_1, z_2)$. ▲

This last example clearly shows that the limit distribution $H(\mathbf{z})$ is not determined by the univariate marginals.

The following concept will prove useful in the investigation of limit distributions for normalized extremes.

Definition 5.2.1. Let $F(\mathbf{x})$ be an m-dimensional distribution function with univariate marginals $F_t(x_t)$, $1 \leq t \leq m$. Let $D(\mathbf{y})$ be an m-dimensional function over the unit cube $0 \leq y_t \leq 1$, $1 \leq t \leq m$, and such that it increases in each of its variables and

$$F(x_1, x_2, \ldots, x_m) = D[F_1(x_1), F_2(x_2), \ldots, F_m(x_m)]. \qquad (9)$$

Then the function $D(\mathbf{y})$ is called a dependence function of $F(\mathbf{x})$. When needed, we shall emphasize the relation of $D(\mathbf{y})$ to $F(\mathbf{x})$ by writing $D_F = D_F(\mathbf{y}) = D(\mathbf{y})$.

Remark 5.2.1. If each of the functions $u_j(x)$, $1 \leq j \leq m$, is increasing, then the dependence function of the distribution of a vector $\mathbf{X} = (X^{(1)}, X^{(2)}, \ldots, X^{(m)})$ is the same as that of the distribution of $\mathbf{Y} = (Y^{(1)}, Y^{(2)}, \ldots, Y^{(m)})$, where $Y^{(j)} = u_j(X^{(j)})$. Hence, if each marginal of $F(\mathbf{x})$ is continuous, then, by the choice $u_j(x_j) = F_j(x_j)$, we can conclude that $D_F(\mathbf{y})$ is a distribution function whose marginals are uniform on the interval $(0, 1)$.

Example 5.2.3. The dependence function $D(y_1, y_2)$ of the Morgenstern distribution (Example 5.1.1) is

$$D(y_1, y_2) = y_1 y_2 [1 + \alpha(1 - y_1)(1 - y_2)].$$

For Mardia's distribution

$$D(y_1, y_2) = y_1 + y_2 - 1 + \left[\frac{1}{1 - y_1} + \frac{1}{1 - y_2} - 1\right]^{-1}.$$

Both formulas are obtained from the definition by a simple substitution.

▲

5.2 WEAK CONVERGENCE OF EXTREMES: BASIC RESULTS

From the definition of dependence functions we have

$$D_{F^n}(\mathbf{y}) = D_F^n\left(y_1^{1/n}, y_2^{1/n}, \ldots, y_m^{1/n}\right). \tag{10}$$

Namely, the tth marginal of $F^n(\mathbf{x})$ is $F_t^n(x_t)$, and thus

$$F^n(x_1, x_2, \ldots, x_m) = D_{F^n}\left(F_1^n(x_1), F_2^n(x_2), \ldots, F_m^n(x_m)\right).$$

On the other hand, if we take the nth power of (9), we get

$$F^n(x_1, x_2, \ldots, x_m) = D_F^n\left(F_1(x_1), F_2(x_2), \ldots, F_m(x_m)\right).$$

A comparison of these last two equations leads to (10).

We can now prove several important results.

Theorem 5.2.1. *If $F(\mathbf{x})$ is such that, with some sequences $\mathbf{a}_n$ and $\mathbf{b}_n$, (6) holds, then the dependence function D_H of the limit $H(\mathbf{x})$ satisfies*

$$D_H^k\left(y_1^{1/k}, y_2^{1/k}, \ldots, y_m^{1/k}\right) = D_H(y_1, y_2, \ldots, y_m),$$

where $k \geq 1$ is an arbitrary integer.

Proof. Let $k \geq 1$ be a fixed integer. Then, by (5) and (6),

$$H_{nk}(\mathbf{a}_{nk} + \mathbf{b}_{nk}\mathbf{x}) = F^{nk}(\mathbf{a}_{nk} + \mathbf{b}_{nk}\mathbf{x}) \to H(\mathbf{x})$$

as $n \to +\infty$. This can also be written as

$$F^n(\mathbf{a}_{nk} + \mathbf{b}_{nk}\mathbf{x}) \to H^{1/k}(\mathbf{x}) \qquad (n \to +\infty). \tag{11}$$

We notice that Lemma 2.2.3 can be extended to multivariate distributions (no change is required in the proof if we adopt that an inequality between vectors is considered componentwise). Therefore, (6) and (11) imply that there are vectors $\mathbf{A}_k$ and $\mathbf{B}_k$, where each component of $\mathbf{B}_k$ is positive, such that

$$H^k(\mathbf{A}_k + \mathbf{B}_k \mathbf{x}) = H(\mathbf{x}). \tag{12}$$

Since the dependence function of $H^k(\mathbf{A}_k + \mathbf{B}_k\mathbf{x})$ is the same as that of $H^k(\mathbf{x})$ (choose each $u_j(x)$ in Remark 5.2.1 a linear function and recall that the components of $\mathbf{B}_k$ are positive), (10) and (12) establish the theorem. ▲

Theorem 5.2.2. *Any limit distribution function $H(\mathbf{x})$ in (6) is continuous. Its univariate marginals belong to the types $H_{1,\gamma}(x)$, $H_{2,\gamma}(x)$, and $H_{3,0}(x)$.*

Proof. In the preceding proof we obtained that $H(\mathbf{x})$ satisfies (12) for all $\mathbf{x}$. If we let each component x_j of $\mathbf{x}$, except x_t, tend to infinity, we obtain (12) for the tth marginal of $H(\mathbf{x})$. We have determined in Section 2.4 all univariate solutions of (12). These are the types of the distributions stated in the theorem. Since the marginals are differentiable, elementary results of calculus imply that $H(\mathbf{x})$ is continuous. The proof is complete. ▲

Theorem 5.2.3. *Let* $\mathbf{X}_1, \mathbf{X}_2, \ldots, \mathbf{X}_n$ *be i.i.d. m-dimensional vectors with common distribution function* $F(\mathbf{x})$. *Then there are vectors* $\mathbf{a}_n$ *and* $\mathbf{b}_n > 0$ *such that* $(\mathbf{Z}_n - \mathbf{a}_n)/\mathbf{b}_n$ *converges weakly to a nondegenerate distribution function* $H(\mathbf{x})$ *if, and only if, each marginal belongs to the domain of attraction of one of the distributions* $H_{1,\gamma}(x)$, $H_{2,\gamma}(x)$, *and* $H_{3,0}(x)$ *and if, as* $n \to +\infty$,

$$D_F^n(y_1^{1/n}, y_2^{1/n}, \ldots, y_m^{1/n}) \to D_H(y_1, y_2, \ldots, y_m). \tag{13}$$

Proof. First, let us assume that, with some vectors $\mathbf{a}_n$ and $\mathbf{b}_n > 0$, $(\mathbf{Z}_n - \mathbf{a}_n)/\mathbf{b}_n$ converges weakly to a nondegenerate distribution $H(\mathbf{x})$. Then, by Theorem 5.2.2, $H(\mathbf{x})$ is continuous. Consequently, we can apply Lemma 5.2.1, which yields that the univariate marginals of $F(\mathbf{x})$ belong to the domain of attraction of one of the mentioned distributions. Furthermore, if (6) holds, then, on account of (5), (9), and (10),

$$D_F^n(z_1, z_2, \ldots, z_m) \to D_H(y_1, y_2, \ldots, y_m), \tag{14}$$

where $z_t = F_t(a_{t,n} + b_{t,n} x_t)$ and $y_t = H_t(x_t)$, where the subscript t refers to the tth component or marginal distribution as appropriate. But $z_t^n \to y_t$ for each t and $D_H(\mathbf{y})$ is continuous by $H(\mathbf{x})$'s being continuous (Theorem 5.2.2). Therefore, (14) implies (13).

Let us now turn to the converse. With the notations of the previous paragraph, we assume that (13) holds and that $z_t^n \to y_t$ for $1 \leq t \leq m$, as $n \to +\infty$. Applying again that $D_H(\mathbf{y})$ is continuous, we get the validity of (14), which, in view of (5), (9), and (10), yields (6). The theorem is established. ▲

For completing the discussion, we add the following simple result.

Theorem 5.2.4. *An m-variate distribution function* $H(\mathbf{x})$ *is a limit distribution in* (6) *if, and only if, its univariate marginals are of the same type as one of the functions* $H_{1,\gamma}(x)$, $H_{2,\gamma}(x)$, *and* $H_{3,0}(x)$ *and if its dependence function* D_H *satisfies the condition of Theorem 5.2.1.*

Proof. If $H(\mathbf{x})$ is a limit in (6), then Theorems 5.2.1 and 5.2.2 imply the conclusion of the theorem. Conversely, let the univariate marginals of $H(\mathbf{x})$ and D_H be as stated. Then, for all $n \geq 1$, and for each $1 \leq t \leq m$, there are numbers $a_{t,n}$ and $b_{t,n} > 0$ such that the marginals $H_t(x)$ satisfy

$$H_t^n(a_{t,n} + b_{t,n} x) = H_t(x). \tag{15}$$

5.2 WEAK CONVERGENCE OF EXTREMES: BASIC RESULTS

Choose $H(\mathbf{x})$ as the population distribution $F(\mathbf{x})$. Let $\mathbf{a}_n = (a_{1,n}, a_{2,n}, \ldots, a_{m,n})$ and $\mathbf{b}_n = (b_{1,n}, b_{2,n}, \ldots, b_{m,n})$. We now apply previous relations in the following order: first (5), then the definition (9). It will be followed by (15), (10), and finally by the condition of Theorem 5.2.1. We get

$$P(\mathbf{Z}_n < \mathbf{a}_n + \mathbf{b}_n \mathbf{x}) = H^n(\mathbf{x})$$

$$= D_{H^n}\left[H_1^n(a_{1,n} + b_{1,n}x_1), \ldots, H_m^n(a_{m,n} + b_{m,n}x_m)\right]$$

$$= D_{H^n}\left[H_1(x_1), \ldots, H_m(x_m)\right]$$

$$= D_H^n\left[H_1^{1/n}(x_1), \ldots, H_m^{1/n}(x_m)\right]$$

$$= D_H\left[H_1(x_1), \ldots, H_m(x_m)\right] = H(\mathbf{x}).$$

Thus, $(\mathbf{Z}_n - \mathbf{a}_n)/\mathbf{b}_n$ converges weakly to $H(\mathbf{x})$; that is, $H(\mathbf{x})$ is a limit in (6). The proof is completed. ▲

In principle, we have completed the investigation. Given an m-variate distribution function $F(\mathbf{x})$, we check if its marginals belong to the domain of attraction of one of $H_{1,\gamma}(x), H_{2,\gamma}(x)$, and $H_{3,0}(x)$. If yes, then we use the methods of Chapter 2 to determine the components of the normalizing vectors $\mathbf{a}_n$ and $\mathbf{b}_n$. Furthermore, we determine $D_F(\mathbf{y})$ by its definition (9) and check if $D_F^n(\mathbf{y}^{1/n})$ converges. If this limit exists, then we check the condition of Theorem 5.2.1. If it holds and if it is a dependence function, then we have got the actual limit distribution.

In several practical problems, the question is only if a given distribution is a limiting form for $\mathbf{Z}_n$, when suitably normalized. If the assumption of the observations' being i.i.d. vectors is justified, then the answer is quite simple: one has to check the types of its marginals and the validity of the condition of Theorem 5.2.1.

Example 5.2.4. Let $H_1(x), H_2(x), \ldots, H_m(x)$ be of the same type as one of $H_{1,\gamma}(x), H_{2,\gamma}(x)$, and $H_{3,0}(x)$. Then

$$H(\mathbf{x}) = H_1(x_1)H_2(x_2)\cdots H_m(x_m)$$

is a possible limit in (6).

By assumption, the condition of Theorem 5.2.3 on the marginals is satisfied. Furthermore, by definition,

$$D_H(\mathbf{y}) = y_1 y_2 \cdots y_m,$$

for which the condition of Theorem 5.2.1 is evident. An appeal to Theorem 5.2.4 yields the claim. ▲

In this example, of course, it was not necessary to use Theorem 5.2.4. One can easily get the conclusion by starting with a basic vector whose components are independent.

Example 5.2.5. The distribution function

$$H(x_1, x_2, \ldots, x_m) = \exp\{-\exp[-\min(x_1, x_2, \ldots, x_m)]\}$$

is a possible limit in (6).

We use Theorem 5.2.4. The marginal distributions $H_t(x_t) = \exp(-e^{-x_t}) = H_{3,0}(x_t)$. Therefore, it remains to check the validity of the condition of Theorem 5.2.1. From the definition, since

$$\exp\{-\exp[-\min(x_1, x_2, \ldots, x_m)]\} = \min\{\exp[-\exp(-x_j)]: 1 \leq j \leq m\},$$

$$D_H(y_1, y_2, \ldots, y_m) = \min(y_1, y_2, \ldots, y_m).$$

Hence, for $k \geq 1$,

$$D_H^k(y_1^{1/k}, \ldots, y_m^{1/k}) = [\min(y_1^{1/k}, \ldots, y_m^{1/k})]^k = \min(y_1, \ldots, y_m),$$

which was to be shown. ▲

Notice that $H(x_1, x_2, \ldots, x_m)$ above is the Fréchet upper bound (Theorem 5.1.1) of all m-variate distribution functions $F(\mathbf{x})$ whose univariate marginals $F_t(x_t) = H_{3,0}(x_t)$ for each $1 \leq t \leq m$.

Example 5.2.6. The distribution

$$H(x_1, x_2) = H_{3,0}(x_1) H_{3,0}(x_2) \left[1 + \tfrac{1}{2}(1 - H_{3,0}(x_1))(1 - H_{3,0}(x_2)) \right]$$

does not occur as a limit in (6).

As a matter of fact, even though the marginals $H_1(x_1) = H_{3,0}(x)$ and $H_2(x_2) = H_{3,0}(x_2)$, the dependence function

$$D_H(y_1, y_2) = y_1 y_2 \left[1 + \tfrac{1}{2}(1 - y_1)(1 - y_2) \right]$$

fails to satisfy the condition of Theorem 5.2.1. ▲

In the next section we give an equivalent result to Theorem 5.2.3, which is simpler to apply to certain distributions.

5.3. FURTHER CRITERIA FOR THE I.I.D. CASE

Let $\mathbf{X} = (X^{(1)}, X^{(2)}, \ldots, X^{(m)})$ be a vector with distribution function $F(\mathbf{x})$. Let $\mathbf{j}(k) = (j_1, j_2, \ldots, j_k)$, $1 \leq k \leq m$ be a vector with components $1 \leq j_1 < j_2 < \cdots < j_k \leq m$. The distribution function $F_{\mathbf{j}(k)}(x_{j_1}, \ldots, x_{j_k})$ of the vector

$(X^{(j_1)},\ldots,X^{(j_k)})$ is called a k-dimensional marginal distribution, which is obtained from $F(\mathbf{x})$ by letting $x_t \to +\infty$ for all $t \neq j_1, j_2,\ldots, j_k$. We also use the previously introduced notation

$$G_{\mathbf{j}(k)}(x_{j_1}, x_{j_2},\ldots, x_{j_k}) = P\left(X^{(j_1)} \geq x_{j_1}, X^{(j_2)} \geq x_{j_2},\ldots, X^{(j_k)} \geq x_{j_k}\right).$$

For $\mathbf{j}(m) = (1, 2,\ldots, m)$ we drop the subscript and we write $G(x_1, x_2,\ldots, x_m)$. We call $G(\mathbf{x})$ the survival function and $G_{\mathbf{j}(k)}(x_{j_1},\ldots, x_{j_k})$ marginal survival functions. If we consider a sequence of distributions, then, when turning to marginals, we indicate the sequence by a second subscript. If we change F into another letter, then this new letter with a subscript $\mathbf{j}(k)$ will denote its corresponding marginal. For example, a limit distribution in (6) is denoted by $H(\mathbf{x})$, and thus $H_{\mathbf{j}(k)}(x_{j_1},\ldots, x_{j_k})$ denotes its marginal corresponding to the components $\mathbf{j}(k)$.

Now let $\mathbf{X}_1, \mathbf{X}_2,\ldots, \mathbf{X}_n$ be i.i.d. vectors which are distributed as $\mathbf{X}$. We assume that $F(\mathbf{x})$ is such that each of its univariate marginals belongs to the domain of attraction of one of $H_{1,\gamma}(x)$, $H_{2,\gamma}(x)$, and $H_{3,0}(x)$. Therefore, there are constants $a_{t,n}$ and $b_{t,n} > 0$ such that, as $n \to +\infty$,

$$\lim F_{t,n}^n (a_{t,n} + b_{t,n} x) = H_t(x), \quad 1 \leq t \leq m, \tag{16}$$

where $H_t(x)$ is of the same type as one of the above mentioned three distributions. We assume that $a_{t,n}$ and $b_{t,n}$ have been determined and we put $\mathbf{a}_n = (a_{1,n}, a_{2,n},\ldots, a_{m,n})$ and $\mathbf{b}_n = (b_{1,n}, b_{2,n},\ldots, b_{m,n})$. We now prove the following result.

Theorem 5.3.1. *With the notations of the preceding paragraph, $(\mathbf{Z}_n - \mathbf{a}_n)/\mathbf{b}_n$ converges weakly to a nondegenerate distribution $H(\mathbf{x})$ if, and only if, for each fixed vector $\mathbf{j}(k)$ and for all $\mathbf{x}$ for which $H_t(x_t), 1 \leq t \leq m$, of (16) are positive, the limits, as $n \to +\infty$*

$$\lim n G_{\mathbf{j}(k),n}(a_{j_1,n} + b_{j_1,n} x_{j_1,n},\ldots, a_{j_k,n} + b_{j_k,n} x_{j_k,n}) = h_{\mathbf{j}(k)}(x_{j_1},\ldots, x_{j_k}) \tag{17}$$

are finite, and the function

$$H(\mathbf{x}) = \exp\left\{\sum_{k=1}^m (-1)^k \sum_{1 \leq j_1 < \cdots < j_k \leq m} h_{\mathbf{j}(k)}(x_{j_1},\ldots, x_{j_k})\right\} \tag{18}$$

is a nondegenerate distribution function. The actual limit distribution function of $(\mathbf{Z}_n - \mathbf{a}_n)/\mathbf{b}_n$ is the one given in (18).

When the limit distribution $H(\mathbf{x})$ exists, then the following inequalities hold. Let $s \geq 0$ be an integer. Then

$$H(\mathbf{x}; 2s + 1) \leq H(\mathbf{x}) \leq H(\mathbf{x}; 2s), \tag{19}$$

where

$$H(\mathbf{x};r) = \exp\left\{\sum_{k=1}^{r}(-1)^k \sum_{1 \leq j_1 < \cdots < j_k \leq m} h_{\mathbf{j}(k)}(x_{j_1},\ldots,x_{j_k})\right\}. \quad (20)$$

Proof. We first prove that if (17) holds, then $(\mathbf{Z}_n - \mathbf{a}_n)/\mathbf{b}_n$ converges weakly and the limit distribution $H(\mathbf{x})$ satisfies (18) and (19). In view of the basic relations (5) and (6), we thus have to prove

$$F^n(\mathbf{a}_n + \mathbf{b}_n\mathbf{x}) \to H(\mathbf{x}). \quad (21)$$

First, note that if $\mathbf{x}$ is such that, for at least one t, $H_t(x_t) = 0$ in (16), then $F^n(\mathbf{a}_n + \mathbf{b}_n\mathbf{x}) \to 0$. Namely, if $\mathbf{x}$ is any vector whose tth component is x_t, then the inequality

$$F(\mathbf{a}_n + \mathbf{b}_n\mathbf{x}) \leq F_t(a_{t,n} + b_{t,n}x_t)$$

implies that the limit in (21) is zero. Therefore, let $\mathbf{x}$ be such that, for all t, $H_t(x_t) > 0$. Then, in view of (17), $F(\mathbf{a}_n + \mathbf{b}_n\mathbf{x}) \to 1$ as $n \to +\infty$ and thus, for large n, $F(\mathbf{a}_n + \mathbf{b}_n\mathbf{x}) > 0$. We can therefore turn to logarithms as well, as we can apply the asymptotic relation (Taylor's formula)

$$\lim_{n = +\infty} \frac{\log F(\mathbf{a}_n + \mathbf{b}_n\mathbf{x})}{1 - F(\mathbf{a}_n + \mathbf{b}_n\mathbf{x})} = -1.$$

Hence, as $n \to +\infty$,

$$F^n(\mathbf{a}_n + \mathbf{b}_n\mathbf{x}) = \exp\left[n \log F(\mathbf{a}_n + \mathbf{b}_n\mathbf{x})\right] \sim \exp\{-n[1 - F(\mathbf{a}_n + \mathbf{b}_n\mathbf{x})]\}. \quad (22)$$

We thus get from (2) and (17) that (21) holds and that the limit $H(\mathbf{x})$ satisfies (18). Applying (3) and (17), we arrive at (19). The sufficiency of the theorem has been proved.

Turning to the converse, we assume (21). Let $\mathbf{x}$ be such that, for all t, $H_t(x_t) > 0$. We shall prove the validity of (17). Since (16) implies (17) for $k = 1$ (apply (22) with one component), the elementary inequality

$$G_{\mathbf{j}(k),n}(y_{j_1},\ldots,y_{j_k}) \leq 1 - F_{j_t}(y_{j_t}), \quad 1 \leq t \leq k,$$

yields that, for any k,

$$nG_{\mathbf{j}(k),n}(a_{j_1,n} + b_{j_1,n}x_{j_1,n},\ldots,a_{j_k,n} + b_{j_k,n}x_{j_k,n})$$

are bounded. Therefore, we can select a subsequence n^* of n on which (17) holds. Let us repeat the first part of the proof for this subsequence. We get that the limit $H(\mathbf{x})$ of (21) satisfies (18) where the limits in (17) may depend on the actual subsequence n^*. Observing, however, that (21) implies that all bivariate marginals of $F^n(\mathbf{a}_n + \mathbf{b}_n\mathbf{x})$ converge to the corresponding bivariate marginals of $H(\mathbf{x})$ (the proof is similar to Lemma 5.2.1, where the univariate marginals were treated), we conclude from the representation (18) with $m=2$ that (17) cannot depend on n^* for $k=2$. Considering trivariate marginals, one gets the validity of (17) for $k=3$. Proceeding this way, (17) follows for all $k \leq m$. The first part of the proof now gives that the limit $H(\mathbf{x})$ of (21) satisfies (18). Therefore, the limits of (17) are such that the expression in (18) is a nondegenerate distribution function. This completes the proof. ▲

Notice that the special case $s=0$ of (19) yields the inequality $H(\mathbf{x}; 1) \leq H(\mathbf{x})$. This, when written in detail, shows that an arbitrary $H(\mathbf{x})$ is never exceeded by the product of its univariate marginals.

Corollary 5.3.1. *Assume that* $(\mathbf{Z}_n - \mathbf{a}_n)/\mathbf{b}_n$ *has a nondegenerate asymptotic distribution* $H(\mathbf{x})$. *Then the components of* $(\mathbf{Z}_n - \mathbf{a}_n)/\mathbf{b}_n$ *are asymptotically independent if, and only if, the limits in* (17) *are identically zero for* $k=2$.

Proof. Let $1 \leq j_1 < j_2 \leq m$ be arbitrary integers. Since $(\mathbf{Z}_n - \mathbf{a}_n)/\mathbf{b}_n$ has an asymptotic distribution, so do its bivariate components. By Theorem 5.3.1, the asymptotic distribution of the bivariate vector $(\{Z_{j_1,n} - a_{j_1,n}\}/b_{j_1,n}, \{Z_{j_2,n} - a_{j_2,n}\}/b_{j_2,n})$ is

$$\exp\{-h_{j_1}(x_{j_1}) - h_{j_2}(x_{j_2}) + h_{j_1,j_2}(x_{j_1}, x_{j_2})\}.$$

Now, if the components of $(\mathbf{Z}_n - \mathbf{a}_n)/\mathbf{b}_n$ are asymptotically independent, then they are asymptotically pairwise independent. Consequently, $h_{j_1,j_2}(x_{j_1}, x_{j_2}) = 0$. Conversely, if $h_{i,j}(x_i, x_j) = 0$ for all $1 \leq i < j \leq m$, then all limits in (17) are identically zero. Namely,

$$0 \leq h_{\mathbf{j}(k)}(x_{j_1}, x_{j_2}, \ldots, x_{j_k}) \leq h_{\mathbf{j}(2)}(x_{j_1}, x_{j_2}),$$

for any $k \geq 2$. Thus, the representation (18) yields

$$H(\mathbf{x}) = \exp\{-h_1(x_1) - h_2(x_2) - \cdots - h_m(x_m)\};$$

that is, the components of $(\mathbf{Z}_n - \mathbf{a}_n)/\mathbf{b}_n$ are asymptotically independent. The proof is completed. ▲

Example 5.3.1. Let $\mathbf{X}$ be an m-dimensional normal vector. Let each of the components of $\mathbf{X}$ have zero expectation and unit variance. Let $\mathbf{X}_1, \mathbf{X}_2, \ldots, \mathbf{X}_n$ be independent observations on $\mathbf{X}$. Then the components of

$(\mathbf{Z}_n - \mathbf{a}_n)/\mathbf{b}_n$ are asymptotically independent, where each component of $\mathbf{a}_n$ and $\mathbf{b}_n$ is an appropriate normalizing constant for the standard normal distribution (see Section 2.3.2). In other words, the asymptotic distribution $H(\mathbf{x})$ of $(\mathbf{Z}_n - \mathbf{a}_n)/\mathbf{b}_n$ is $H_{3,0}(x_1) H_{3,0}(x_2) \cdots H_{3,0}(x_m)$.

For showing this claim, we appeal to Corollary 5.3.1. In view of its conclusion, we have to investigate (17) with $k=2$. Because a bivariate marginal of normal vectors is normal, it suffices to show that if (X, Y) has a bivariate normal distribution then, as $n \to +\infty$,

$$nP(X \geq a_n + b_n x, Y \geq a_n + b_n y) \to 0, \qquad (23)$$

where a_n and $b_n > 0$ are chosen as in Section 2.3.2. This choice implies that, as $n \to +\infty$,

$$nP(X \geq a_n + b_n x) \to e^{-x}, \qquad nP(Y \geq a_n + b_n y) \to e^{-y}.$$

Thus, if one writes $(u_n = a_n + b_n x, v_n = a_n + b_n y)$,

$$P(X \geq u_n, Y \geq v_n) = P(X \geq u_n) \frac{P(X \geq u_n, Y \geq v_n)}{P(X \geq u_n)}, \qquad (24)$$

we get (23) by the well-known property of the bivariate normal distribution (which is obtained by an easy calculation) that the last fraction in (24) tends to zero as both u_n and v_n tend to infinity. ▲

Notice that, although Corollary 5.3.1 is stated in terms of the normalizing constants $\mathbf{a}_n$ and $\mathbf{b}_n$, the criterion (23) can always be reduced to the last fraction in (24) tending to zero. In this latter limit, the actual form of u_n and v_n is not essential. In other words, one does not have to compute $\mathbf{a}_n$ and $\mathbf{b}_n$ for applying Corollary 5.3.1. A similar remark also applies to Theorem 5.3.1 (see the discussion after Example 5.4.2).

5.4. ON THE PROPERTIES OF $H(\mathbf{x})$

We have given two characterizations of the possible asymptotic distribution $H(\mathbf{x})$ of $(\mathbf{Z}_n - \mathbf{a}_n)/\mathbf{b}_n$. One was given in Section 5.2 in terms of the univariate marginals and the dependence function. The other was obtained in Theorem 5.3.1, and a specific representation (18) was given. We now return to the first case.

We have seen (Theorem 5.2.4) that a function $H(\mathbf{x})$ can occur as a limit in (6) if, and only if, its univariate marginals $H_t(x_t)$, $1 \leq t \leq m$, belong to one of the types of $H_{1,\gamma}(x)$, $H_{2,\gamma}(x)$, and $H_{3,0}(x)$ and its dependence function satisfies, for $k \geq 2$,

$$D_H^k\left(y_1^{1/k}, y_2^{1/k}, \ldots, y_m^{1/k}\right) = D_H(y_1, y_2, \ldots, y_m). \qquad (25)$$

We know that a monotonic transformation does not affect a dependence function. We know also that if $T(x)$ is a distribution function which belongs to one of the types of $H_{1,\gamma}(x)$, $H_{2,\gamma}(x)$, and $H_{3,0}(x)$, then $T(x)$ can be transformed to $H_{3,0}(x)$ by a monotonic transformation. Therefore, we can assume, without loss of generality, that $H(\mathbf{x})$ is such that, for each t, $1 \leq t \leq m$, $H_t(x_t) = H_{3,0}(x_t)$. In addition, (25) is assumed to hold. For the present section, $H(\mathbf{x})$ will always denote a multivariate distribution function with the mentioned univariate marginals and which satisfies (25).

Theorem 5.4.1. *For all $H(\mathbf{x})$,*

$$\exp\left[-\sum_{t=1}^{m} \exp(-x_t)\right] \leq H(\mathbf{x}) \leq \exp\{-\exp[-\min(x_1, x_2, \ldots, x_t)]\}.$$

Both bounds are sharp.

Proof. The upper inequality is the Fréchet bound of Theorem 5.1.1. On the other hand, the lower inequality is the special case $s=0$ of (19) (for its meaning, see the remark after the end of the proof of Theorem 5.3.1). The fact that both bounds are sharp follows by observing that both bounds are actually H-functions. Indeed, their univariate marginals are $H_{3,0}(x_t)$, $1 \leq t \leq m$ (let $x_j \to +\infty$ for all $j \neq t$). In addition, a simple calculation shows that their dependence functions $y_1 y_2 \cdots y_m$ and $\min(y_1, y_2, \ldots, y_m)$, respectively, satisfy (25). This completes the proof. ▲

If $H(\mathbf{x})$ does not split into the product of its univariate marginals, then more restrictive bounds are provided by (19). Of course, if marginals in all dimensions are known, then (18) gives an exact expression for $H(\mathbf{x})$. We now deduce another representation for $H(\mathbf{x})$. We first prove the following extension of (25).

Lemma 5.4.1. *If $H(\mathbf{x})$ satisfies (25) for all integers $k \geq 1$, then it satisfies (25) when k is replaced by any real number $s > 0$.*

Proof. Let $s \geq 1$ be a real number and let k be its integer part. Since $D_H(y_1, y_2, \ldots, y_m)$ is nondecreasing in each of its variables,

$$D_H\left(y_1^{1/k}, \ldots, y_m^{1/k}\right) \leq D_H\left(y_1^{1/s}, \ldots, y_m^{1/s}\right) \leq D_H\left(y_1^{1/(k+1)}, \ldots, y_m^{1/(k+1)}\right).$$

Therefore (abbreviating $(y_1^t, \ldots, y_m^t)$ to $\mathbf{y}^t$)

$$D_H^{k+1}\left(\mathbf{y}^{1/k}\right) \leq D_H^s\left(\mathbf{y}^{1/s}\right) \leq D_H^k\left(\mathbf{y}^{1/(k+1)}\right). \tag{26}$$

By assumption, (25) holds for integer k. Thus

$$D_H^k\left(\mathbf{y}^{1/k}\right) = D_H^{k+1}\left(\mathbf{y}^{1/(k+1)}\right) = D_H(\mathbf{y}).$$

Substituting these last identities into (26) and letting $s \to +\infty$, we get

$$\lim D_H^s(\mathbf{y}^{1/s}) = D_H(\mathbf{y}). \tag{26a}$$

Namely, $D_H(\mathbf{y})$ is continuous (Theorem 5.2.2 and Remark 5.2.1) and thus $\lim D_H(\mathbf{y}^{1/k}) = \lim D_H(\mathbf{y}^{1/(k+1)}) = D_H(1,1,\ldots,1) = 1$. From (26a), for any $s > 0$,

$$D_H(\mathbf{y}) = \lim_{n=+\infty} D_H^{sn}(\mathbf{y}^{1/ns})$$

$$= \lim_{n=+\infty} \left\{ D_H^n\left[(\mathbf{y}^{1/s})^{1/n}\right]\right\}^s$$

$$= D_H^s(\mathbf{y}^{1/s}),$$

which was to be proved. ▲

Let us introduce the function

$$d_H(y_1, y_2, \ldots, y_m) = -\log D_H(e^{-y_1}, e^{-y_2}, \ldots, e^{-y_m}),$$

where $0 \leq y_t < +\infty$, $1 \leq t \leq m$. In view of Lemma 5.4.1,

$$s d_H(\mathbf{y}) = d_H(s\mathbf{y}), \qquad s > 0. \tag{27}$$

A function that satisfies (27) is known as Euler's homogeneous function (of order one). This equation has drawn much attention in the mathematical literature and comes up in different contexts. Its best-known solution, stated for our form of $H(\mathbf{x})$, says that there is a function $v(u_1, u_2, \ldots, u_{m-1})$ of $m-1$ variables such that

$$H(\mathbf{x}) = \left[H_{3,0}(x_1) H_{3,0}(x_2) \cdots H_{3,0}(x_m)\right]^{v[x_2 - x_1, \ldots, x_m - x_1]}. \tag{28}$$

Since marginals of $H(\mathbf{x})$ in any dimension are also possible limits in (6) with a reduced number of variables, all marginals of $H(\mathbf{x})$ have a representation similar to (28). From this fact, together with inequalities on $H(\mathbf{x})$, one can easily deduce several restrictions on $v(u_1, u_2, \ldots, u_{m-1})$. However, the notations become very complicated for arbitrary m. We therefore restrict ourselves to $m=2$, the bivariate case, for some further discussion. In this case, (28) reduces to

$$H(x_1, x_2) = \left[H_{3,0}(x_1) H_{3,0}(x_2)\right]^{v(x_2 - x_1)}. \tag{28a}$$

Evaluating the marginals by letting separately x_1 or x_2 tend to infinity, we get $v(+\infty) = v(-\infty) = 1$. The inequalities of Theorem 5.4.1 yield

$$\frac{\max(1, e^{-y})}{1 + e^{-y}} \leq v(y) \leq 1. \tag{29}$$

Further restriction on $v(y)$ is obtained by the following consideration. Let $(Z^{(1)}, Z^{(2)})$ be a random vector with distribution function $H(x_1, x_2)$. Then, with $\Delta x_1 > 0$ and $\Delta x_2 > 0$,

$$0 \leq P(x_1 \leq Z^{(1)} < x_1 + \Delta x_1, x_2 \leq Z^{(2)} < x_2 + \Delta x_2)$$
$$= H(x_1 + \Delta x_1, x_2 + \Delta x_2) - H(x_1, x_2 + \Delta x_2) - H(x_1 + \Delta x_1, x_2) + H(x_1, x_2).$$

This inequality, together with the monotonicity of the marginals, yields for $v(y)$ the properties listed below.

$$(1 + e^y)v(y) \text{ is nondecreasing} \tag{30}$$

$$(1 + e^{-y})v(y) \text{ is nonincreasing} \tag{31}$$

and, for $z > x$, $s > y$,

$$(e^{-z} + e^{-s})v(s-z) + (e^{-x} + e^{-y})v(y-x)$$
$$\leq (e^{-x} + e^{-s})v(s-x) + (e^{-z} + e^{-y})v(y-z). \tag{32}$$

Finally, in view of Theorem 5.2.2, $v(y)$ is continuous. The reader can easily verify that the listed properties of $v(y)$—that is, $v(+\infty) = v(-\infty) = 1$, $v(y)$ is continuous and satisfies (29)–(32)—are also sufficient for $H(x_1, x_2)$ of (28a) to be a bivariate limit in (6). One simply has to check that $H(x_1, x_2)$ of (28a) is a distribution function, its marginals are $H_{3,0}(x_1)$ and $H_{3,0}(x_2)$, respectively, and its dependence function satisfies (25).

If $H(x_1, x_2)$ has a density, then a neat representation holds.

Theorem 5.4.2. *If $H(x_1, x_2)$ has density $\partial^2 H / (\partial x_1 \partial x_2)$, then $v(y)$ in the representation (28a) is of the form*

$$v(y) = 1 - \frac{e^y \int_y^{+\infty} g(u) \, du + \int_{-\infty}^y e^u g(u) \, du}{1 + e^y}, \tag{33}$$

where $g(y) \geq 0$ is an arbitrary function with

$$\int_{-\infty}^{+\infty} g(u) \, du \leq 1, \qquad \int_{-\infty}^{+\infty} e^u g(u) \, du \leq 1. \tag{34}$$

Conversely, any function $H(x_1, x_2)$ of (28a) with $v(y)$ satisfying (33) and (34) is a limiting distribution in (6).

Proof. We first observe that $v(y)$ of (33) satisfies $v(+\infty) = v(-\infty) = 1$. Next, we conclude from (28a) that if $H(x_1, x_2)$ has density, then $v(y)$ is twice differentiable. Therefore, collecting all terms to the right hand side in

(32), dividing by $z-x$, and letting $z \to x$ results in the inequality

$$e^{-x}v(s-x)+e^{-x}v'(s-x)+e^{-s}v'(s-x)-e^{-y}v'(y-x)$$
$$-e^{-x}v'(y-x)-e^{-x}v(y-x) \geq 0.$$

If we now divide this inequality by $s-y$ and let $s \to y$, we get

$$(e^{-x}-e^{-y})v'(y-x)+(e^{-x}+e^{-y})v''(y-x) \geq 0,$$

or, dividing by e^{-x},

$$(1-e^{-u})v'(u)+(1+e^{-u})v''(u) \geq 0. \qquad (35)$$

Let us put

$$g(u)=(1-e^{-u})v'(u)+(1+e^{-u})v''(u). \qquad (36)$$

Then, on account of (35), $g(u) \geq 0$. Further, the solution of the differential equation (36), under the condition $v(+\infty)=v(-\infty)=1$ as well as (30) and (31), is the function (33), which should satisfy (34). This proves the claimed representation.

As to the converse, one has to check that the function $H(x_1,x_2)$ of (28a) with $v(y)$ defined in (33) and (34) is a distribution function. This is a simple and routine calculation. We thus omit the details. The theorem is established. ▲

Example 5.4.1. Let $g(u)=e^{-2u}$ for $u \geq 0$ and zero otherwise. We then get

$$(1+e^y)v(y) = \begin{cases} 1+\tfrac{1}{2}e^y & \text{if } y<0, \\ e^y+\tfrac{1}{2}e^{-y} & \text{if } y \geq 0. \end{cases}$$

Consequently, by (28a), the function

$$H(x_1,x_2) = \exp\left[(-e^{-x_1}-e^{-x_2})v(x_2-x_1)\right]$$
$$= \exp\left(-e^{-x_2}-\tfrac{1}{2}e^{-x_1}\right) \qquad \text{if } x_1 > x_2$$
$$= \exp\left(-e^{-x_1}-\tfrac{1}{2}e^{x_1-2x_2}\right) \qquad \text{if } x_1 \leq x_2,$$

is a possible limit distribution in (6).

The claim of the example is obtained by a straight substitution into Theorem 5.4.2. Starting with a function $g(u) \geq 0$, one first has to check the validity of (34). If it holds, then $v(y)$ can be evaluated with (33) and $H(x_1,x_2)$ by (28a). ▲

Example 5.4.2. The distribution $F(x,y)$ of a vector (X,Y) is in the domain of attraction of $H(x_1,x_2)$ of the preceding example if, and only if, its marginals $F(x,+\infty)$ and $F(+\infty,y)$ are in the domain of attraction of $H_{3,0}(x)$ and if $G(x,y) = P(X \geqslant x, Y \geqslant y)$ satisfies

$$\lim_{n=+\infty} nG(a_{1,n}+b_{1,n}x_1, a_{2,n}+b_{2,n}x_2) = e^{-x_2}u(x_2-x_1),$$

where $u(x_2-x_1) = \frac{1}{2}e^{x_2-x_1}$ or $1 - \frac{1}{2}e^{x_1-x_2}$ according as $x_1 > x_2$ or $x_1 \leqslant x_2$, respectively. Here, $(a_{1,n}, a_{2,n})$ and $(b_{1,n}, b_{2,n})$ are the normalizing constants for the univariate marginals.

We now arrive at the conclusion of the example by an appeal to Theorem 5.3.1. The function $H(x_1,x_2)$ of the previous example is given in the form (18), where $h_1(x_1) = e^{-x_1}, h_2(x_2) = e^{-x_2}$, and $h_{1,2}(x_1,x_2) = e^{-x_2}u(x_2-x_1)$, where $u(y)$ is as specified above. Thus, the criterion expressed in (17) is exemplified above. ▲

We can avoid the necessity of actually calculating $(a_{1,n}, a_{2,n})$ and $(b_{1,n}, b_{2,n})$ before applying Theorem 5.3.1. This can be done in the same manner in which we avoided the computation of these constants in Example 5.3.1. For simplicity, we carry out the necessary steps for the case when the marginals of $F(x,y)$ are identical, $F(x,+\infty) = F(+\infty,x) = F(x)$. Then, for $F(x)$ to be in the domain of attraction of $H_{3,0}(x)$, it is necessary and sufficient that, with some $h(t)$, as $t \to \omega(F)$,

$$\lim \frac{1-F[t+xh(t)]}{1-F(t)} = e^{-x}, \quad x \text{ real.}$$

Here, $h(t)$ can be chosen so that $h(a_n) = b_n$, where a_n and b_n denote the common values of $a_{1,n}, a_{2,n}$ and $b_{1,n}, b_{2,n}$, respectively (see Section 2.5). We have also seen that

$$n[1-F(a_n)] \to 1$$

as $n \to +\infty$. Thus, if

$$\frac{G(t+h(t)x_1, t+h(t)x_2)}{1-F(t)} \to e^{-x_2}u(x_2-x_1) \tag{37}$$

as $t \to \omega(F)$, then, with $t = a_n$, $h(t) = b_n$,

$$nG(a_n+b_nx_1, a_n+b_nx_2) = n[1-F(t)]\frac{G(t+h(t)x_1, t+h(t)x_2)}{1-F(t)}$$

also converges to $e^{-x_2}u(x_2-x_1)$. The converse statement is also true (which is immediate by the methods of Sections 2.4 or 2.5). Thus, the criterion of Example 5.4.2 is equivalent to (37). The general statement of the transformation of Theorem 5.3.1 into a parametric form like (37) is asked of the reader in Exercise 13.

Not all limiting distributions $H(x_1,x_2)$ have density, and thus Theorem 5.4.2 does not apply to arbitrary $H(x_1,x_2)$. For example, $H(x_1,x_2)= \exp[-\max(e^{-x_1},e^{-x_2})]$, obtained in Theorem 5.4.1 as the upper bound of all possible limiting distributions (see the convention stated just before Theorem 5.4.1 concerning the form of $H(\mathbf{x})$), does not have a density. Some further analysis of this distribution, however, can lead to another representation that is applicable to any limiting distribution in (6).

Let us first introduce a parameter in the above $H(x_1,x_2)$. Namely, let us consider

$$H(x_1,x_2;p) = \exp\{-\max[pe^{-x_1},(1-p)e^{-x_2}]\}, \qquad (38)$$

where $0<p<1$. This imposes a parameter on the margins as well, but they remain of the same type as $H_{3,0}(x)$. Next, we enlarge the above parametric family to

$$H^*(x_1,x_2) = \exp\left\{-\int_0^1 \max(pe^{-x_1},(1-p)e^{-x_2})dU(p)\right\}, \qquad (39)$$

where $U(p)$ is a distribution function concentrated on the interval $[0,1]$. Then we get back (38) with a degenerate distribution function $U(p)$, which is degenerate at some point in $(0,1)$. It is immediate from Theorem 5.2.4 that the function in (39) is a possible limit in (6). As a matter of fact, its margins are of the same type as $H_{3,0}(x)$, and its dependence function

$$D_{H^*}(y_1,y_2) = \exp\left\{-\int_0^1 \max(-p\log y_1, -(1-p)\log y_2)dU(p)\right\}$$

evidently satisfies (25). Now the fact is that the class $H^*(x_1,x_2)$ coincides with the class of all limiting distributions in (6) whose margins are of the type of $H_{3,0}(x)$ (by the nature of (39), we actually get $H_{3,0}(x+A)$, $A<0$, for the margins). The additional advantage of (39) is that it can easily be extended to a representation of the limits in (6) for arbitrary dimension. Before formulating the exact statement in higher dimensions, we introduce a concept.

Definition 5.4.1. The m-dimensional unit simplex S is the set of vectors $\mathbf{p}$ with nonnegative components p_t such that $\sum_{t=1}^m p_t = 1$.

We now state, without proof, a very recent result.

Theorem 5.4.3 (The Pickands Representation). *Let $H(\mathbf{x})$ be a function with univariate margins $H_t(x_t) = H_{3,0}(x_t + A_t)$, $A_t < 0$. Let $U(\mathbf{p})$ be a finite measure on the m-dimensional unit simplex S. Then $H(\mathbf{x})$ is a limiting distribution in (6) if, and only if,*

$$H(\mathbf{x}) = \exp\left\{-\int_S \left[\max_{1 \leq t \leq m}(p_t e^{-x_t})\right] dU(\mathbf{p})\right\}. \tag{40}$$

In the same manner as for (39), it follows that $H(\mathbf{x})$ of (40) is a possible limit in (6). Its converse is the emphasis here, namely, that any limit in (6) for which the restriction on the margins is satisfied can be written in the form (40).

It is not an easy task to give the representation (40) for a given $H(\mathbf{x})$. But one usually does not aim at giving another form of $H(\mathbf{x})$ when it is already known. The value of (40) lies in its possibility of generating functions which are limits in (6). Such a problem of producing limiting distributions $H(\mathbf{x})$ is faced by the statistician who wants to make a decision on the form of $H(\mathbf{x})$ by a goodness of fit test, based on a given set of data. The comments made on this difficult practical question in the univariate case apply to multivariate situations as well. An additional difficulty here is that when a reasonable decision has been made on the univariate marginal distributions, there still remain a large number of possibilities for the actual multivariate distribution of the population. This is further complicated by the fact that the theory of multivariate distributions is far from being thorough. And finally, even if one is convinced that the appropriate distribution $F(\mathbf{x})$ of the population has been found, its actual computation in dimension $m \geq 5$ can be tremendously difficult (this is why probably most readers have not seen a table even for four-dimensional normal distributions). Some of these difficulties, however, can be avoided if the interest is $H(\mathbf{x})$. This is illustrated in the following numerical examples.

Example 5.4.3. Assume that $X_1, X_2, \ldots, X_{250}$ are independent observations on $\mathbf{X} = (X^{(1)}, X^{(2)}, X^{(3)}, X^{(4)})$, where $\mathbf{X}$ is a normal vector with $E(X^{(j)}) = 0$ and $V(X^{(j)}) = 1$, $1 \leq j \leq 4$. Let us find $P(\mathbf{Z}_{250} < \mathbf{x})$, where $\mathbf{x} = (3, 3.5, 2.8, 4)$.

If we wanted the exact distribution of $\mathbf{Z}_{250}$, we would need a table for four-dimensional normal distributions with actually arbitrary covariances. This can, however, be avoided by an appeal to asymptotic theory. In Example 5.3.1, we have seen that the components of $(\mathbf{Z}_n - \mathbf{a}_n)/\mathbf{b}_n$ are asymptotically independent, where the components of both $\mathbf{a}_n$ and $\mathbf{b}_n$ are identical and they can be computed by the formulas of Section 2.3.2. We get

$$a_{250} = 2.685, \quad b_{250} = 0.301.$$

Thus, in view of $(\mathbf{x} - \mathbf{a}_{250})/\mathbf{b}_{250} = (1.047, 2.708, 0.382, 4.369)$,

$$P(\mathbf{Z}_{250} < \mathbf{x}) = P\left(\frac{\mathbf{Z}_{250} - \mathbf{a}_{250}}{\mathbf{b}_{250}} < \frac{\mathbf{x} - \mathbf{a}_{250}}{\mathbf{b}_{250}}\right)$$

$$\sim H_{3,0}(1.047) H_{3,0}(2.708) H_{3,0}(0.382) H_{3,0}(4.369)$$

$$= 0.329. \qquad \blacktriangle$$

Example 5.4.4. Let $\mathbf{X}_1, \mathbf{X}_2, \ldots, \mathbf{X}_{100}$ be independent observations on $\mathbf{X} = (X^{(1)}, X^{(2)}, X^{(3)}, X^{(4)})$, where each component is unit exponential variate and each bivariate marginal distribution is a Morgenstern distribution (see Example 5.1.1). Find $P(\mathbf{Z}_{100} < \mathbf{x})$ with $\mathbf{x} = (6, 6.5, 6.2, 5.8)$.

Notice that we did not specify the distribution of $\mathbf{X}$. Hence, the answer could not be given without the asymptotic theory. With the asymptotic theory, however, we know that $\mathbf{Z}_n - \mathbf{a}_n$ has an asymptotic distribution, where each component of $\mathbf{a}_n$ is $\log n$. As a matter of fact, by the result of Example 5.2.1 and by Theorem 5.3.1 and Corollary 5.3.1, the components of $\mathbf{Z}_n - \mathbf{a}_n$ are asymptotically independent, with univariate margins $H_{3,0}(x_t)$, $1 \leq t \leq 4$. One has only to observe that

$$nG(x + \log n, y + \log n) \to 0$$

as $n \to +\infty$. Hence, all limits of (17) exist (and they are zero). Thus ($\log 100 = 4.605$)

$$P(\mathbf{Z}_{100} < \mathbf{x}) = P(\mathbf{Z}_{100} - \mathbf{a}_{100} < \mathbf{x} - \mathbf{a}_{100})$$

$$\sim H_{3,0}(1.395) H_{3,0}(1.895) H_{3,0}(1.595) H_{3,0}(1.195)$$

$$= 0.405. \qquad \blacktriangle$$

Example 5.4.5. Let us change the Morgenstern distributions to Mardia's distribution in the previous example. Let us estimate $P(\mathbf{Z}_{100} < \mathbf{x})$, where again $\mathbf{x} = (6, 6.5, 6.2, 5.8)$.

Again, the distribution of $\mathbf{X}$ is not specified, and thus $P(\mathbf{Z}_{100} < \mathbf{x})$ cannot be computed exactly. Neither can the asymptotic theory be applied to compute the above probability, because the components are not independent as $n \to +\infty$ (see Example 5.2.2). Therefore, formula (18) shows that the asymptotic distribution of $\mathbf{Z}_n - \mathbf{a}_n$ (each component of $\mathbf{a}_n$ is again $\log n$, as shown in Example 5.2.1) depends, for example, on the trivariate margins, which we do not know. We can, however, give estimates. The inequalities of Theorem 5.4.1 yield, when asymptotic theory is applied,

$$0.405 \leq P(\mathbf{Z}_{100} < \mathbf{x}) \leq H_{3,0}(1.195) = 0.739.$$

The lower bound can be improved considerably by the inequality of Exercise 1. That an improvement is possible can be expected, since in the estimates above we did not use bivatiate margins. We found in Example 5.2.2 that the bivariate margin of the asymptotic distribution of $\mathbf{Z}_n - \mathbf{a}_n$ that corresponds to the first and the jth component is

$$H_{1j} = H_{3,0}(z_1) H_{3,0}(z_j) \exp\left[(e^{z_1} + e^{z_j})^{-1} \right],$$

where $\mathbf{z} = (z_1, z_2, z_3, z_4) = \mathbf{x} - \mathbf{a}_{100} = (1.395, 1.895, 1.595, 1.195)$. Thus, by Exercise 1,

$$P(\mathbf{Z}_{100} < \mathbf{x}) \geq H_{12} + H_{13} + H_{14} - 2H_{3,0}(z_1) = 0.550.$$

Further improvement is possible by interchanging the roles of the first component and the fourth one. The reader is advised to carry out the calculations. ▲

5.5. CONCOMITANTS OF ORDER STATISTICS

Let (X_j, Y_j), $j = 1, 2, \ldots, n$, be independent and identically distributed random vectors. We consider the order statistics $X_{1:n} \leq X_{2:n} \leq \cdots \leq X_{n:n}$ of the first component and we denote the corresponding Y's by $Y_{[1:n]}, Y_{[2:n]}, \ldots, Y_{[n:n]}$. That is, if $X_j = X_{r:n}$, then $Y_{[r:n]} = Y_j$. The sequence $Y_{[r:n]}$, $1 \leq r \leq n$, is called the concomitants of order statistics.

The $Y_{[r:n]}$ are of interest in selection problems, where selection is based on the $X_{r:n}$. That is, we select those m individuals who had the highest X-scores and we wish to know something about the behavior of the concomitant Y-scores. For example, the X's may refer to a first test and the Y's to a later test, or the X's to a characteristic in a parent and the Y's to the same characteristic in an offspring. In more general terms we can say that a theory of the concomitants is needed whenever we want to judge individuals whose characteristics are measured by the Y's but we can observe only some related measurements X.

Some asymptotic results are easily obtained in certain cases. One possibility is illustrated below.

Example 5.5.1. Let (X_j, Y_j) be i.i.d. normal vectors with $E(X_j) = E(Y_j) = 0$ and $V(X_j) = V(Y_j) = 1$. Then, as $n \to +\infty$,

$$\lim P\left(Y_{[n:n]} < \rho (2 \log n)^{1/2} + y \right) = \Phi\left[(1 - \rho^2)^{-1/2} y \right],$$

where $\Phi(y)$ is the standard normal distribution function and ρ is the correlation coefficient of X_1 and Y_1.

The limit relation above can be deduced from the following representation. For $1 \leq j \leq n$,

$$Y_j = \rho X_j + (1-\rho^2)^{1/2} U_j, \qquad (41)$$

where the X's and U's are independent standard normal variates. Thus

$$Y_{[n:n]} = \rho X_{n:n} + (1-\rho^2)^{1/2} U_{[n]}.$$

We know that $X_{n:n} - (2\log n)^{1/2}$ converges to zero in probability (see Example 3.3.2). In addition, $U_{[n]}$ is independent of the X's with distribution

$$P(U_{[n]} < x) = \sum_{j=1}^{n} P(U_{[n]} < x | X_{n:n} = X_j) P(X_{n:n} = X_j)$$

$$= \sum_{j=1}^{n} P(U_j < x) P(X_j = X_{n:n}) = \Phi(x), \qquad (42)$$

because, for each j, $P(U_j < x) = \Phi(x)$. Lemma 2.2.1 now leads to the claimed limit. ▲

There is a notable difference between $Y_{[n:n]}$ and $Y_{n:n}$ both in the normalizing constants for a nondegenerate limiting distribution and in the actual limit law (see Section 2.3.2 for $Y_{n:n}$). The argument can be repeated, without any change, for $Y_{[n-k:n]}$ with k fixed as $n \to +\infty$. We thus get that, for any k, as $n \to +\infty$,

$$(2\log n)^{-1/2} Y_{n-k:n} \to 1, \qquad (2\log n)^{-1/2} Y_{[n-k:n]} \to \rho$$

in probability. A particular consequence of this result is that, in a large population, offspring of the top individuals will not be among the top members of the next generation in terms of the measurement represented by X and Y.

The normality of X and Y was not essential. The aspect of major importance was the decomposition (41) of Y_j into $aX_j + bU_j$, where U_j is independent of the X's.

We can, of course, evaluate the exact distribution of $Y_{[r:n]}$ without any structural assumption. For simplicity of analysis, let us assume that the common distribution function $F(x,y)$ of (X_j, Y_j) is absolutely continuous with density function $f(x,y)$. Let $f(y|x)$ denote the density function of Y_j given $X_j = x$. Since the vectors (X_j, Y_j) are i.i.d., the conditional density of

$Y_{[r:n]}$ given $X_{r:n}=x$ also equals $f(y|x)$ (apply the argument of (42)). Therefore

$$P(Y_{[r:n]}<y|X_{r:n}=x)=P(Y_1<y|X_1=x),$$

and thus, by the continuous version of the total probability rule (Appendix I),

$$P(Y_{[r:n]}<y)=\int_{-\infty}^{+\infty}P(Y_1<y|X_1=x)f_{r:n}(x)dx, \tag{43}$$

where $f_{r:n}(x)$ is the density function of $X_{r:n}$. Denoting by $F_1(x)=F(x,+\infty)$ and by $f_1(x)=F_1'(x)$ the special case $r=n$ of (43) yields

$$P(Y_{[n:n]}<y)=n\int_{-\infty}^{+\infty}P(Y_1<y|X_1=x)F_1^{n-1}(x)f_1(x)dx. \tag{44}$$

We now prove the following general result.

Theorem 5.5.1. *Let (X_j, Y_j), $1\leq j\leq n$, be i.i.d. vectors with absolutely continuous distribution function $F(x,y)$. Let the marginal distribution $F_1(x) = F(x,+\infty)$ be such that $\omega(F_1)=+\infty$, $F_1''(x)$ exists for all large x, and $F_1'(x)=f_1(x)\neq 0$. Furthermore, let*

$$\lim_{x=+\infty}\frac{d}{dx}\left[\frac{1-F_1(x)}{f_1(x)}\right]=0.$$

If the sequences a_n, $b_n>0$, A_n and $B_n>0$ are such that, as $n\to+\infty$,

$$\lim F_1^n(a_n+b_nz)=H_{3,0}(x) \tag{45}$$

and

$$\lim P(Y_1<A_n+B_nu|X_1=a_n+b_nz)=T(u,z), \tag{46}$$

a nondegenerate distribution function, then

$$\lim_{n=+\infty}P(Y_{[n:n]}<A_n+B_nu)=T(u), \tag{47}$$

where

$$T(u)=\int_{-\infty}^{+\infty}T(u,z)H_{3,0}(z)e^{-z}dz. \tag{48}$$

Proof. The conditions on $F_1(x)$ are such that all conditions of Theorem 2.7.2 are satisfied. Therefore, there are sequences a_n and $b_n>0$ for which

(45) holds. In addition, the reader was asked in Exercise 12 of Chapter 2 to show that, with these same a_n and b_n,

$$nb_n f_1(a_n + b_n z) \to e^{-z}. \tag{49}$$

If we now substitute $x = a_n + b_n z$ in (44), the conclusion (47) and (48) follows from (45) and (49) by the dominated convergence theorem (Appendix I). ▲

Notice that if the marginal $F_1(x)$ is smooth, the sole condition of the theorem is (46). For example, this is the case for all bivariate exponential distributions as well as for logistic, gamma, and the limit laws $H(x,y)$ in (6). Therefore, the theorem has a very wide appeal. Evidently, the normal case is also covered.

As was pointed out for the normal distribution, the concomitants of the extremes among the X's are not extremes among the Y's (with high probability). It is therefore an interesting question to investigate the rank $\lambda(r)$ of $Y_{[r:n]}$. For defining $\lambda(r)$, let us first introduce the function

$$I(x) = \begin{cases} 1 & \text{if } x \geq 0, \\ 0 & \text{otherwise}. \end{cases}$$

Then we define for continuous marginals $F_2(y) = F(+\infty, y)$

$$\lambda(r) = \sum_{j=1}^{n} I(Y_{[r:n]} - Y_j).$$

Thus $Y_{[r:n]} = Y_{\lambda(r):n}$ and

$$P(\lambda(r) = s) = \sum_{j=1}^{n} P(Y_j = Y_{s:n}, X_j = X_{r:n})$$

$$= nP(Y_1 = Y_{s:n}, X_1 = X_{r:n}),$$

the last equation being due to the fact that the vectors (X_j, Y_j) are identically distributed. Now, the event $\{Y_1 = Y_{s:n}, X_1 = X_{r:n}\}$ means that out of Y_j, $2 \leq j \leq n$, exactly $s-1$ are smaller than Y_1 and out of the X_j, $2 \leq j \leq n$, exactly $r-1$ do not exceed X_1. Collecting the terms according as $\{X_i < X_1, Y_i < Y_1\}$, or $\{X_i < X_1, Y_i > Y_1\}$, or $\{X_i > X_1, Y_i < Y_1\}$, or $\{X_i > X_1, Y_i > Y_1\}$, we get, by conditioning on (X_1, Y_1),

$$P(\lambda(r) = s) = n \sum_{k=0}^{s-1} \binom{n-1}{s-1} \binom{s-1}{k} \binom{n-s}{r-1-k} \int_{-\infty}^{+\infty} \int_{-\infty}^{+\infty} g(x, y; k) \, dx \, dy,$$

where

$$g(x,y;k) = u_1^k u_2^{s-1-k} u_3^{r-1-k} u_4^{n-r-(s-k-1)} f(x,y)$$

with

$u_1 = P(X_1 < x, Y_1 < y)$, $u_2 = P(X_1 > x, Y_1 < y)$, $u_3 = P(X_1 < x, Y_1 > y)$,

$u_4 = P(X_1 > x, Y_1 > y)$ and $f(x,y) = \dfrac{\partial^2 F(x,y)}{\partial x \, \partial y}$.

For a given distribution $F(x,y)$, $P(\lambda(r) = s)$ can easily be given by a computer. From the exact distribution, the expected rank $E[\lambda(r)]$ can also be computed.

The theory of concomitants of order statistics is at a very early stage, in particular as it concerns extremes. It is hoped that the early results will induce further research. It should be of great interest for sociologists, psychologists, and medical researchers.

5.6. SURVEY OF THE LITERATURE

The asymptotic theory of the extremes for bivariate distributions started with the short announcement of results by Finkelstein (1953). His work was not followed by details. Several years later, approximately at the same time and independently of each other, three basic works appeared on bivariate extremes: J. Geffroy (1958/1959), J. Tiago de Oliveira (1958), and M. Sibuya (1960). Each of these papers arrives at a representation equivalent to (28a), and actually the multivariate case (28) is also obtained in the first two. Geffroy and Sibuya obtain conditions for the asymptotic independence of the components. Geffroy's criterion is equivalent to the form expressed at (24) (see also Exercise 20), and Sibuya's result is as given in Exercise 21. Although each of the above three works discusses several properties of $v(y)$ occurring in the representation (28a), the method leading to formulas (29)–(32) and, in particular, to the representation (33) and (34), when the density exists, is due to Tiago de Oliveira (1962/1963). Here he also establishes that the components of the normalized maxima are independent in arbitrary dimension if, and only if, the bivariate marginals are asymptotically independent. This is obtained by setting bounds on $H(\mathbf{x})$. His bounds include the important lower bound of Theorem 5.4.1. The special case of $m = 2$ of this bound is supported by the observation that the correlation coefficient ρ of the two components of a vector with distribution (28a) is always nonnegative (see Exercise 29). The formula of Exercise 30 gives a further insight into the structure of the bivariate distribution (28a).

Formula (28a) greatly enriched the set of easily accessible bivariate distributions. One should realize that, by monotonic transformations, one immediately gets distributions with uniform, exponential, logistic, Pareto or Weibull marginals from (28a) and thus, starting with (33) and (34), arbitrary $g(u)$ leads to a bivariate distribution (28a) whose density function exists and is easy to handle. Of course, not all bivariate distributions can be generated this way because of the restriction of the dependence function of (28a). Some methods of generating multivariate distributions became well known and are mentioned in bivariate exponential forms in Example 5.1.1. The works treating those methods are by D. Morgenstern (1956), E. J. Gumbel (1960), A. W. Marshall and I. Olkin (1967), and K. V. Mardia (1964a and 1970). A basic work on the theory of multivariate distributions is that of M. Fréchet (1951). A good collection of material and references is found in N. L. Johnson and S. Kotz (1972).

The asymptotic independence of the components of $(\mathbf{Z}_n - \mathbf{a}_n)/\mathbf{b}_n$ has drawn much further attention. S. M. Berman (1961) continued the work of Geffroy (1958/1959) along this line. This was then extended to a condition guaranteeing the asymptotic independence of other extremes (not just maxima) of the components in the bivariate case in the works of K. V. Mardia (1964b) and O. P. Srivastava (1967). Finally, independently of each other, V. G. Mikhailov (1974) and J. Galambos (1975d) obtained a necessary and sufficient condition for the asymptotic independence of arbitrary extremes in any dimension. This condition, which is the same for all extremes, is formulated for maxima in Corollary 5.3.1.

The paper by J. Galambos (1975d) is the only one which gives the asymptotic distribution of all extremes (for the definition in the bivariate case, see Exercise 24). This distribution is given in terms of the limits in (17). Its wide applicability lies in combining the two representations (28a) and (18). For example, in the bivariate case, let us generate a distribution $H(x_1, x_2)$ by (28a). Then, writing this in the form of (18), one gets $h_1(x_1)$, $h_2(x_2)$, and $h_{1,2}(x_1, x_2)$. With these functions known, the asymptotic distribution of $(X_{n-i:n}, Y_{n-j:n})$, when normalized, can be computed by the formula of Galambos (1975d) (see Exercise 24). It should be observed that, just as in the univariate case, the same constants are used for normalization of all upper extremes (fixed i and j above) as for maxima. On the other hand, if the population distribution is available (which is rare, and a statistical choice is as difficult, or even more so, as in the univariate case), then one directly computes the limits (17), and thus (18) becomes the basic representation. The inequalities (19) appear for the first time in such generality here.

In a recent paper J. Pickands III (1977) gave a valuable representation of multivariate asymptotic distributions of the maxima. It will serve as a basic tool for generating these distributions in any dimension—in particular, when the statistical properties of these distributions become fully understood. See also A. A. Balkema and S. I. Resnick (1977), whose work serves as basis for another approach to multivariate extremes by L. de Haan and S. I. Resnick (1977).

The possibility of combining the two representations (18) and (28a) was overlooked in the literature. Neither was the possibility realized of transforming Theorem 5.3.1 into a parametric form as in Example 5.4.2. Hence, these results are being reobtained, or some are mentioned in recent surveys as problems to be settled. However, Theorem 5.3.1 was not yet available to K. A. Nair (1976). For specific classes of population distributions, extensive studies were made by J. Villasenor (1976). He also extended the investigations to exchangeable sequences of multivariate distributions along the line of the treatment of the univariate case in Section 3.6. Another special class, the density of which admits a special series expansion, is treated by J. W. Campbell and C. Tsokos (1973).

The number of published papers dealing with applications of the asymptotic theory of multivariate extremes is very small. The major reason for this should be a general lack of multivariate applications when the underlying distribution is not normal. However, interesting applications are discussed in E. J. Gumbel and N. Goldstein (1964), E. J. Gumbel and C. K. Mustafi (1967), and E. C. Posner et al. (1969). Other papers dealing with the statistical aspects of multivariate extreme value distributions are by B. Arnold (1968) and by Tiago de Oliveira (1970, 1971, and 1974). Eve Bofinger and V. J. Bofinger (1965) analyze the accuracy of the approximation of bivariate extremes for normal populations through the correlation of the extremes. This is extended to some nonnormal cases in V. J. Bofinger (1970). It should be noted that for high correlations of the population, the extremes do not show independence for sample sizes as high as $n=50$. T. Cacoullos and H. DeCicco (1967) determine the distribution of the bivariate range. Also on the bivariate range, see K. V. Mardia (1967). The surveys by Tiago de Oliveira (1975) and E. J. Gumbel (1962) are of interest.

The influence of the maximum in the sum of multivariate observations is investigated by N. Kalinauskaite (1973 and 1976).

The theory of concomitants of order statistics is comparatively new. It was initiated by H. A. David (1973). More detail developed on the asymptotic theory by H. A. David and J. Galambos (1974), was extended by M. J. O'Connell and H. A. David (1976), H. A. David, M. J. O'Connell, and S. S. Yang (1977), and S. S. Yang (1977). Another aspect of the theory, namely, functional limit laws, under the term of induced order statistics, was developed by P. K. Bhattacharyya (1974) and P. K. Sen (1976).

For other kinds of ordering of multivariate data, see the survey by V. Barnett (1976).

5.7. EXERCISES

1. Let $\mathbf{X}$ be an m-dimensional vector with distribution function $F(\mathbf{x})$, whose univariate and bivariate marginals are $F_i(x_i)$ and $F_{i,j}(x_i, x_j)$, respectively, where $1 \leq i, j \leq m, i < j$. Show that

$$F(\mathbf{x}) \geq \sum_{j=2}^{m} F_{1,j}(x_1, x_j) - (m-2)F_1(x_1).$$

[Hint: Apply Exercise 18 of Chapter 1 and Theorem 5.1.2 with $m = 2$.]

2. Using the notation of the preceding exercise, show that, for any integer $1 \leq k \leq m - 1$,

$$F(\mathbf{x}) \leq \frac{2}{k(k+1)} \left\{ \sum_{1 \leq i < j \leq m} F_{i,j}(x_i, x_j) - (m-k+1) \sum_{i=1}^{m} F_i(x_i) + k^2 - k + 2 \right\}.$$

(Apply Theorem 1.4.3.)

3. Evaluate the two bounds above for $m = 5$, if $F_{i,j}(x_i, x_j)$, for all $1 \leq i < j \leq 5$, equals (i) Gumbel's type I distribution with $\theta = \frac{1}{2}$ and (ii) Mardia's distribution (see Example 5.1.1). Choose numerical values for $\mathbf{x} = (x_1, x_2, x_3, x_4, x_5)$ and compare the bounds obtained. Also compare the results with the Fréchet bounds (Theorem 5.1.1).

4. Let (X, Y) have a bivariate normal distribution with $E(X) = 0$, $E(Y) = -2$, $V(X) = 1$, $V(Y) = 4$ and with correlation coefficient $\rho = .3$. For an independent sample (X_j, Y_j) of size $n = 40$ on (X, Y), evaluate $P(Z_{1,40} < 2.8, Z_{2,40} < 1.8)$. Compare the exact value with the appropriate asymptotic expression. Make this same comparison if n is increased to 100.

5. (i) Let $\mathbf{X}_j$, $1 \leq j \leq n$, be i.i.d. normal vectors such that each correlation coefficient of the components is positive and less than one. Show that the asymptotic distribution of $(\mathbf{Z}_n - \mathbf{a}_n)/\mathbf{b}_n$ is a lower estimate of the exact distribution.

(ii) Let ρ be the largest correlation coefficient of the components of $\mathbf{X}_1$. Let $\mathbf{Z}_n^*$ be the $\mathbf{Z}$-vector of i.i.d. normal variates whose components are equally correlated with coefficient ρ. Assume that n is such that the distribution of $(\mathbf{Z}_n^* - \mathbf{a}_n)/\mathbf{b}_n$ is accurately obtained up to five decimal digits by its limit distribution. Show that then the same is true for $(\mathbf{Z}_n - \mathbf{a}_n)/\mathbf{b}_n$. [Hint: Apply Lemma 3.8.1.]

6. Let $F_j(\mathbf{x})$, $1 \leq j \leq k$, be m-dimensional distribution functions whose univariate marginals do not depend on j. Let $\mathbf{u} = (u_1, u_2, \ldots, u_k)$ be such that $0 \leq u_j \leq 1$ and $u_1 + u_2 + \cdots + u_k = 1$. Show the following relations for

5.7 EXERCISES

dependence functions:

(i)
$$D_T(\mathbf{y}) = \sum_{i=1}^{k} u_i D_{F_i}(\mathbf{y}) \quad \text{for } T(\mathbf{x}) = \sum_{i=1}^{k} u_i F_i(\mathbf{x}),$$

and

(ii)
$$D_R(\mathbf{y}) = \prod_{i=1}^{k} D_{F_i}^{u_i}(\mathbf{y}) \quad \text{for } R(\mathbf{x}) = F_1^{u_1}(\mathbf{x}) \cdots F_k^{u_k}(\mathbf{x}).$$

7. Show that if each of the m-dimensional distribution functions $H_j(\mathbf{x})$, $1 \leq j \leq k$, is a limit in (6), then so is

$$R(\mathbf{x}) = H_1^{u_1}(\mathbf{x}) H_2^{u_2}(\mathbf{x}) \cdots H_k^{u_k}(\mathbf{x}),$$

where $0 \leq u_j \leq 1$ and $u_1 + u_2 + \cdots + u_k = 1$. [Hint: Apply the preceding exercise and Theorem 5.2.4.]

8. Let $L(\mathbf{x})$ be the limit distribution of $(\mathbf{W}_n - \mathbf{c}_n)/\mathbf{d}_n$ for a given population distribution $F(\mathbf{x})$ and for the appropriate $\mathbf{c}_n$ and $\mathbf{d}_n$. Give a characterization of the dependence function $D_L(\mathbf{y})$, based on Theorem 5.2.1.

9. Give a necessary and sufficient condition on $L(\mathbf{x})$ of the preceding exercise based on Theorem 5.2.4.

10. Let $L(\mathbf{x})$ be as in Exercise 8. Let $T(\mathbf{x})$ be the survival function of a vector whose distribution function is $L(\mathbf{x})$. Give a representation of $T(\mathbf{x})$ in the form of (18).

11. With the notations of the preceding exercise, show that $T(\mathbf{x})$ is never smaller than the product of its univariate marginals.

12. Prove that (37) is also necessary for the conclusion of Example 5.4.2.

13. Extend the method discussed after Example 5.4.2 without assuming that the univariate marginals are identical. Also extend this method to dimensions higher than two.

14. With the appropriate transformation of the marginals, transform the representation (28a) to obtain a form for an arbitrary limit occurring in (6).

15. Let $v(y_1, y_2)$ be the exponent in (28) for $m = 3$. Rewrite the inequalities of Theorem 5.4.1 to obtain bounds on $v(y_1, y_2)$.

16. Let the distribution function of the vector (X, Y) be $H(x_1, x_2)$ of (28a) with $v(y) = \max(1, e^{-y})/(1 + e^{-y})$. Show that $P(X = Y) = 1$.

17. Evaluate $v(y)$ of (33) and $H(x_1,x_2)$ of (28a) if $g(u)=\exp(-3|u|)$ for all u.

18. Let X and Y be independent random variables with common distribution function $H_{3,0}(x)$. Put $X_1 = X$ and $X_2 = \max[X + \log \nu, Y + \log(1-\nu)]$. Find the distribution function $H(x_1,x_2)$ of (X_1,X_2). Show that $H(x_1,x_2)$ is a possible limit in (6), and give its representation in the form of (28a).

19. Let $g(u) = Cu^s$ for $0 \leq u \leq A < +\infty$ and zero otherwise, where $C > 0$ and $s > 0$ are given numbers. Determine $v(y)$ of (33) and $H(x_1,x_2)$ of (28a).

20. Let $F(x,y)$ be a bivariate distribution function with identical marginals $F(x) = F(x, +\infty) = F(+\infty, x)$. It was shown that in (24) if u_n and v_n tend to $\omega(F)$ as $n \to +\infty$ and if

$$\frac{P(X \geq u_n, Y \geq v_n)}{P(X \geq u_n)} \to 0 \quad \text{as } n \to +\infty,$$

where (X,Y) is a vector with distribution $F(x,y)$, then, for i.i.d. observations on (X,Y), the components of $(\mathbf{Z}_n - \mathbf{a}_n)/\mathbf{b}_n$ are asymptotically independent whenever it has an asymptotic distribution. Reobtain this result from Corollary 3.7.1 by using the following property. If (X_j, Y_j) are i.i.d. random vectors with common distribution $F(x,y)$, then the sequence $X_1, Y_1, X_2, Y_2 \ldots$ is a two-dependent sequence (see Section 3.7).

21. Let (X_j, Y_j), $1 \leq j \leq n$, be i.i.d. with common distribution function $F(x,y)$. Let the marginals of $F(x,y)$ be such that, with suitable vectors $\mathbf{a}_n$ and $\mathbf{b}_n$, the components of $(\mathbf{Z}_n - \mathbf{a}_n)/\mathbf{b}_n$ have limiting distributions. Then $(\mathbf{Z}_n - \mathbf{a}_n)/\mathbf{b}_n$ itself converges weakly, and its components are asymptotically independent, whenever the dependence function $D_F(y_1, y_2)$ of $F(x,y)$ satisfies the asymptotic property

$$D_F(1-s, 1-s) = 1 - 2s + o(s), \quad s \to 0.$$

[M. Sibuya (1960)]

22. Let (X,Y) be a random vector with survival function $P(X \geq x, Y \geq y) = G(x,y)$ whose univariate marginals are denoted by $G_1(x)$ and $G_2(y)$. Let (X_t, Y_t), $1 \leq t \leq n$, be n independent observations on (X,Y). Put $A_{1t} = \{X_t \geq x\}$ and $A_{2t} = \{Y_t \geq y\}$. Show that $S(u,v)$ of Exercise 19 of Chapter 1 becomes

$$S(u,v) = \sum_{d=0}^{\min(u,v)} T_n(u,v;d),$$

where

$$T_n(u,v;d) = \binom{n}{d}\binom{n-d}{u-d}\binom{n-u}{v-d} G^d(x,y) G_1^{u-d}(x) G_2^{v-d}(y).$$

23. With the notation of the preceding problem, we introduce the following additional quantities. $X_{r:n}$ and $Y_{r:n}$ denote the rth order statistic of the X's and the Y's, respectively. Let

$$V(k_1,k_2;t) = (-1)^{t-k_1-k_2} \sum_u \binom{u_1}{k_1}\binom{u_2}{k_2} S(u_1,u_2),$$

where, in $\sum_u$, summation is for (u_1,u_2) with $u_i \geq 0$ and $u_1+u_2 = t$. Finally, let

$$\tau_n(i,j;s) = \sum_{k_1=0}^{i} \sum_{k_2=0}^{j} \sum_{t=k_1+k_2}^{k_1+k_2+s} V(k_1,k_2;t).$$

Use the inequalities of Exercises 19 and 20 of Chapter 1 to establish, for any $s \geq 0$,

$$\tau_n(i,j;2s+1) \leq P(X_{n-i:n} < x, Y_{n-j:n} < y) \leq \tau_n(i,j;2s). \qquad (50)$$

[J. Galambos (1975d)]

24. We use the notation of Exercise 22. Assume that $G(x,y)$, $G_1(x)$, and $G_2(y)$ are such that, with suitable constants $\mathbf{a}_n = (a_{1,n}, a_{2,n})$ and $\mathbf{b}_n = (b_{1,n}, b_{2,n})$, (17) holds. Show that, as $n \to +\infty$,

$$\lim S(u,v) = \sum_{d=0}^{\min(u,v)} \frac{1}{d!(u-d)!(v-d)!} h_{1,2}^d(x_1,x_2) h_1^{u-d}(x_1) h_2^{v-d}(x_2).$$

Hence, find the limit of the bounds in (50) for fixed i, j, and s. By letting $s \to +\infty$, conclude that, under (17), the bivariate extremes $(X_{n-i:n}, Y_{n-j:n})$, i,j fixed, have limiting distributions, when normalized by $\mathbf{a}_n$ and $\mathbf{b}_n$.

[J. Galambos (1975d)]

25. Prove the conclusion of Exercise 24 for m-dimensional vectors.

[J. Galambos (1975d)]

26. Show that the limiting distribution obtained in Exercise 24 for $\{(X_{n-i:n} - a_{1,n})/b_{1,n}, (Y_{n-j:n} - a_{2,n})/b_{2,n}\}$ is the Cauchy product of the functions

$$\sum_{k_1=0}^{i} \frac{h_1^{k_1}(x_1)}{k_1!} \sum_{d=0}^{+\infty} (-1)^d \frac{h_1^d(x_1)}{d!} = \sum_{k_1=0}^{i} \frac{h_1^{k_1}(x_1)}{k_1!} \exp\{-h_1(x_1)\}$$

and a similar function, where i is replaced by j and $h_1(x_1)$ by $h_2(x_2)$.

[O. P. Srivastava (1967)]

27. Extend the preceding result to dimension m.

[J. Galambos (1975d)]

28. Rework Exercises 22–27 for the lower extremes $(X_{i:n}, Y_{j:n})$, where i and j do not vary with n.

29. Let the random vector (X, Y) have distribution function $H(x_1, x_2)$ of the form (28a). Show the formula

$$\rho = -\frac{6}{\pi^2} \int_{-\infty}^{+\infty} \log v(y) dy,$$

where ρ is the correlation coefficient of X and Y. Hence conclude that X and Y are independent if, and only if, they are uncorrelated (that is, $\rho = 0$). (Note that $\rho \geq 0$. Why?)

[J. Tiago de Oliveira (1962–1963)]

30. In the preceding exercise, let $H(x_1, x_2)$ have a bivariate density. Evaluate the distribution function $T(x)$ of the difference $Y - X$. Establish the relation

$$v(y) = (1 + e^y)^{-1} \exp\left\{\int_{-\infty}^{y} T(x) dx\right\}.$$

[J. Tiago de Oliveira (1962–1963)]

31. (i) Let the distribution function of (X, Y) be

$$F(x, y) = \int_0^1 (1 - e^{-ux})(1 - e^{-uy}) du.$$

Let (X_j, Y_j), $1 \leq j \leq n$, be independent observations on (X, Y). Show that, with suitable vectors $\mathbf{a}_n$ and $\mathbf{b}_n$, $(\mathbf{Z}_n - \mathbf{a}_n)/\mathbf{b}_n$ converges weakly to $\exp\{-x^{-1} - y^{-1} + (x+y)^{-1}\}$.

(ii) If $1 - e^{-ux}$ and $1 - e^{-uy}$ are replaced by some distribution functions $G_1(u, x)$ and $G_2(y, u)$ in the definition of $F(x, y)$, and furthermore if the integration is over the whole real line with respect to a distribution function $T(u)$, find sufficient conditions when the components of $(\mathbf{Z}_n - \mathbf{a}_n)/\mathbf{b}_n$ are asymptotically independent.

[J. A. Villasenor (1976)]

32. (i) Let U, V, and S be independent random variables with common distribution function $F(x) = 1 - 1/x$, $x \geq 1$. Set $X = U + S$ and $Y = V + S$. For independent observations on (X, Y), show that there are vectors $\mathbf{a}_n$ and $\mathbf{b}_n$ such that $(\mathbf{Z}_n - \mathbf{a}_n)/\mathbf{b}_n$ converges weakly to $\exp(-1/x - 1/y + \frac{1}{2}\max(x,y))\}$.

(ii) Let U and V be independent random variables with nondegenerate distribution functions $F(x)$ and $G(y)$, respectively. Show that if $\omega(F) < +\infty$, then, for independent observations on $(X^{(1)}, X^{(2)}) = (U, U + V)$, the components of $(\mathbf{Z}_n - \mathbf{a}_n)/\mathbf{b}_n$ are asymptotically independent, whenever it has an asymptotic distribution.

[J. A. Villasenor (1976)]

CHAPTER 6

Miscellaneous Results

The present chapter is devoted to three major topics. We deal first with the weak convergence of extremes when the sample size itself is a random variable. Then we turn to a special case of random sample sizes, when sampling stops at a maximum or minimum. These stoppings are known as record times and the actual values at time of termination of experimentation as records. Finally, we shall present the foundations of the so-called extremal processes. These are continuous time stochastic processes constructed from extremes by a limiting procedure. The theory is similar to the better-known extension of the central limit theorem to approximating specially constructed piecewise linear random functions by the Wiener process.

The section dealing with the extremal processes requires more than a basic knowledge of probability theory, but the other sections are at the level of previous chapters. The combination of these three topics into one chapter, then, should not hinder the reader from going through the other sections if he, or she, has decided not to prepare for the investigation of a continuous process.

6.1. THE MAXIMUM QUEUE LENGTH IN A STABLE QUEUE

We shall consider the following simple model of a one-server system. The system starts at time $t_0 = 0$ when the first customer arrives who is ready to be served. Additional customers arrive at time $t_n, n \geq 1$, where the intervals $t_n - t_{n-1}$ are assumed to be i.i.d. random variables with distribution function $U(x)$ such that

$$U(0+) = 0 \quad \text{and} \quad 1 < a = \int_0^{+\infty} x \, dU(x) < +\infty.$$

If there is no customer in the system at time t_n, then individual $n+1$ starts

6.1 THE MAXIMUM QUEUE LENGTH IN A STABLE QUEUE

being served. Otherwise, he joins the queue and awaits his turn. The service times, $s_n, n \geq 1$, of the successive customers are assumed to be independent unit exponential variates and independent of the interarrivals $t_n - t_{n-1}$.

Notice that our assumptions include the choice of the time unit to be the expected service time of an individual. Hence, the meaning of $a > 1$ is that service is expected to be shorter than the intervals between arrivals. It is thus immediate from the strong law of large numbers that, with probability one, there is a finite time t^* when the server becomes idle. The period $(0, t^*)$ is called the busy period. With the arrival of the first customer after t^* the process starts again, and we can speak of the second busy period, and so on. With the interarrivals being independent, the busy periods are i.i.d. random variables.

Our interest is the maximum queue length. That is, let Y_k represent the number of customers present in the system just prior to the kth arrival. We wish to investigate $Q_n = \max\{Y_k : 1 \leq k \leq n\}$. The problem would belong to Chapter 3, because the Y's are strongly dependent. However, a simpler approach presents itself by the following observation. Let $N(n)$ be the number of busy periods completed just prior to the arrival of the nth customer. Furthermore, let $X_j + 1$ be the maximum queue length in the jth busy period. Then, as pointed out above, the X_j are i.i.d. random variables and, evidently,

$$Z_{N(n)} \leq Q_n \leq Z_{N(n)+1},$$

where, as usual, $Z_n = \max(X_1, X_2, \ldots, X_n)$. We thus arrive at a new problem, namely, finding the asymptotic behavior of $Z_{N(n)}$, where $N(n)$ is a random variable.

The subsequent sections will provide asymptotic results for Q_n via $Z_{N(n)}$. In particular, it will follow that there is a finite number $c > 0$ such that, as $n \to +\infty$,

$$\lim \frac{Q_n}{\log n} = c \quad \text{in probability.} \tag{1}$$

In this conclusion, we do not need anything about the interrelation of $N(n)$ and the sequence X_j, $j \geq 1$. However, the following property is essential.

Lemma 6.1.1. *As $n \to +\infty$, $N(n)/n$ converges (almost surely) to a finite, positive constant.*

Proof. Let U_j be the number of customers served in the jth busy period and put $T_n = \sum_{j=1}^{n} U_j$. We have seen that random variables associated with different busy periods are i.i.d. Let $\mu = E(U_j)$. Evidently $\mu > 0$. On the other hand, it can be shown that $\mu < +\infty$ (see, e.g., Prabhu (1965), p. 168). Thus, by the strong law of large numbers, for sufficiently large m,

$$(\mu - \varepsilon)m \leq T_m \leq (\mu + \varepsilon)m,$$

where $\varepsilon > 0$ is arbitrary with $\varepsilon < \mu$. Applying the above inequalities with $m = N(n)$ and with $m = N(n) + 1$, we get

$$n(\mu + \varepsilon)^{-1} - 1 \leq N(n) \leq n(\mu - \varepsilon)^{-1}.$$

Since $\varepsilon > 0$ is arbitrary, $0 < \mu < +\infty$, and all statements above are valid with probability one, the lemma follows. ▲

6.2. EXTREMES WITH RANDOM SAMPLE SIZE

Partially guided by the discussion in the preceding section, we now investigate the following problem. Let $X_1, X_2, \ldots$ be i.i.d. random variables with common distribution function $F(x)$. Let $N(n)$ be a positive integer valued random variable. What can be said about $Z_{N(n)}$ or $W_{N(n)}$, or the other extremes if the sample size is $N(n)$? In the first part of the present section we shall develop a technique for the following result.

Theorem 6.2.1. *Let, as $n \to +\infty$, $N(n)/n \to \tau$ in probability, where τ is a positive random variable. Assume that there are sequences a_n and $b_n > 0$ such that $(Z_n - a_n)/b_n$ converges weakly to a nondegenerate distribution function $H(x)$. Then, as $n \to +\infty$,*

$$\lim P(Z_{N(n)} < a_n + b_n x) = \int_{-\infty}^{+\infty} H^y(x) dP(\tau < y). \qquad (2)$$

The theorem above will follow from a sequence of lemmas which individually express very interesting facts. They can be applied to solving a great variety of problems which are not necessarily related to extremes.

Lemma 6.2.1. *If a sequence U_n of random variables satisfies the limit relations*

$$\lim_{n = +\infty} P(U_n < x) = T(x) \qquad (3)$$

and, for x's for which $P(U_k < x) > 0$,

$$\lim_{n = +\infty} P(U_n < x | U_k < x) = T(x), \qquad k = 1, 2, \ldots, \qquad (4)$$

where $T(x)$ is a distribution function and convergence is for continuity points of $T(x)$, then, for any event B,

$$\lim_{n=+\infty} P(\{U_n < x\} \cap B) = T(x)P(B). \tag{5}$$

Proof. Let $I_k(x)$ and $I(B)$ be the indicator variables of the events $\{U_k < x\}$ and B, respectively. Then (4) can be rewritten as

$$\lim_{n=+\infty} E[I_n(x)I_k(x)] = T(x)E[I_k(x)].$$

We thus have for any fixed m and for constants c_j, $1 \leq j \leq m$,

$$\lim_{n=+\infty} E[Y_0 I_n(x)] = T(x)E(Y_0), \tag{6}$$

where $Y_0 = c_1 I_1(x) + c_2 I_2(x) + \cdots + c_m I_m(x)$. Now let Y be a random variable with finite variance and such that there is a sequence $Y_{0,m}, m \geq 1$, of the form occurring in (6) with $E[(Y - Y_{0,m})^2] \to 0$ as $m \to +\infty$. Then, by the Cauchy-Schwarz inequality,

$$\{E[YI_n(x)] - E[Y_{0,m}I_n(x)]\}^2 \leq E[(Y - Y_{0,m})^2] \to 0$$

and

$$[E(Y) - E(Y_{0,m})]^2 \leq E[(Y - Y_{0,m})^2] \to 0$$

as $m \to +\infty$. We thus get from (6) that, by first letting $n \to +\infty$ and then $m \to +\infty$,

$$\lim_{n=+\infty} E[YI_n(x)] = T(x)E(Y). \tag{7}$$

Write $I(B) = Y + R$, where Y is of the above property and R is such that for each n, $E[RI_n(x)] = 0$. Such a representation is possible (see Appendix II). Since

$$E[I(B)I_n(x)] = E[YI_n(x)],$$

the limit relation (7) implies (5), which was to be proved. ▲

Lemma 6.2.2. *Let $X_1, X_2, \ldots$ be i.i.d. random variables with distribution function $F(x)$. Let there exist constants a_n and $b_n > 0$ such that, as $n \to +\infty$,*

$$\lim P(Z_n < a_n + b_n x) = H(x), \tag{8}$$

where $H(x)$ is nondegenerate. Then, for any event B,

$$\lim_{n=+\infty} P(\{Z_n < a_n + b_n x\} \cap B) = H(x)P(B).$$

Proof. We apply Lemma 6.2.1 with $U_n = (Z_n - a_n)/b_n$. Since (8) corresponds to the assumption (3), the lemma will be proved if we establish the validity of (4). For this purpose, let us write $Z_n = \max(Z_k, Z_{k,n})$, where $Z_{k,n} = \max(X_{k+1}, \ldots, X_n)$. Then, by the independence of the X's,

$$P(U_n < x | U_k < x) = P(Z_k < a_n + b_n x | U_k < x) P(Z_{k,n} < a_n + b_n x).$$

Because $a_n + b_n x \to \omega(F)$ as $n \to +\infty$ (which is evident from $F^n(a_n + b_n x) \to H(x)$), for any fixed k, as $n \to +\infty$,

$$\lim P(Z_k < a_n + b_n x | U_k < x) = 1.$$

On the other hand,

$$P(Z_{k,n} < a_n + b_n x) = F^{n-k}(a_n + b_n x) \to H(x)$$

for fixed k. This completes the proof. ▲

Lemma 6.2.3. *Let $X_1, X_2, \ldots$ be i.i.d. random variables for which (8) holds. Let $N(n)$ be a positive integer-valued random variable such that $N(n)/n \to \tau$, where $\tau > 0$ is a random variable. Then, for any event B, as $n \to +\infty$,*

$$\lim P(\{Z_{N(n)} < a_{N(n)} + b_{N(n)} x\} \cap B) = H(x)P(B).$$

Proof. Let $\varepsilon > 0$ be arbitrary. Choose $y_1 < y_2$ such that $P(y_1 \leq \tau < y_2) \geq 1 - \varepsilon$. From the assumptions it follows that there is an integer n^* such that for $n \geq n^*$, $P(y_1 \leq N(n)/n < y_2) \geq 1 - 2\varepsilon$. Let us fix y_1, y_2, and n^*. Let us divide the interval $[y_1, y_2]$ by the points $y_1 = s_0 < s_1 < \cdots < s_m = y_2$. Let us put $n(j)$ for the integer part of ns_j. We now have

$$-2\varepsilon + \sum_{j=1}^{m} P\left(\{Z_{n(j)} < a_{N(n)} + b_{N(n)} x\} \cap B \cap \left\{s_{j-1} \leq \frac{N(n)}{n} < s_j\right\}\right)$$

$$\leq P(Z_{N(n)} < a_{N(n)} + b_{N(n)} x)$$

$$\leq \sum_{j=1}^{m} P\left(\{Z_{n(j-1)} < a_{N(n)} + b_{N(n)} x\} \cap B \cap \left\{s_{j-1} \leq \frac{N(n)}{n} < s_j\right\}\right) + 2\varepsilon.$$

We make two modifications in the above inequalities. First, we replace $N(n)/n$ by its limit τ. Because m does not depend on n, the effect of this change is arbitrarily small if n^* is suitably chosen. Thus, the above inequalities hold if 2ε is replaced by 3ε, say, and $N(n)/n$ by τ. Next, let us write

$$\{Z_{n(j)} < a_{N(n)} + b_{N(n)}x\} = \left\{\frac{Z_{n(j)} - a_{n(j)}}{b_{n(j)}} \frac{b_{n(j)}}{b_{N(n)}} + \frac{a_{n(j)} - a_{N(n)}}{b_{N(n)}} < x\right\},$$

and similarly the event in the upper inequality. If we choose the points s_j, $0 \leq j \leq m$, sufficiently close, then, by Theorem 2.2.1, for large n,

$$\left|\frac{b_{n(j)}}{b_{N(n)}} - 1\right| < \delta \quad \text{and} \quad \left|\frac{a_{n(j)} - a_{N(n)}}{b_{N(n)}}\right| < \delta,$$

where $\delta > 0$ is again arbitrary. The same argument applies if $n(j)$ is replaced by $n(j-1)$, and thus we can conclude that if the s_j are sufficiently close and if n is large,

$$-3\varepsilon + \sum_{j=1}^{m} P\left(\left\{\frac{Z_{n(j)} - a_{n(j)}}{b_{n(j)}} < \frac{x-\delta}{1+\delta}\right\} \cap B \cap \{s_{j-1} \leq \tau < s_j\}\right)$$

$$\leq P(Z_{N(n)} < a_{N(n)} + b_{N(n)}x)$$

$$\leq \sum_{j=1}^{m} P\left(\left\{\frac{Z_{n(j-1)} - a_{n(j-1)}}{b_{n(j-1)}} < \frac{x+\delta}{1-\delta}\right\} \cap B \cap \{s_{j-1} \leq \tau < s_j\}\right) + 3\varepsilon.$$

An application of Lemma 6.2.2 thus yields

$$-3\varepsilon + H\left(\frac{x-\delta}{1+\delta}\right) P(B \cap \{y_1 \leq \tau < y_2\}) \leq \liminf_{n = +\infty} P(Z_{N(n)} < a_{N(n)} + b_{N(n)}x)$$

$$\leq \limsup_{n = +\infty} P(Z_{N(n)} < a_{N(n)} + b_{N(n)}x) \leq H\left(\frac{x+\delta}{1-\delta}\right) P(B \cap \{y_1 \leq \tau < y_2\}) + 3\varepsilon.$$

In view of the choice of y_1 and y_2,

$$|P(B \cap \{y_1 \leq \tau < y_2\}) - P(B)| \leq P(\tau < y_1 \text{ or } \tau \geq y_2) < \varepsilon.$$

Thus, because $\varepsilon > 0$ and $\delta > 0$ were arbitrary and $H(x)$ is continuous (Section 2.4), the proof is completed. ▲

Lemma 6.2.4. *Under the assumptions of Lemma 6.2.3, as $n \to +\infty$,*

$$\lim \frac{b_n}{b_{N(n)}} = B_\tau, \quad \lim \frac{a_{N(n)} - a_n}{b_{N(n)}} = -A_\tau \tag{9}$$

in probability, where A_t and B_t are defined by the relation

$$H^t(x) = H(A_t + B_t x).$$

Proof. We first prove (9) for the following special sequence $N(n)$, when convergence in probability becomes the convergence of a sequence. Let $t_n > 0$ be a sequence of numbers which converges to t with $0 < t < 1$. Define $N(n)$ as the integer part of nt_n. Then $N(n)/n \to t$. Hence, by (8), as $n \to +\infty$,

$$F^{N(n)}(a_n + b_n x) \to H^t(x).$$

On the other hand, by Lemma 6.2.3,

$$F^{N(n)}(a_{N(n)} + b_{N(n)} x) \to H(x).$$

Lemma 2.2.3 thus yields (9) for this special sequence.

For arbitrary $N(n)$ satisfying our assumptions, we can now proceed as follows. Because there are only three possibilities as limits in (8), a quick check shows that B_t and A_t are continuous and monotonic functions of t (in fact, A_t is either zero or $\log t$ and $B_t = t^s$ with $s = 0$ for $H_{3,0}(x)$). By a standard argument of calculus it follows that, for any point of the probability space for which $s \leq N(n)/n \leq s^*$,

$$B_s - \delta \leq \frac{b_n}{b_{N(n)}} \leq B_{s^*} + \delta$$

for all large n, where $\delta > 0$ is arbitrary. Therefore, choosing again $y_1 < y_2$ with $P(y_1 \leq \tau < y_2) \geq 1 - \varepsilon$ and a division $y_1 = s_0 < s_1 < \cdots < s_m = y_2$ such that

$$|B_{s_i} - B_{s_{i-1}}| < \delta,$$

we get

$$P\left(\left|\frac{b_n}{b_{N(n)}} - B_\tau\right| > 3\delta\right)$$

$$\leq \sum_{i=1}^m P\left(\left|\frac{b_n}{b_{N(n)}} - B_\tau\right| > 3\delta, s_{i-1} \leq \tau < s_i\right) + \varepsilon$$

$$\leq \sum_{i=1}^m P\left(\left|\frac{b_n}{b_{N(n)}} - B_\tau\right| > 3\delta, s_{i-1} \leq \tau < s_i, s_{i-1} \leq \frac{N(n)}{n} < s_i\right)$$

$$+ \sum_{i=1}^m P\left\{s_{i-1} \leq \tau < s_i \text{ but } \frac{N(n)}{n} \notin [s_{i-1}, s_i)\right\} + \varepsilon.$$

6.2 EXTREMES WITH RANDOM SAMPLE SIZE

The terms of the second sum evidently tend to zero as $n\to +\infty$. On the other hand, each term of the first sum becomes zero for all large n (not only in limit). Namely, the monotonicity of B_t and the choice of $s_i, 0 \leq i \leq m$, yield that, for all large n,

$$\left|\frac{b_n}{b_{N(n)}} - B_\tau\right| \leq |B_{s_i} - B_{s_{i-1}}| + 2\delta < 3\delta,$$

whenever $s_{i-1} \leq N(n)/n \leq s_i$. This completes the proof of the first limit in (9). The proof of the second limit is similar, and thus the details are not repeated once more. The lemma is established. ▲

We now turn to the proof of Theorem 6.2.1.

Proof of Theorem 6.2.1. Let $s_0 < s_1 < \cdots < s_m$ be given real numbers. Define the events $D_k = \{s_{k-1} \leq \tau < s_k\}, 1 \leq k \leq m$, and let $D_0 = \{\tau < s_0\}$ and $D_{m+1} = \{\tau \geq s_m\}$. Then, starting with the result of Lemma 6.2.3, we have, for $0 \leq k \leq m+1$,

$$P(\{Z_{N(n)} < a_{N(n)} + b_{N(n)}x\} \cap D_k) \to H(x)P(D_k). \tag{10}$$

Next, we rewrite the fraction

$$\frac{Z_{N(n)} - a_{N(n)}}{b_{N(n)}} = \frac{Z_{N(n)} - a_n}{b_n} \frac{b_n}{b_{N(n)}} + \frac{a_n - a_{N(n)}}{b_{N(n)}}$$

$$= \frac{Z_{N(n)} - a_n}{b_n} B_\tau + A_\tau + \left(\frac{b_n}{b_{N(n)}} - B_\tau\right)\frac{Z_{N(n)} - a_n}{b_n} + \left(\frac{a_n - a_{N(n)}}{b_{N(n)}} - A_\tau\right). \tag{11}$$

By (9), the last term in (11) tends to zero in probability. We now deduce that so does the last but one term. Namely,

$$P\left(\left|\left(\frac{b_n}{b_{N(n)}} - B_\tau\right)\frac{Z_{N(n)} - a_n}{b_n}\right| \geq \varepsilon\right)$$

$$\leq P\left(\left|\frac{Z_{N(n)} - a_n}{b_n}\right| \geq r\right) + P\left(\left|\frac{b_n}{b_{N(n)}} - B_\tau\right| > \frac{\varepsilon}{r}\right),$$

where r is arbitrary. The last term here tends to zero by (9) again, while the first equality in (11), together with Lemma 6.2.3 and with (9), implies

$$P\left(\left|\frac{Z_{N(n)} - a_n}{b_n}\right| \geq r\right) \to 0 \quad \text{as } r \to +\infty.$$

Thus, we can conclude from (10) and Lemma 2.2.1 that, as $n \to +\infty$,

$$\lim P\left(\left\{\frac{Z_{N(n)} - a_n}{b_n} B_\tau + A_\tau < x\right\} \cap D_k\right) = H(x)P(D_k). \qquad (12)$$

For $1 \leqslant k \leqslant m$, let $s_{k-1} \leqslant s(k) \leqslant s_k$ be fixed points. Then by the basic equation for A_t and B_t in Lemma 6.2.4 and by (12), as $n \to +\infty$,

$$P\left(\left\{\frac{Z_{N(n)} - a_n}{b_n} B_\tau + A_\tau < A_{s(k)} + B_{s(k)}x\right\} \cap D_k\right) \to H^{s(k)}(x)P(D_k).$$

Consequently, the continuity of the functions $H(x)$, A_t and $B_t > 0$, imply that, if s_{k-1} and s_k are sufficiently close, then, for $1 \leqslant k \leqslant m$,

$$\left|P\left(\left\{\frac{Z_{N(n)} - a_n}{b_n} < x\right\} \cap D_k\right) - H^{s(k)}(x)P(D_k)\right| < \frac{\varepsilon}{m}$$

for all large n. If we choose s_0 and s_m so that

$$P(D_0) + P(D_{m+1}) < \varepsilon,$$

then, with the choice of s_k required above,

$$P\left(\frac{Z_{N(n)} - a_n}{b_n} < x\right) = \sum_{k=0}^{m+1} P\left(\left\{\frac{Z_{N(n)} - a_n}{b_n} < x\right\} \cap D_k\right)$$

would deviate from

$$\sum_{k=1}^{m} H^{s(k)}(x)P(D_k)$$

by less than 2ε for all large n. But this latter sum is a Riemann sum of the integral

$$\int_{s_0}^{s_m} H^t(x)\,dP(\tau < t).$$

Therefore, for all large n,

$$\left|P(Z_{N(n)} < a_n + b_n x) - \int_{-\infty}^{+\infty} H^t(x)\,dP(\tau < t)\right| < 3\varepsilon.$$

Because $\varepsilon > 0$ is arbitrary, passing to infinity with n, we obtain (2), which was to be proved. ▲

6.2 EXTREMES WITH RANDOM SAMPLE SIZE

The difficulty of the proof is due to the facts that the interrelation of the basic variables $X_1, X_2, \ldots$ and $N(n)$ was not restricted and that a_n and b_n do not contain the random size $N(n)$. Had we been satisfied with a random normalization $a_{N(n)}$ and $b_{N(n)}$, we could have stopped at Lemma 6.2.3. Such a theorem, however, is unsatisfactory, because it does not show clearly the behavior of $Z_{N(n)}$.

When the sequence $X_1, X_2, \ldots$ and $N(n)$ are independent, then a more complete statement is possible about $Z_{N(n)}$. In fact, a family of limit theorems can be combined into one, as described in the following model.

Let $X_{j,n}, 1 \leq j \leq N(n)$, be independent random variables with common distribution function $F_n(x)$. Let $N(n)$ itself be a positive integer-valued random variable which is distributed independently of the $X_{j,n}$.

Let $\nu_n(x)$ be the number of $j \leq N(n)$ such that $\{X_{j,n} \geq x\}$. Then, by the total probability rule,

$$P_n(\nu_n(x) = t) = \sum_{j=1}^{+\infty} \binom{j}{t} [1 - F_n(x)]^t F_n^{j-t}(x) P_n(N(n) = j)$$

$$= \int_0^{+\infty} \binom{y}{t} [1 - F_n(x)]^t F_n^{y-t}(x) dP_n(N(n) < y).$$

We denote by $X_{r:N(n)}^{(n)}$ the rth order statistic of $X_{j,n}, 1 \leq j \leq N(n)$. Now, Theorem 3.4.2.b and the frequently used relation

$$P_n(X_{N(n)-k:N(n)}^{(n)} < x) = \sum_{t=0}^{k} P_n(\nu_n(x) = t),$$

immediately yield the following result.

Theorem 6.2.2. *Let $X_{j,n}, 1 \leq j \leq N(n)$, be independent random variables with common distribution function $F_n(x)$. Let $N(n)$ be a positive integer-valued random variable which is distributed independently of the sequence $X_{j,n}, 1 \leq j \leq N(n)$. Let a_n and $b_n > 0$ be sequences of real numbers such that $F_n(a_n + b_n x) \to 1$ for any x as $n \to +\infty$. Then, for each k,*

$$\lim_{n = +\infty} P_n(X_{N(n)-k:n}^{(n)} < a_n + b_n x) = E_k(x)$$

exists if, and only if, as $n \to +\infty$,

$$\lim P_n \left[N(n) < \frac{u}{1 - F_n(a_n + b_n x)} \right] = U(u; x)$$

exists. The limits $E_k(x)$ and $U(u;x)$ are related by the formula

$$E_k(x) = \sum_{t=0}^{k} \frac{1}{t!} \int_0^{+\infty} u^t e^{-u} dU(u;x). \tag{13}$$

If the sequences a_n and b_n are characteristic to the extremes in the sense of (39) and (40) of Chapter 3, then $U(u;x)$ is a proper distribution function.

Let us evaluate $U(u;x)$ when $F_n = F$ for each n and $N(n)/n \to \tau$, where τ is a positive random variable. Let a_n and b_n be such that, as $n \to +\infty$,

$$n[1 - F(a_n + b_n x)] \to h(x), \tag{14}$$

where $0 < h(x) < +\infty$ on some interval (α, ω) (possibly infinite). Then, on account of Lemma 2.2.1,

$$U(u;x) = \lim_{n=+\infty} P\left\{ \frac{N(n)}{n} < \frac{u}{n[1 - F(a_n + b_n x)]} \right\}$$

$$= \lim_{n=+\infty} P\left(\frac{N(n)}{n} < \frac{u}{h(x)} \right) = P\left(\tau < \frac{u}{h(x)} \right).$$

Thus

$$E_0(x) = \int_0^{+\infty} e^{-u} dU(u;x) = \int_0^{+\infty} e^{-yh(x)} dP(\tau < y),$$

which is the result of Theorem 6.2.1 with $H(x) = e^{-h(x)}$ (recall from Chapter 2 that (14) is equivalent to $(Z_n - a_n)/b_n$ converging weakly to $H(x) = e^{-h(x)}$). If $P(\tau = 1) = 1$, we then, of course, get back $E_0(x) = H(x)$.

6.3. RECORD TIMES

We turn to the investigation of a specific sequence $N(n), n \geq 1$, of random sample size and of the corresponding random maximum $Z_{N(n)}$. Let $X_1, X_2, \ldots$ be independent random variables with common continuous distribution function $F(x)$. Let $N(1) = 1$ and, for $n \geq 2$, let

$$N(n) = \min\{ j : j > N(n-1), X_j > X_{N(n-1)} \}. \tag{15}$$

It is immediate that $P(N(n) < +\infty) = 1$, and thus the sequence $p_{k,n} = P(N(n) = k)$ is a proper distribution for each n.

Remark 6.3.1. The sequences $N(n)$ and $X_{N(n)}$ can be interpreted as follows. Consider an infinite sequence $X_1, X_2, \ldots$ of i.i.d. random variables whose distribution function is $F(x)$ (which is assumed to be continuous). Then let us go through the sequence $X_1, X_2, \ldots$ with the aim of picking out larger and larger terms. Obviously, the first largest is X_1. Then, for any m, if $Z_m = X_1$, we ignore $X_2, \ldots, X_m$, and we take that X as the next one, i.e. $X_{N(2)}$, when, for the first time, $Z_m > X_1$. We then continue the process. In other words, the investigation of $N(n)$ gives an insight into the positions of those observations that change Z_m (by the assumption of continuity, ties can be neglected). The values $X_{N(n)} = Z_{N(n)}$ are thus the increasing values $Z_1 < Z_{N(2)} < \cdots$.

As an example, let us consider a concrete case. Let X_j be the amount of water added to a given river at spring of the jth year by the melting of snow. Since the times are a year apart, we can assume that the X_j are i.i.d. If records have been kept since 1900, say, then the amount of water measured in the above manner in 1900 was $X_1 = X_{N(1)}$. If the records show that up to 1936 this amount of water was always less than in 1900, but in 1936 the melting of snow resulted in a big flood, then $N(2) = 37$, etc.

The theorems that follow will reveal that the sequence Z_m changes very rarely as m increases.

We shall use the following terms for the sequences $N(n)$ and $X_{N(n)}$.

Definition 6.3.1. The sequence $N(n), n \geq 1$, defined at (15) is called the sequence of record times. The corresponding X-value, that is, $X_{N(n)} = Z_{N(n)}$, is called a record.

One could define records and record times by reversing the inequality in (15). In such cases, we speak of lower records, and, for comparison, the previous definition of records is termed upper records. Because we deal with upper records only (the theory would be the same for lower records), we drop this qualification and use the concept records and record times as specified in Definition 6.3.1.

Lemma 6.3.1. *The value of $N(n)$ does not depend on $F(x)$.*

Proof. The lemma is evident; it is formulated for easier reference only. As a matter of fact, if $X_j \geq X_t$ then $F(X_j) \geq F(X_t)$. But the sequence $F(X_j)$ is a sequence of independent uniform variates. Hence, for arbitrary (continuous) $F(x)$, $N(n)$ can be defined in (15) by the additional assumption that the variables X_j are independent and uniformly distributed on $(0, 1)$. The lemma is established. ▲

Theorem 6.3.1. *The distribution of $N(2)$ is given by*

$$P(N(2) = j) = \frac{1}{j(j-1)}, \quad j \geq 2.$$

Consequently, $E[N(n)] = +\infty$ for $n \geq 2$.

Proof. Let $X_1, X_2, \ldots$ be independent random variables with uniform distribution on the interval $(0, 1)$. Then, by the continuous version of the total probability rule (Appendix I), for $j \geq 2$,

$$P(N(2) = j) = \int_0^1 P(N(2) = j | X_1 = x) dx$$

$$= \int_0^1 x^{j-2}(1-x) dx$$

$$= \frac{1}{j-1} - \frac{1}{j} = \frac{1}{j(j-1)}.$$

Hence,

$$E[N(2)] = \sum_{j=2}^{+\infty} j P(N(2) = j) = +\infty.$$

Because $N(n) \geq N(2)$, $n \geq 2$, the proof is completed. ▲

Notice the meaning of Theorem 6.3.1. If a disaster has been recorded by the value of X_1, then both of the following statements are valid. The value X_2 that will bring an even larger disaster has a probability $1/2$, but the actual expected waiting time to a larger disaster is infinity.

Theorem 6.3.2. *The sequence $N(n)$, $n \geq 2$, forms a Markov chain. That is,*

$$P(N(n) = k | N(t) = j_t, 2 \leq t < n) = P(N(n) = k | N(n-1) = j_{n-1})$$

for all vectors $(j_2, \ldots, j_{n-1})$ for which the condition of the left hand side has positive probability. The transitions

$$P(N(n) = k | N(n-1) = j) = \frac{j}{k(k-1)} \quad \text{for } k > j \geq n-1 \geq 2,$$

and the conditional probabilities above are equal to zero for any other values of j and k.

Proof. We again appeal to Lemma 6.3.1 and consider the sequence $X_1, X_2, \ldots$, which are i.i.d. with common distribution function $F(x) = x$ for $0 \leq x \leq 1$. For uniformity of notation, put $j_1 = 1$. Thus, by the total probability rule, for $j_t > j_{t-1}, t \geq 2$,

$$P(N(t) = j_t, 2 \leq t \leq n)$$

$$= \int_0^1 \cdots \int_0^1 P(N(t) = j_t, 2 \leq t \leq n | X_{j_t} = x_t, 1 \leq t \leq n-1) dx_1 \cdots dx_{n-1}$$

$$= \int \cdots \int_{0 < x_1 < \cdots < x_{n-1} < 1} x_1^{j_2 - 2} x_2^{j_3 - j_2 - 1} \cdots x_{n-1}^{j_n - j_{n-1} - 1} (1 - x_{n-1}) dx_1 \cdots dx_{n-1}$$

$$= [(j_2 - 1)(j_3 - 1) \cdots (j_{n-1} - 1)(j_n - 1) j_n]^{-1}.$$

Hence,

$$P(N(n) = k | N(t) = j_t, 2 \leq t < n) = \frac{P(N(n) = k, N(t) = j_t, 2 \leq t < n)}{P(N(t) = j_t, 2 \leq t < n)}$$

$$= \frac{j_{n-1}}{j_n(j_n - 1)}, \quad j_t > j_{t-1}, t \geq 2,$$

from which the statement of the theorem is immediate. ▲

From Theorems 6.3.1 and 6.3.2, the distribution of $N(n)$ can be determined by the total probability rule. For example,

$$P(N(3) = k) = \sum_{j=2}^{+\infty} P(N(3) = k | N(2) = j) P(N(2) = j)$$

$$= \sum_{j=2}^{k-1} \frac{j}{k(k-1)} \frac{1}{j(j-1)} = \frac{1}{k(k-1)} \sum_{t=1}^{k-2} \frac{1}{t}.$$

Theorem 6.3.3. *For $n \geq 2$, let $M(n) = N(n)/N(n-1)$. Let us define the integers $T(n)$ by the inequalities*

$$T(n) - 1 < M(n) \leq T(n). \tag{16}$$

Then the random variables $T(n), n \geq 2$, are i.i.d. with

$$P(T(n) = j) = \frac{1}{j(j-1)}, \quad j \geq 2.$$

Proof. In view of (16), $\{T(n) > j\} = \{M(n) > j\}$ for any integer. Thus, for integers $j_k \geq 1$,

$$P(T(k)>j_k, 2 \leq k \leq n) = P(M(k)>j_k, 2 \leq k \leq n)$$
$$= P(N(k)>j_k N(k-1), 2 \leq k \leq n). \quad (17)$$

If we decompose the extreme right hand side of (17) as

$$\sum P(N(k)=t_k, 2 \leq k \leq n-1, N(n)>j_n t_{n-1}), \quad (18)$$

where the summation is over all $(t_2, t_3, \ldots, t_{n-1})$ such that $t_2 > j_2$ and $t_k > j_k t_{k-1}, 3 \leq k \leq n-1$, we can arrive at a recursive formula for the extreme left hand side of (17) by applying Theorem 6.3.2 to the general term of (18). We have

$$P(N(k)=t_k, 2 \leq k \leq n-1, N(n)>j_n t_{n-1})$$
$$= \sum_{s=j_n t_{n-1}+1}^{+\infty} P(N(k)=t_k, 2 \leq k \leq n-1, N(n)=s)$$
$$= P(N(k)=t_k, 2 \leq k \leq n-1) \sum_{s=j_n t_{n-1}+1}^{+\infty} \frac{t_{n-1}}{s(s-1)}$$
$$= \frac{P(N(k)=t_k, 2 \leq k \leq n-1)}{j_n}.$$

If we substitute this into (18), (17) yields

$$P(T(k)>j_k, 2 \leq k \leq n) = \frac{1}{j_n} P(T(k)>j_k, 2 \leq k \leq n-1).$$

Hence, by induction,

$$P(T(k)>j_k, 2 \leq k \leq n) = \frac{1}{j_2 j_3 \cdots j_n}. \quad (19)$$

The special case $j_t = 1$ for $t \neq k$, $2 \leq t \leq n$, gives

$$P(T(k)>j_k) = \frac{1}{j_k}, \quad j_k \geq 1. \quad (20)$$

The expressions (19) and (20) are equivalent to what was to be proved. ▲

A number of interesting results follow from Theorem 6.3.3.

Corollary 6.3.1. *Let $\Delta(n) = N(n) - N(n-1), n \geqslant 2$. Then*

$$E\left[\frac{\Delta(n)}{N(n-1)}\right] = +\infty.$$

Proof. Because

$$\frac{\Delta(n)}{N(n-1)} = M(n) - 1 > T(n) - 2, \tag{21}$$

Theorem 6.3.3 immediately yields the statement. ▲

Corollary 6.3.2. *With probability one, as $n \to +\infty$,*

$$\limsup \frac{\log \Delta(n) - \log N(n-1)}{\log n} = 1. \tag{22}$$

Proof. We start again with (21). Because, for integer j, $\{M(n) > j\} = \{T(n) > j\}$,

$$P(M(n) > j(n) \text{ infinitely often}) = P(T(n) > j(n) \text{ infinitely often}),$$

for any sequence $j(n)$ of integers. But, because the variables $T(n)$ are completely independent (Theorem 6.3.3), the Borel-Cantelli lemmas of Section 4.2 (Lemmas 4.2.1 and 4.2.3) are applicable. Since

$$\sum_{n=2}^{+\infty} P(T(n) > n) = +\infty, \quad \sum_{n=2}^{+\infty} P(T(n) > n(\log n)^2) < +\infty,$$

we have, with probability one,

$$T(n) > n \text{ infinitely often and } T(n) \leqslant n(\log n)^2 \text{ ultimately}.$$

A change of $n(\log n)^2$ into its integer part has no effect on the upper inequalities; therefore these same inequalities hold for $M(n)$. But these inequalities are somewhat stronger than the fact expressed in (22). The corollary is proved. ▲

In order to determine the liminf of the fraction occurring in (22), we establish a simple inequality.

Lemma 6.3.2. *For $n \geqslant 3$, and for any real number $s > 0$,*

$$P(M(n) \leqslant 1 + s) \leqslant 2s.$$

Proof. By the total probability rule and by the fact that $N(n) \geq n$,

$$P(M(n) \leq 1+s) = \sum_{j=n-1}^{+\infty} P(M(n) \leq 1+s | N(n-1)=j) P(N(n-1)=j)$$

$$= \sum_{j=n-1}^{+\infty} \sum_{k=j+1}^{[(1+s)j]} P(N(n)=k | N(n-1)=j) P(N(n-1)=j),$$

where $[y]$ signifies the integer part of y. With an appeal to Theorem 6.3.2, we thus get

$$P(M(n) \leq 1+s) = \sum_{j=n-1}^{+\infty} jP(N(n-1)=j) \left\{ \frac{1}{j} - \frac{1}{[(1+s)j]} \right\}.$$

It remains to observe that

$$j \left\{ \frac{1}{j} - \frac{1}{[(1+s)j]} \right\} \leq \frac{(1+s)j-j}{(1+s)j-1} = \frac{s}{1+s-1/j}$$

$$\leq \frac{s}{1+s-1/(n-1)} \leq 2s,$$

the last but one step being justified by $j \geq n-1$. The proof is completed. ▲

Corollary 6.3.3. *With probability one, as $n \to +\infty$,*

$$\limsup \frac{|\log \Delta(n) - \log N(n-1)|}{\log n} = 1. \tag{23}$$

Proof. By (21) and by Lemma 6.3.2,

$$P\left(\frac{\Delta(n)}{N(n-1)} \leq \frac{1}{n(\log n)^2} \right) \leq \frac{2}{n(\log n)^2}, \quad n \geq 3.$$

If we sum the above inequality for $n \geq 3$, we get a convergent series. In view of Lemma 4.2.1,

$$P\left(\frac{\Delta(n)}{N(n-1)} > \frac{1}{n(\log n)^2} \text{ ultimately} \right) = 1.$$

Hence, with probability one,

$$\liminf_{n=+\infty} \frac{\log \Delta(n) - \log N(n-1)}{\log n} \geq -1.$$

On account of (22), this suffices for (23). The corollary is established. ▲

The relation (23) will be a basic tool to show that similar limit relations hold for $\log \Delta(n)$ and $\log N(n-1)$. For investigating the latter, we need the following elementary lemma.

Lemma 6.3.3. *Let the random variables e_k, $k \geq 2$, be defined as follows. Let $e_k = 1$ if there is an $n \geq 2$ such that $N(n) = k$, and $e_k = 0$ otherwise. Then $e_2, e_3, \ldots$ are independent with*

$$P(e_k = 1) = \frac{1}{k}.$$

Proof. Let us first consider the event $e_k = 1$. This means that $X_j < X_k$ for all $j < k$. Thus, assuming that the basic random variables $X_1, X_2, \ldots$ are uniformly distributed on $(0, 1)$ (see Lemma 6.3.1),

$$P(e_k = 1) = \int_0^1 P(X_j < X_k, 1 \leq j < k | X_k = x) \, dx$$

$$= \int_0^1 x^{k-1} \, dx$$

$$= \frac{1}{k}.$$

Let now $2 \leq k_1 < k_2 < \cdots < k_m$ be integers. Then the event $e_{k_t} = 1$, $1 \leq t \leq m$, means that $X_j < X_{k_t}$ whenever $j < k_t$. Hence

$$P(e_{k_t} = 1, 1 \leq t \leq m)$$

$$= \int_0^1 \cdots \int_0^1 P(e_{k_t} = 1, 1 \leq t \leq m | X_{k_t} = x_t, 1 \leq t \leq m) \, dx_1 \cdots dx_t$$

$$= \int \cdots \int_{0 < x_1 < \cdots < x_t < 1} x_1^{k_1 - 1} x_2^{k_2 - k_1 - 1} \cdots x_t^{k_t - k_{t-1} - 1} \, dx_1 \, dx_2 \cdots dx_t$$

$$= \frac{1}{k_1 k_2 \cdots k_t}.$$

The proof is completed. ▲

In view of Lemma 6.3.3, the relation

$$P(N(n) > t) = P(e_1 + e_2 + \cdots + e_t < n) \tag{24}$$

and elementary probability theory give a large number of limit theorems for $N(n)$. As a matter of fact, since

$$E(e_k) = \frac{1}{k}, \qquad V(e_k) = \frac{1}{k}\left(1 - \frac{1}{k}\right)$$

and

$$\sum_{k=1}^{n} \frac{1}{k} \sim \log n \qquad \text{as } n \to +\infty,$$

we have, as $t \to +\infty$,

$$P\left(\lim \frac{e_1 + e_2 + \cdots + e_t}{\log t} = 1\right) = 1, \tag{25}$$

$$P\left(\limsup \frac{e_1 + e_2 + \cdots + e_t - \log t}{(2\log t \log\log\log t)^{1/2}} = 1\right) = 1, \tag{26}$$

and

$$\lim P\left(e_1 + e_2 + \cdots + e_t - \log t < x\sqrt{\log t}\,\right) = \Phi(x), \tag{27}$$

where $\Phi(x)$ is the standard normal distribution. Furthermore, (26) can be supplemented by taking the lim inf of the fraction there, which becomes (-1). (The limits (25) and (27) are routine in elementary probability theory, and (26) is well known, although its proof is somewhat complicated—see Appendix I.) We thus have the following results.

Theorem 6.3.4. *With probability one, as $n \to +\infty$,*

$$\lim \frac{\log N(n)}{n} = \lim \frac{\log \Delta(n)}{n} = 1.$$

Proof. By (24) and (25), with probability one, ultimately,

$$(1-\varepsilon)n < \log N(n) < (1+\varepsilon)n,$$

where $\varepsilon > 0$ is arbitrary. We thus have the limit for $N(n)$. This, in turn, yields the stated limit for $\Delta(n)$ in view of (23). The proof is completed. ▲

Theorem 6.3.5. *As $n \to +\infty$,*

$$\lim P\left(\frac{\log N(n) - n}{\sqrt{n}} < x\right) = \lim P\left(\frac{\log \Delta(n) - n}{\sqrt{n}} < x\right) = \Phi(x).$$

Proof. It again suffices to prove the statement for $N(n)$—namely,

$$\frac{\log \Delta(n) - n}{\sqrt{n}} = \frac{\log \Delta(n) - \log N(n-1)}{\sqrt{n}} + \frac{\log N(n-1) - n}{\sqrt{n}}$$

where the first term on the right hand side tends to zero on account of (23). Therefore, by Lemma 2.2.1, this term has no effect on an asymptotic distribution.

On the other hand, for proving the stated limit relation for $N(n)$, choose t in (24) as the integer part of $\exp(n + x\sqrt{n})$. Then, as $n \to +\infty$,

$$n = \log t - x\sqrt{\log t} + O\left[(\log t)^{-1/2}\right].$$

Therefore, (27) and Lemma 2.2.1 give

$$\lim_{n = +\infty} P\left(\log N(n) > n + x\sqrt{n}\right) = \Phi(-x) = 1 - \Phi(x),$$

as stated. The proof is completed. ▲

A similar simple transformation of (26) results in

$$\limsup_{n = +\infty} \frac{\log N(n) - n}{(2n \log \log n)^{1/2}} = 1,$$

which is valid with probability one. In view of Corollary 6.3.3, the same relation holds for $\log \Delta(n)$ as well.

6.4. RECORDS

We now turn to the investigation of records as defined in Definition 6.3.1. Throughout this section it is assumed that $X_1, X_2, \ldots$ are independent random variables with common continuous distribution function $F(x)$. We denote by $N(n)$, $n \geq 1$, the record times and by $X_{N(n)}$ the records themselves.

In view of the results on $N(n)$ in the previous section, $X_{N(n)}$ is rarely observed with large n. Therefore, an asymptotic theory of $X_{N(n)}$ is not of much value to the applied scientist. However, this theory is interesting from the mathematical point of view, which justifies its detailed discussion.

Clearly from its definition, $X_{N(n)} = Z_{N(n)}$. Hence, one would think that when it converges, $(X_{N(n)} - A_n)/B_n$, where A_n and $B_n > 0$ are suitable constants, converges weakly to a mixture of distributions as at (2). The surprising result is that this is not the case. In sharp contrast to Section 6.2, the limiting distribution of $(X_{N(n)} - A_n)/B_n$ is closely related to the normal distribution. The reason for this is best demonstrated if we consider first a special case: $F(x)=1-e^{-x}, x>0$. For this distribution the following result holds.

Theorem 6.4.1. *Let* $F(x)=1-e^{-x}$, $x>0$. *Let* $X_{N(n)}, n\geq 1$, *be the records and let* $Y_1=X_{N(1)}=X_1, Y_j=X_{N(j)}-X_{N(j-1)}, j\geq 2$. *Then* $Y_1, Y_2,\ldots$ *are i.i.d. random variables and their common distribution function is* $F(x)$ *itself.*

Proof. Let us first evaluate the joint distribution of Y_1 and Y_2. We apply the decomposition

$$P(Y_1<x, Y_2<y) = \sum_{j=2}^{+\infty} P(Y_1<x, Y_2<y, N(2)=j)$$

$$= \sum_{j=2}^{+\infty} P(X_1<x, X_j-X_1<y, X_j \geq X_1, X_t<X_1, 2\leq t<j).$$

The general term on the right hand side can easily be evaluated by the formula

$$P(A) = \int_0^{+\infty} P(A|X_1=z) e^{-z} dz, \qquad (28)$$

where A is an arbitrary event (see Appendix I). When conditioned on $X_1=z$, the mentioned term becomes zero if $x<z$ and, for $x>z$, it is

$$[F(y+z)-F(z)]F^{j-2}(z) = e^{-z}(1-e^{-y})(1-e^{-z})^{j-2}.$$

Hence,

$$P(Y<x, Y_2<y) = \sum_{j=2}^{+\infty} \int_0^x e^{-z}(1-e^{-y})(1-e^{-z})^{j-2} e^{-z} dz$$

$$= (1-e^{-y}) \int_0^x \sum_{j=2}^{+\infty} (1-e^{-z})^{j-2} e^{-2z} dz$$

$$= (1-e^{-y}) \int_0^x e^{-z} dz$$

$$= (1-e^{-y})(1-e^{-x}) = F(x)F(y).$$

We thus have proved the theorem for Y_1 and Y_2. The proof in the general case is similar, only the notations become more complicated. We again decompose

$$P(Y_t<y_t, 1\leq t\leq n) = \sum P(Y_t<y_t, N(t)=j_t, 1\leq t\leq n),$$

where the summation is for all integers $1=j_1<j_2<\cdots<j_n$. We then apply the formula (28) in a vector form, namely, conditioning will be on $\{X_{j_t}=z_t, 1\leq t<n\}$ with $0<z_1<z_2<\cdots<z_{n-1}$. Under this condition, the general term on the right hand side above becomes zero if $y_1<z_1$ or if $y_t+z_{t-1}<z_t$. Otherwise, it equals

$$\prod_{t=2}^{n}[F(y_t+z_{t-1})-F(z_{t-1})]F^{j_t-j_{t-1}-1}(z_{t-1})$$

$$=\prod_{t=2}^{n}(1-e^{-y_t})\prod_{t=2}^{n}F^{j_t-j_{t-1}-1}(z_{t-1})\exp(-z_{t-1}).$$

If we sum now these terms for $1=j_1<j_2<\cdots<j_n$ and then multiply by the density $\exp(-z_1-z_2-\cdots-z_{n-1})$ and integrate over $0<z_1<z_2<\cdots<z_{n-1}$ and $z_t<z_{t-1}+y_t, 2\leq t\leq n-1$, we get the desired result. The proof is completed. ▲

Using the result of Theorem 6.4.1, we get that, for unit exponential variates,

$$X_{N(n)} = Y_1+Y_2+\cdots+Y_n,$$

where the Y_j are i.i.d. random variables. The classical theory of probability theory thus gives the asymptotic distribution of, and almost sure results for, $X_{N(n)}$, when suitably normalized. For example, as $n\to+\infty$,

$$\lim P\left(\frac{X_{N(n)}-n}{\sqrt{n}}<x\right)=\Phi(x). \tag{29}$$

While Theorem 6.4.1 is a characteristic property of the exponential distribution (see Exercise 7), its conclusion is a basic tool for arbitrary population distributions. Namely, if $F(x)$ is an arbitrary continuous distribution function, then the random variables

$$X_j^* = -\log[1-F(X_j)], \quad j\geq 1,$$

are i.i.d. with unit exponential distribution. Furthermore, the above transformation transforms a record among the X_j into a record among X_j^*. More precisely, if $U_F(x)=-\log[1-F(x)]$, then $X_{N(n)}^* = U_F(X_{N(n)})$. We thus have the following general result.

Corollary 6.4.1. *Let $F(x)$ be an arbitrary continuous distribution. Then there are constants A_n and $B_n > 0$ such that, as $n \to +\infty$,*

$$\lim P(X_{N(n)} < A_n + B_n x) = T(x)$$

exists and it is nondegenerate if, and only if,

$$\lim_{n = +\infty} \frac{U_F(A_n + B_n x) - n}{\sqrt{n}} = g(x) \tag{30}$$

exists and is finite on an interval, where $g(x)$ has at least two points of increase. When it exists, $T(x) = \Phi[g(x)]$.

Proof. Because of the relation of $X^*_{N(n)}$ and $X_{N(n)}$, we have

$$\{X_{N(n)} < A_n + B_n x\} = \{X^*_{N(n)} < U_F(A_n + B_n x)\}$$

$$= \left\{ \frac{X^*_{N(n)} - n}{\sqrt{n}} < \frac{U_F(A_n + B_n x) - n}{\sqrt{n}} \right\}.$$

Applying (29) to $X^*_{N(n)}$, we obtain the corollary. ▲

The limit $g(x)$ in (30) can have only three forms.

Theorem 6.4.2. *The limit $g(x)$ in (30) is one of the following three functions (i) $g(x) = x$ for all x, (ii) $g(x) = \gamma \log x$ with some $\gamma > 0$ for $x > 0$ and $g(x) = -\infty$ for $x < 0$, and, finally, (iii) $g(x) = +\infty$ for $x > 0$ and $g(x) = -\gamma \log(-x)$ for $x < 0$, where $\gamma > 0$ is an arbitrary constant.*

Proof. It follows from (30) that, as $n \to +\infty$,

$$U_F(A_n + B_n x) \sim n. \tag{31}$$

Next, using the elementary identity $a^2 - b^2 = (a-b)(a+b)$, we observe that, since

$$\frac{U_F(A_n + B_n x) - n}{\sqrt{n}} = \left[U_F^{1/2}(A_n + B_n x) - \sqrt{n} \right] \frac{U_F^{1/2}(A_n + B_n x) + \sqrt{n}}{\sqrt{n}}$$

converges to $g(x)$, and since the last fraction, by (31), converges to 2,

$$\lim_{n = +\infty} U_F^{1/2}(A_n + B_n x) - \sqrt{n} = \tfrac{1}{2} g(x). \tag{32}$$

Let $M = M(n)$ be the closest integer to $\exp(\sqrt{n})$. Then (32) can be rewritten as

$$\lim_{M=+\infty} M[1-F^*(A_M^* + B_M^* x)] = \exp[-\tfrac{1}{2} g(x)], \qquad (33)$$

where

$$F^*(x) = 1 - \exp[-U_F^{1/2}(x)] \qquad (34)$$

and $A_M^* = A_n$, $B_M^* = B_n$, where M and n uniquely determine each other by the definition of $M = M(n)$. We now apply Corollary 1.3.1 and conclude that $F^*(y)$ of (34) is in the domain of attraction of one of the limiting distributions $H(x)$ for the maxima and that $\exp[-\tfrac{1}{2} g(x)]$ is necessarily $-\log H(x)$. It thus follows that apart from a linear transformation, $g(x)$ equals $-\log[-\log H(x)]$, where $H(x)$ is one of the functions $H_{1,\gamma}(x)$, $H_{2,\gamma}(x)$, and $H_{3,0}(x)$. This completes the proof. ▲

As a side result, we also obtained that the type of $g(x)$ and thus of $T(x)$ in Corollary 6.4.1 depends on the distribution function (34). Let us call $\Phi(\gamma \log x)$, $x > 0$, $\gamma > 0$, positive lognormal and $\Phi[-\gamma \log(-x)]$, $x > 0, \gamma > 0$, negative lognormal distribution. Then the limiting distribution $T(x) = \Phi[g(x)]$ is normal, positive lognormal, or negative lognormal according as $F^*(x)$ of (34) is in the domain of attraction of $H_{3,0}(x)$, $H_{1,\gamma}(x)$, or $H_{2,\gamma}(x)$, respectively.

From the dependence of $X_{N(n)}$ on $F^*(x)$ of (34), weak laws of large numbers (both additive and multiplicative) can be deduced. This relation can also be used to obtain almost sure results. For these, the technique of Chapter 4 suffices.

If one compares the results on the logarithm $\log N(n)$ of record times in the preceding section with those on the records $X_{N(n)}$ themselves under the additional assumption that $F(x) = 1 - e^{-x}$, $x > 0$ (Theorem 6.4.1 and (29)), a striking similarity is observed. This is not an accident; rather, the following result is true.

Theorem 6.4.3. *Let the population distribution* $F(x) = 1 - e^{-x}$, $x > 0$. *Then, with probability one,*

$$\limsup_{n=+\infty} \frac{|\log N(n) - X_{N(n)}|}{\log n} = 1.$$

Because we have analyzed the two sequences $\log N(n)$ and $X_{N(n)}$ separately and the approach applied in each case was very straightforward, we do not prove this theorem in detail. The basis of establishing the relation above is the property that the random variables $N(n)$ are conditionally independent, given the sequence $X_{N(k)}, k \geq 1$. The conditional distribution of $N(n)$

$$P(N(n)>t|X_{N(k)},k\geqslant 1)=[1-\exp(-X_{N(n)})]^t.$$

From this fact, one gets at once by a Borel-Cantelli type of argument that

$$\limsup_{n=+\infty}\frac{|\log[N(n)\exp(-X_{N(n)})]|}{\log n}=1$$

almost surely, given the sequence $\{X_{N(k)},k\geqslant 1\}$. But, since the lim sup is a constant, the same result is obtained if the condition is dropped.

6.5. EXTREMAL PROCESSES

Deviating from the basic assumption of the present book, we shall devote this last section to a special continuous time stochastic process called an extremal process. However, these processes are obtained directly from a sequence of extremes which warrants the inclusion of their theory in the present treatment.

Let $X_1, X_2, \ldots$ be a sequence of random variables and, as usual, put $Z_n = \max(X_1, X_2, \ldots, X_n)$. Let a_n and $b_n > 0$ be sequences of real numbers such that $(Z_n - a_n)/b_n$ converges weakly to a nondegenerate distribution $H(x)$. Let now $t > 0$ and define ($[y]$ signifies the integer part of y)

$$z_n(t) = \frac{Z_{[nt]} - a_n}{b_n} \quad \text{if } t \geqslant \frac{1}{n},$$

$$= \frac{X_1 - a_n}{b_n} \quad \text{if } 0 < t < \frac{1}{n}.$$

We call $z_n(t)$ a partial maxima process. When it exists, a limit $z(t)$ of $z_n(t)$ is called an extremal process, where the limit is in the following sense.

Definition 6.5.1. We say that $z_n(t)$ converges to $z(t)$ if, for any finite number k of reals $0 < t_1 < t_2 < \cdots < t_k$, as $n \to +\infty$,

$$\lim P(z_n(t_j) < x_j, 1 \leqslant j \leqslant k) = P(z(t_j) < x_j, 1 \leqslant j \leqslant k),$$

whenever the right hand side is continuous at x_j, $1 \leqslant j \leqslant k$.

Notice that, by assumption, the distribution of $z_n(1)$ always converges to $H(x)$. Thus, for any extremal process, $P(z(1) < x) = H(x)$. The so-called marginal distributions $P(z(t) < x)$ can also be determined for all structures investigated in Chapter 3. However, since our aim is to lay down the foundations of the theory, we restrict ourselves to independent variables. For several models, no change is required to carry out the corresponding details.

Let first $X_1, X_2,\ldots$ be i.i.d. random variables with distribution function $F(x)$. Then, because, for large n, a fixed t exceeds $1/n$ and the ratio $[nt]/nt$ approaches one, we get

$$P(z_n(t)<x)\to H^t(x).$$

That is, the marginal distribution $P(z(t)<x)=H^t(x)$. With the same argument, for $0<t_1<t_2<\cdots<t_k$, we get the distributions

$$P(z(t_j)<x_j, 1\leqslant j\leqslant k)=H^{t_1}(y_1)H^{t_2-t_1}(y_2)\cdots H^{t_k-t_{k-1}}(y_k), \quad (35)$$

where $y_j=\min(x_j, x_{j+1},\ldots,x_k)$. Indeed, putting $s_j=[nt_j]$, then

$$P(z_n(t_j)<x_j, 1\leqslant j\leqslant k)$$
$$=F^{s_1}(a_n+b_n y_1)F^{s_2-s_1}(a_n+b_n y_2)\cdots F^{s_k-s_{k-1}}(a_n+b_n y_k).$$

From this, the limiting form is immediate on account of $s_j/n\to t_j$ and $F^n(a_n+b_n y)\to H(y)$.

Now the question arises whether in fact there is a stochastic process $\{z(t), t>0\}$, whose marginal distributions are those given in (35). The answer is yes. Its existence can be proved by an appeal to the general theory of stochastic processes. It can also be established in a constructive manner, which course we shall follow.

Before we construct a representative of $z(t)$, $t>0$, let us study further its basic properties. It will make clearer the reason for our method of construction.

In view of (35), there are three types of extremal processes $z(t)$; that is, there are three types of $H(x)$ (Section 2.4). In the discussion below, we choose $H(x)=H_{3,0}(x)$. All statements can be transformed to an equivalent form for the other two types.

Because for $H(x)=H_{3,0}(x)$, $H^t(x)=H(x-\log t)$, the bivariate marginals of $z(t)$ are

$$H(x_1, x_2)=H(\min(x_1, x_2)-\log t_1]H[x_2-\log(t_2-t_1)].$$

This is a bivariate extreme value distribution (see Section 5.4—in particular, formulas (28a)–(32) there). If we introduce the new variables $u_1=x_1\log t_1$ and $u_2=x_2\log t_2$, we can write $H(u_1, u_2)$ in the form

$$H(u_1, u_2)=[H_{3,0}(u_1)H_{3,0}(u_2)]^{v(u_2-u_1)}$$

where

$$v(r) = \frac{1 - t_1/t_2 + \max(e^r, t_1/t_2)}{1 + e^r}.$$

In this form, one immediately obtains from Exercise 29 of Chapter 5 for the correlation coefficient $\rho(t_1, t_2)$ of $z(t_1)$ and $z(t_2)$,

$$\rho(t_1, t_2) = -\frac{6}{\pi^2} \int_0^{t_1/t_2} \frac{\log r}{1 - r} dr.$$

Therefore, the process, $z(t), t > 0$, is continuous and integrable in mean square.

We return to the general case; that is, $H(x)$ is any of the three possible types.

The process $z(t), t > 0$, is a Markov process with transition probabilities

$$P(z(t_2) < y | z(t_1) = x) = H^{t_2 - t_1}(y) \quad \text{if } x \leq y$$

and it equals zero for $x > y$. One immediately gets from this that $z(t)$ is a step function. We now show that, for $0 < t_1 < t_2$,

$$P(z(t_1) < z(t_2)) = 1 - \frac{t_1}{t_2}.$$

We prove this via the definition of $z(t)$. Namely, we start with a sequence $X_1, X_2, \ldots$ of i.i.d. random variables with common distribution function $F(x)$, evaluate $P(z_n(t_1) < z_n(t_2))$, and let $n \to +\infty$. For evaluating this latter probability, we observe that $z_n(t_1) < z_n(t_2)$ if, and only if, an integer k with $[nt_1] < k \leq [nt_2]$ is a record time. Let the random variable $e_k = 1$ or 0 according as $X_k = Z_k$, or not. Lemma 6.3.3 tells us that the $\{e_k\}$ are independent with $P(e_k = 1) = 1/k$. Therefore,

$$P(z_n(t_1) < z_n(t_2)) = P\left[\sum_{k=[nt_1]+1}^{[nt_2]} e_k \geq 1\right]$$

$$= 1 - P(e_k = 0 \text{ for } [nt_1] < k \leq [nt_2])$$

$$= 1 - \prod_{k=[nt_1]+1}^{[nt_2]} \left(1 - \frac{1}{k}\right),$$

6.5 EXTREMAL PROCESSES

which is easily seen to tend to $1-t_1/t_2$ by applying the relations

$$\sum_{k=T_1}^{T_2} \frac{1}{k} \sim \log\left(\frac{T_2}{T_1}\right) \quad \text{as } T_1, T_2 \to +\infty$$

and

$$1-x \sim e^{-x} \quad \text{as } x \to 0.$$

Because of the independence of the e_k, it is clear from the above discussion that, for $0 < t_1 < t_2 < \cdots < t_k$, the events $\{z(t_j) > z(t_{j-1})\}$, $2 \leq j \leq k$, are independent, the jth term having probability $1 - t_{j-1}/t_j$. Let us choose $t_j = T_1 + (T_2 - T_1)(j-1)/k$, $1 \leq j \leq k$, where $T_1 < T_2$ are fixed positive real numbers. Then the number ν_k of $1 \leq j \leq k$ for which $z(t_j) > z(t_{j-1})$ tends to the number of discontinuities of $z(t)$ in (T_1, T_2) as $k \to +\infty$, which limit can also be infinite. On the other hand, ν_k is the sum of the indicators I_j of the events $\{z(t_j) > z(t_{j-1})\}$, which indicators, as was just shown, are independent. By construction, as $k \to +\infty$,

$$\max_{1 \leq j \leq k} P(I_j = 1) \to 0, \quad \sum_{j=1}^{k} P(I_j = 1) \to \log \frac{T_2}{T_1}.$$

Therefore, as $k \to +\infty$,

$$\lim P(\nu_k = m) = \frac{T_1}{T_2} \frac{[\log(T_2/T_1)]^m}{m!}.$$

Summarizing, we established that the number of jumps in any finite interval (T_1, T_2) is asymptotically a Poisson variate with parameter $\log(T_2/T_1)$. Furthermore, a similar argument leads to the following result. If $\tau_1 > \tau_2 > \cdots$ are the successive points of discontinuity of $z(t)$ in $0 < t < T$, counting the points from T backward, then τ_1/T and τ_j/τ_{j-1}, $j \geq 2$, are independent random variables, each uniformly distributed on the interval $(0, 1)$. This observation, together with the Markovian property of $z(t)$, leads to the following representation of $z(t)$, when the marginal distribution $P(z(t) < x) = H_{3,0}^t(x)$. Let Y_s have distribution function $H_{3,0}(x - \log s)$, $s > 0$. Let U_j and V_j, $j \geq 1$, be i.i.d. unit exponential variates, which are also independent of Y_s. Define

$$s_0 = s, \quad s_j = s_{j-1} + U_j \exp(Y_s + V_1 + \cdots + V_{j-1}), \quad j \geq 1,$$

where an empty sum is taken as zero. Then

$$z(t) = Y_s + V_1 + V_2 + \cdots + V_{j-1}, \quad s_{j-1} \leq t < s_j. \tag{36}$$

The above representation automatically transforms to a representation for $z(t)$ with the other two marginals by the proper transformation of $H_{3,0}(x)$ into the other types of extreme value distributions.

The description given here was intended to be an introduction to the theory of extremal processes. Out of a number of possible attacks, we also selected those which are closest to the approaches of the previous sections and chapters.

The advantages of the theory of extremal processes are yet to be developed. It would become very valuable if the so-called invariance principle were available, by which one could reduce the proof of a limit theorem to a specific population distribution $F(x)$. While such a principle is not yet available, the representations such as (36) and its equivalents for the other types do provide new analytical tools for the investigation of limit theorems.

6.6. SURVEY OF THE LITERATURE

In order to introduce the problem of extremes for random sample sizes, we first described a model from queueing theory, an approach due to C. C. Heyde (1971). We do not plan to review the theory of queues. Its foundations can be found in the book by N. U. Prabhu (1965), and many new references are given in the survey by L. Takács (see Chapter 3). We mention, however, the work by J. W. Cohen (1967), where the property assumed in Exercise 1 is established for the model of Section 6.1. Consequently, we have the limit relation (1). Another type of application in reliability theory is given by R. V. Canfield and L. E. Borgman (1975).

The method of proof of Section 6.2 is due to J. Mogyoródi (1967), but the result itself was first established by O. Barndorff-Nielsen (1964) (Theorem 6.2.1). Mogyoródi obtains an extension as well, in that he can handle $(Z_{N(n)} - a_{M(n)})/b_{M(n)}$, where $M(n)$ is another random variable. This is not of interest to us directly, because $M(n)$ may have too strong an influence on the whole expression, but the method is of value because a wide class of statistics fall into such a group of expressions. Mogyoródi's method is based on a result of A. Rényi (1963), which has a wide field of potential applications. A special case was established by W. Richter (1965). Mogyoródi's method and result are extended by J. Galambos (1973b) to a class of dependent samples, which is a special class of E_n-sequences introduced in Section 3.9. One of the theorems of Galambos is extended by H. Rootzén (1974) for a larger class of dependent systems.

H. Cohn and A. G. Pakes (1977) establish an interesting relation between certain limit theorems of a simple Galton-Watson process and the theory of maxima with random sample size.

6.6 SURVEY OF THE LITERATURE

For the case when $N(n)$ is assumed to be independent of the variables $X_1, X_2, \ldots$, the first general result was obtained by S. M. Berman (1964). He realized that the same technique works in such a case as for extremes of exchangeable variables, but he did not unify the two theories. This unification came in a paper by J. Galambos (1975e), who deduced both results from a single theorem on limits for mixtures. It should be noted that the result is applicable to general sequences (see Chapter 3, in particular the survey section). The above paper extends results of Berman as well as those of D. I. Thomas (1972). Several theorems can be deduced from the results of S. Guiasu (1971). Although these would not extend the conclusions for Z_n, the advantage of his approach is that one can handle all extremes in one theorem. For random kth extremes, see W. Dziubdziela (1972).

The theory of record times started with the works of K. N. Chandler (1952) and F. G. Foster and A. Stuart (1954). These papers establish classical types of theorems on the sequence $N(n)$, but the fact that $E[N(2)] = +\infty$ dissuaded statisticians from continuing the work. While the disinterest of statisticians is understandable, the theoretical implications of the results justify continued work on this subject. This theory shows how extremes actually change (see Remark 6.3.1). In addition, it induced much work and produced neat techniques that are applicable in other branches of probability as well. Development of the theory began with Rényi's (1962) discovery (which is actually contained in the work of M. Dwass (1960) as well) of Lemma 6.3.3. All previous results follow from it, and several new ones were obtained with very little effort. M. Neuts (1967) established the close relation of $\Delta(n)$ and $N(n-1)$—quite a surprising result. The theory then developed very rapidly. The works of R. W. Shorrock (1972a,b and 1973), P. T. Holmes and W. Strawderman (1969), W. Strawderman and P. Holmes (1970), W. Vervaat (1972), J. Galambos and E. Seneta (1975), and D. Williams (1973) all added new illumination and new results to the theory. We have adopted the approach of Galambos and Seneta, which is given in Theorem 6.3.3. It immediately yields that $\Delta(n)$ and $N(n-1)$ are closely related. This method is more elementary than the others that are known, but it is not claimed to be superior if one allows advanced results of probability theory to enter the investigation. In particular, Shorrock's method and that of Vervaat contributed much to the theory's advance. Another simple method on the relation of $\Delta(n)$ and a transformation of the records themselves can be found in M. M. Siddiqui and R. W. Biondini (1975).

Thorough investigation of the records themselves started with M. N. Tata (1969) and J. Pickands III (1971). Theorem 6.4.1, due to Tata, became a basic theorem for future investigations. Corollary 6.4.1 is also hers. A major development occurred with the works of S. I. Resnick (1973a, b). He established Theorem 6.4.2, and within the framework of extremal processes he gave extensive analysis. Theorem 6.4.3 is due to Shorrock (1972b). Records and record times for discrete populations are discussed by W. Vervaat (1973a) and records on Markov chains by R. Biondini and M. M. Siddiqui (1975). To so-called kth records W. Dziubdziela and B. Kopocinsky (1976b) extend Resnick's theorem (Theorem 6.4.2). This same theorem is extended to random index by W. Freudenberg and D. Szynal (1976). A different theory applies if the records are taken from a finite sequence. For fixed number of elements in the sequence, see D. Haghighi-Talab and C. Wright (1973), and for records in sequences with increasing number of elements see M. C. Yang (1975).

It is very popular at this time to reobtain results that are known for sequences in the form of so-called functional limit theorems. The theory of extremal processes is part of this approach. It started with the works of M. Dwass (1964) and J. Lamperti (1964) and then developed rapidly. M. Dwass (1966 and 1973), J. Tiago de Oliveira (1968, 1971, and 1972b), S. I. Resnick (1973c, 1974, and 1975), S. I. Resnick and M. Rubinovitch (1973), and R. Shorrock (1974) contributed substantially to this field. A new approach, using random difference equations, was proposed by W. Vervaat (1977). See also his survey (1973b) and the papers by M. Wichura (1974) and T. Mori and H. Oodaira (1976). P. K. Sen (1972) obtained results on the line of extremal processes when one starts with a random sample size.

L. de Haan and S. I. Resnick (1973) give results of the type of Chapter 4 for records.

D. P. Gaver's (1976) work is on a different line and can perhaps be related to the approach of R. Biondini and M. M. Siddiqui (1975). Here one considers a point process, and to each occurrence there corresponds a random variable X_j. The records and record times of this sequence are investigated. Its relevance to applied models is also indicated.

Extremal processes obtained as limits of Z_n of independent but not identically distributed variables are introduced and studied by I. Weissman (1975a,b,c). The structure is similar to the process of Section 6.5, but the marginals are in agreement with the results of Section 3.10.

The survey of D. L. Iglehart (1974) points to fields of application of the last section.

In sequential analysis, the following theorem of F. J. Anscombe (1952) plays a central role. Let $y_1, y_2, \ldots$, be i.i.d. random variables with zero expectation and unit variance. Let $M(n)$ be a sequence of positive integer-valued random variables such that, with an increasing sequence $g(n)$ of numbers which tend to $+\infty$ with n, $M(n)/g(n)$ tends in probability to a positive constant. Then $[g(n)]^{-1/2}(y_1 + \cdots + y_{M(n)})$ is asymptotically normal. Let us apply this result to the following case. Let $X_1, X_2, \ldots,$ be i.i.d. unit exponential variates. Put $y_1 = X_1 - 1$ and $y_j = X_{N(j)} - X_{N(j-1)} - 1, j \geq 2$, where $N(n), n \geq 1$, is the sequence of record times. By Theorem 6.4.1, the y's satisfy the requirements in the Anscombe theorem. Let $M(n)$ be the last record time which does not exceed n; that is, $X_{M(n)} = Z_n$. Hence, by Theorem 6.3.4, $M(n)/\log n \to 1$ with probability one. Now, on the one hand,

$$(\log n)^{-1/2}(y_1 + \cdots + y_{M(n)}) = (\log n)^{-1/2}(Z_n - M(n))$$

is asymptotically normal, but on the other hand,

$$P(Z_n - \log n < x) \to H_{3,0}(x).$$

If one writes $Z_n - M(n) = (Z_n - \log n) - [M(n) - \log n]$, it follows that $Z_n - \log n$ does not contribute to the first limit law, only the second (see Theorem 6.3.5). But this in fact says that the y's did not contribute to "their own sum"; that is, all contributions came from $M(n)$. This example, due to J. Galambos (1976), illustrates that much caution is to be taken when random sized limit theorems are applied.

6.7. EXERCISES

1. Let $X_1, X_2, \ldots$ be i.i.d. random variables with common distribution function $F(x)$ such that, as $x \to +\infty$, $1 - F(x) \sim cq^x$, where $c > 0$ and $0 < q < 1$ are constants. Let $N(n), n \geq 1$, be positive integer-valued random variables. Assume that $N(n)/n$ converges in probability to a positive constant, as $n \to +\infty$. Show that $Z_{N(n)}/\log n$ is asymptotically a constant in probability.

2. Use the method of proof of Theorem 6.2.1 to establish the following extension. Let $g(n)$ be an increasing function, tending to $+\infty$ with n. Let $N(n)/g(n) \to \tau$ in probability, where τ is a positive random variable. Show that if $(Z_n - a_n)/b_n$ tends weakly to $H(x)$, then

$$\frac{Z_{N(n)} - a_{[g(n)]}}{b_{[g(n)]}} \to U(x) \quad \text{weakly}$$

where

$$U(x) = \int_0^{+\infty} H^y(x) \, dP(\tau < y),$$

and $[g(n)]$ is the integer part of $g(n)$.

3. Extend the conclusion of Exercise 2 to the kth extremes.

4. Use the total probability rule and Theorem 6.3.2 to show that, for the record times $N(n), n \geq 1, P(N(n) > xN(n-1)) = 1/x$ for $x \geq 1$ integer and, for arbitrary $x \geq 1$, as $n \to +\infty$,

$$P(N(n) > xN(n-1)) \to \frac{1}{x}.$$

5. Extend the result of Exercise 4 and show that the events $N(n+k)/N(n+k-1), 1 \leq k \leq m$, are asymptotically independent for any fixed $m \geq 2$.

6. Show that the limit distribution in (2) is never normal (see also Exercises 18 and 19 of Chapter 3).

7. Show the following converse to Theorem 6.4.1. If, for i.i.d. random variables, the differences $X_{N(n)} - X_{N(n-1)}, n \geq 2$, are independent, then the population is exponential.

[M. N. Tata (1969)]

8. Let $N(n), n \geq 1$, be the sequence of record times for i.i.d. continuous variates. Let $k_n(s)$ be the number of integers $j, 2 \leq j \leq n$, for which $\Delta(j) = N(j) - N(j-1) > sN(j-1)$. Show that, with probability one, $k_n(s)/n \to 1/(s+1)$. Record the special case $s=1$ and compare it with the result $P(N(2)=2)$. Discuss the meaning of this relation as the origin of time is shifted.

[J. Galambos and E. Seneta (1975)]

9. Let again $N(n)$ be the sequence of record times for i.i.d. continuous variates. Put $R(n) = N(n)/N(n-1), n \geq 2$. Determine the asymptotic behavior of $\max\{R(j): 2 \leq j \leq m\}$.

10. Go through the proof of Theorem 6.4.2 and find the asymptotic distribution of the record $X_{N(n)}$ if (i) the population is standard normal, (ii) the population distribution $F(x) = H_{3,0}(x)$, and (iii) $F(x)$ is one of the Weibull distributions.

11. Let $X_1, X_2, \ldots$ be i.i.d. discrete variates taking the nonnegative integers. Let their distribution $F(x)$ be such that $\omega(F) = +\infty$. Let $e_k^* = 1$ or 0 according as $X_k > X_j, 1 \leq j < k$, or not. Show that the events $\{e_k^* = 1\}$ are independent. Evaluate $P(e_k^* = 1)$.

12. Define the record times $N(n), n \geq 1$, by (15) without assuming $F(x)$ to be continuous. Show that the sequence $N(n)$ is always infinite if either $\omega(F) = +\infty$ or $\omega(F) < +\infty$ but $\omega(F)$ is a point of continuity of $F(x)$.

6.7 EXERCISES

13. Let $F(x)$ be a distribution function such that there is a number $x_0 < \omega(F)$ with the property that $F(x)$ is continuous for all $x \geq x_0$. Show that the extended concept of record times of Exercise 12 leads to the same asymptotic properties for $N(n)$ and $\Delta(n)$ as for populations with continuous distributions.

14. Let $F(x)$ be the distribution function of a random variable that takes nonnegative integers. For an integer m, define $F_m(x)$ as $F(x)$ if $a + 1/m \leq x \leq a + 1 - 1/m$, where a is a positive integer and $F_m(x)$ is continuous and linear for $a - 1/m \leq x \leq a + 1/m$. Let $N_F(n)$ be the sequence of record times for $F(x)$ and $N(n)$ for $F_m(x)$ ($F_m(x)$ being continuous, $N(n)$ does not depend on $F_m(x)$). Estimate the difference between $N_F(n)$ and $N(n)$. Extend your conclusion to other discrete distributions.

APPENDIX I

Some Basic Formulas for Probabilities and Expectations

Collected here are a number of results of probability theory which are used in the book and which are not necessarily covered in an introductory course on probability theory. Throughout this appendix the sample space is denoted by Ω, the set of its subsets that are considered as events by $\mathcal{C}$, and the probability measure on $\mathcal{C}$ by P. As usual, the complement of an event is denoted by A^c. The theorems on expectations are stated in the more familiar form of integrals.

Let us first establish two elementary results.

Theorem AI.1. *Let* $A_1 \subset A_2 \subset \cdots$ *be a nondecreasing sequence of events. Then, as* $n \to +\infty$, $P(A_n)$ *converges to* $P(\cup_{j=1}^{+\infty} A_j)$.

Remark. By taking complements, it immediately follows from the above statement that, for a nonincreasing sequence $A_1 \supset A_2 \supset \cdots$ of events, $P(A_n)$ converges to $P(\cap_{j=1}^{+\infty} A_j)$ for $n \to +\infty$.

Proof. Let us put $B = \cup_{j=1}^{+\infty} A_j$. From the monotonicity of the sequence A_j it follows that B can be decomposed as

$$B = A_1 \cup (A_1^c A_2) \cup (A_2^c A_3) \cup \cdots \cup (A_{n-1}^c A_n) \cup \cdots.$$

The terms on the right hand side are disjoint. Thus, by the additive property of P,

$$P(B) = P(A_1) + \sum_{j=1}^{+\infty} P(A_j^c A_{j+1}) = P(A_1) + \lim_{n=+\infty} \sum_{j=1}^{n-1} P(A_j^c A_{j+1}).$$

The axioms of probability imply that, for $A_j \subset A_{j+1}$, $P(A_j^c A_{j+1}) = P(A_{j+1})$

SOME BASIC FORMULAS FOR PROBABILITIES AND EXPECTATIONS   315

$-P(A_j)$. Hence

$$P(B) = \lim_{n = +\infty} \left\{ P(A_1) + \sum_{j=1}^{n-1} \left[P(A_{j+1}) - P(A_j) \right] \right\} = \lim_{n = +\infty} P(A_n),$$

which was to be proved. ▲

The next theorem is an immediate consequence of the axioms and the definition of conditional probability. Its proof is therefore omitted.

Theorem AI.2 (The Total Probability Rule). *Let $A_1, A_2, \ldots$ be a sequence of events satisfying*

(i) $P(A_i A_j) = 0$ *for all $i \neq j$,*
(ii) $\sum P(A_j) = 1$.

Then, for an arbitrary event B,

$$P(B) = \sum P(B|A_j) P(A_j),$$

where $P(B|A_j) P(A_j) = 0$ whenever $P(A_j) = 0$.

We also prove a continuous version of the total probability rule. It will involve the following extended concept of conditional probability, which is, of course, much simpler than the one used in an advanced theory.

Definition. Let A be an event and let X be a random variable. If, as x_1 and x_2 with $x_1 < x_2$ converge to the same value x, the conditional probability $P(A|x_1 \leq X < x_2)$ converges, then this limit is called the conditional probability $P(A|X = x)$ of A given $X = x$. If X is a vector, then inequalities and equations are meant componentwise.

Notice that $P(A|X = x)$ can easily be calculated in the following special case. Let Y be a random vector which is independent of X. Let A be the event that X and Y fall into a (Borel) set S. Then $P(A|X = x) = P(Y \in S(x))$, where $S(x)$ is the set obtained from S by replacing X by x. For example, if Y is one-dimensional and $A = \{X < 2Y\}$, say, then $P(A|X = x) = P(Y > \frac{1}{2}x)$. In the preceding abstract notation $S = \{(x,y): -\infty < x < 2y < +\infty\}$ and $S(x) = \{y: \frac{1}{2}x < y < +\infty\}$.

We can now state a continuous version of Theorem AI.2.

Theorem AI.3 (The Continuous Total Probability Rule). *Let the event A and the random variable X be such that, for almost all x (with respect to $F(x) = P(X < x)$), $P(A|X = x)$ exists. Then*

$$P(A) = \int_{-\infty}^{+\infty} P(A|X = x) \, dF(x). \tag{A1}$$

Before proving this theorem, let us establish some properties of integrals. First recall that the concept of integral is reduced to the limit of a specially constructed finite sum. Namely, the integral of a (measurable) function is defined in three steps as follows. Let $(U, \mathcal{B}, m)$ be a measure space. Let $g(u)$ be a real-valued (measurable) function defined on U. If $g(u)$ takes only a finite number of values $x_1, x_2, \ldots, x_t$, and $B_j = \{u : g(u) = x_j\}$, then

$$\int_U g(u)\, dm = \sum_{j=1}^{t} x_j m(B_j). \tag{A2}$$

Let now $g(u) \geq 0$. Let $g_n(u) \leq g_{n+1}(u)$ be a monotonic sequence of functions, each of which takes a finite number of values, and $g_n(u) \to g(u)$ as $n \to +\infty$. Then, by definition,

$$\int_U g(u)\, dm = \lim_{n = +\infty} \int_U g_n(u)\, dm, \tag{A3}$$

whenever the limit exists. A basic and simple theorem of integration is that this limit does not depend on the actual choice of the sequence $g_n(u)$. The final step in the definition is that if $g(u)$ is an arbitrary (measurable) function, then decompose $g(u) = g^+(u) - g^-(u)$, where $g^+(u) = \max(g(u), 0)$ and $g^-(u) = \max(-g(u), 0)$, and then define

$$\int_U g(u)\, dm = \int_U g^+(u)\, dm - \int_U g^-(u)\, dm, \tag{A4}$$

whenever the right hand side is meaningful. The crucial step is therefore the middle one (A3), which is indeed the limit of finite sums of the type of (A2).

Notice the similarity between the classical Riemann integral and the abstract integral just described. If U is a finite interval and m assigns to an interval its length, then (A2) is in fact a Riemann sum whenever $g(u)$ is constant on t disjoint subintervals of U. On the other hand, if $g(u) \geq 0$ and bounded on U, then (A3) expresses that the integral of $g(u)$ is the limit of the so-called lower Riemann sums (which are monotonically increasing). That is, Riemann integration was built on the same concepts as (A2)–(A4), except that U and m were very specific. The reader who is not very familiar with this abstract concept of integration is advised to go through the definition (A2)–(A4) and several theorems known for Riemann integrals, when U and m are chosen as follows. Let $(\Omega, \mathcal{C}, P)$ be a probability space and let X be a random variable on it with distribution function $F(x)$. Let U be the whole real line and let m be defined for intervals $B = (a, b]$ as $m(B) = F(b) - F(a)$. It is also advisable to relate integrals over Ω with respect to P to those over the whole real line with respect to the above m.

SOME BASIC FORMULAS FOR PROBABILITIES AND EXPECTATIONS

For example, prove that if $g(x)$ is continuous for all x, then

$$\int_\Omega g(X)\,dP = \int_{-\infty}^{+\infty} g(x)\,dF(x),$$

where $dF(x)$ represents dm with the above special choice.

We now prove two theorems for integrals. All integrals are over a set U and with respect to a given measure m on a set $\mathcal{B}$ of subsets of U.

Theorem AI.4. Let $0 \leq g_k(u) \leq g_{k+1}(u)$ and let $g_k(u) \to g(u)$ as $k \to +\infty$. Then

$$\lim_{k=+\infty} \int_U g_k(u)\,dm = \int_U g(u)\,dm. \tag{A5}$$

Proof. Notice that formally (A5) is the same as (A3) except that here $g_k(u)$ is not assumed to take only a finite number of values.

In the proof, we appeal only to the definition. Let $g_{nk}(u)$, $n = 1, 2, \ldots$ be an increasing sequence of functions which tends to $g_k(u)$ as $n \to +\infty$ and each of which takes a finite number of values. Put $G_n(u) = \max_{k \leq n} g_{nk}(u)$. Then $G_n(u)$ also takes only a finite number of values, and $G_n(u) \leq G_{n+1}(u)$. Furthermore, $g_{nk}(u) \leq G_n(u) \leq g_n(u)$ and

$$\int_U g_{nk}(u)\,dm \leq \int_U G_n(u)\,dm \leq \int_U g_n(u)\,dm.$$

Let $n \to +\infty$. We get

$$g_k(u) \leq \lim G_n(u) \leq g(u), \tag{A6}$$

and, by (A3),

$$\int_U g_k(u)\,dm \leq \int_U \lim G_n(u)\,dm \leq \lim \int_U g_n(u)\,dm. \tag{A7}$$

If we now let $k \to +\infty$, (A6) implies that $\lim G_n(u) = g(u)$, and thus (A7) reduces to (A5), which was to be proved. ▲

Theorem AI.5 (The Dominated Convergence Theorem). Let $g_n(u)$ be a sequence of functions which satisfies
  (i) $|g_n(u)| \leq G(u)$ with $\int_U G(u)\,dm < +\infty$;
  (ii) $g_n(u) \to g(u)$ for almost all u as $n \to +\infty$.
Then

$$\lim_{n=+\infty} \int_U g_n(u)\,dm = \int_U g(u)\,dm. \tag{A8}$$

Proof. We first observe that our conditions imply that the functions $g_n(u)$, $n \geq 1$, and $g(u)$ are integrable (the integral of their absolute values is bounded by the integral of $G(u)$). Therefore, so are the functions $G \pm g_n$ and $G \pm g$. These new functions are all nonnegative, and thus so are the functions

$$h_n(u) = \inf_{k \geq n} \left[G(u) + g_k(u) \right] \quad \text{and} \quad s_n(u) = \inf_{k \geq n} \left[G(u) - g_k(u) \right].$$

Both $h_n(u)$ and $s_n(u)$ are nonnegative and nondecreasing sequences and $h_n(u) \to G(u) + g(u)$ and $s_n(u) \to G(u) - g(u)$. Hence, Theorem AI.4 is applicable, which yields, as $n \to +\infty$,

$$\lim \int_U h_n(u) \, dm = \int_U G(u) \, dm + \int_U g(u) \, dm \tag{A9}$$

and

$$\lim \int_U s_n(u) \, dm = \int_U G(u) \, dm - \int_U g(u) \, dm. \tag{A10}$$

On the other hand, from their definitions,

$$h_n(u) \leq G(u) + g_n(u), \quad s_n(u) \leq G(u) - g_n(u),$$

and thus

$$\int_U h_n(u) \, dm \leq \int_U G(u) \, dm + \int_U g_n(u) \, dm,$$

$$\int_U s_n(u) \, dm \leq \int_U G(u) \, dm - \int_U g_n(u) \, dm.$$

Letting $n \to +\infty$ and taking the lim inf of the above inequalities, we get from (A9) and (A10),

$$\int_U g(u) \, dm \leq \liminf \int_U g_n(u) \, dm$$

$$\leq \limsup \int_U g_n(u) \, dm$$

$$\leq \int_U g(u) \, dm,$$

which implies (A8). The theorem is established. ▲

SOME BASIC FORMULAS FOR PROBABILITIES AND EXPECTATIONS 319

We can now prove Theorem AI.3.

Proof of Theorem AI.3. Let $a<b$ be two finite real numbers. Let us divide the interval $[a,b]$ by the points $a=x_0<x_1<\cdots<x_t=b$, which are continuity points of $F(x)$. By Theorem AI.2,

$$P(A)=P(A|X<a)F(a)+\sum_{j=1}^{t}P(A|x_{j-1}\leqslant X<x_j)[F(x_j)-F(x_{j-1})]$$
$$+P(A|X\geqslant b)[1-F(b)].$$

If we define $g_t(x)=P(A|x_{j-1}\leqslant X<x_j)$ for $x_{j-1}\leqslant x<x_j$, then $g_t(x)$ takes t values only and, by assumption, $g_t(x)\to P(A|X=x)$ as $\max(x_j-x_{j-1})\to 0$. The function $g_t(x)$ is bounded by $G(x)\equiv 1$ and

$$\int_a^b g_t(x)\,dF(x)=\sum_{j=1}^{t}P(A|x_{j-1}\leqslant X<x_j)[F(x_j)-F(x_{j-1})].$$

Theorem AI.5 thus yields

$$P(A)=P(A|X<a)F(a)+\int_a^b P(A|X=x)\,dF(x)+P(A|X\geqslant b)[1-F(b)].$$

Letting $a\to-\infty$ and $b\to+\infty$ gives (A1). The proof is completed. ▲

We give one more formula for integrals which is useful when evaluating expectations. We establish that

$$\int_0^{+\infty}x\,dF(x)=\int_0^{+\infty}[1-F(x)]\,dx, \qquad (A11)$$

whenever either side is finite. Indeed, integrating by parts, we get for finite $b>0$,

$$\int_0^b x\,dF(x)=-b[1-F(b)]+\int_0^b[1-F(x)]\,dx. \qquad (A12)$$

We first observe that, by (A12), if the right hand side of (A11) is finite, so is the left hand side (namely, the first term on the right hand side of (A12) is negative). Therefore, we have to establish (A11) under the assumption that its left hand side is finite. This immediately follows from (A12), by letting $b\to+\infty$, if we show that

$$\lim_{b=+\infty}b[1-F(b)]=0, \qquad (A13)$$

whenever the left hand side of (A11) is finite. But, for large b and arbitrary $\varepsilon > 0$,

$$\varepsilon > \int_b^{+\infty} x\, dF(x) \geq b \int_b^{+\infty} dF(x) = b\bigl[1 - F(b)\bigr],$$

from which (A13) follows.

Let us conclude the present appendix by collecting three basic results of probability theory. The first part of each theorem (the i.i.d. case) is assumed to be familiar to the reader, except perhaps in Theorem AI.8. The second part contains a slightly modified version, which can be proved by any of the standard methods of proof for the first parts; we therefore omit proofs. These results are collected here for easier reference.

Let $X_1, X_2, \ldots, X_n$ be independent random variables on a probability space $(\Omega, \mathcal{C}, P)$. Let $E(X_j)$ and $V(X_j)$, $j \geq 1$, denote the expectation and variance of X_j, respectively. We put $S_n = X_1 + \cdots + X_n$, $E_n = E(X_1) + \cdots + E(X_n)$, and $V_n = V(X_1) + \cdots + V(X_n)$.

Theorem AI.6 (The Strong Law of Large Numbers). *Let $X_1, X_2, \ldots, X_n$ be independent and identically distributed with finite expectation $E = E(X_1)$. Then, almost surely, as $n \to +\infty$,*

$$\lim \frac{S_n}{n} = E \quad \text{or} \quad \lim \frac{S_n}{E_n} = 1.$$

The statement in the latter form holds if the X_j are uniformly bounded and independent.

Theorem AI.7 (The Central Limit Theorem). *Let $X_1, X_2, \ldots, X_n$ be independent and identically distributed random variables with finite expectation $E = E(X_1)$ and $V = V(X_1)$. Then, as $n \to +\infty$,*

$$\lim P\bigl(S_n < nE + x\sqrt{nV}\,\bigr) = \frac{1}{\sqrt{2\pi}} \int_{-\infty}^x e^{-t^2/2}\, dt.$$

If we rewrite nE and nV as E_n and V_n, respectively, then the conclusion holds if the X_j are assumed to be independent and uniformly bounded.

Theorem AI.8 (The Iterated Logarithm Theorem). *Let $X_1, X_2, \ldots, X_n$ be independent and identically distributed random variables with finite expectation $E = E(X_1)$ and variance $V = V(X_1)$. Then, almost surely, as $n \to +\infty$,*

$$\limsup \frac{S_n - nE}{(2nV \log \log nV)^{1/2}} = 1.$$

SOME BASIC FORMULAS FOR PROBABILITIES AND EXPECTATIONS 321

If we replace nE and nV by E_n and V_n, respectively, then the conclusion holds under the assumption that the X_j are independent and uniformly bounded, and $V_n \to +\infty$ with n.

There is only one reference to Theorem AI.8 in the book (it occurs in Chapter 6). Therefore, a reader who is not familiar with it will not be hindered. Let us add that its proof is similar to several arguments of Chapter 4.

APPENDIX II

Theorems from Functional Analysis

In this book we have used the following theorems, which are well known in functional analysis. We collect them here, rather than referring to a textbook containing them, for two reasons. First, we formulate them in probabilistic language. Second, they appear in textbooks of functional analysis after a number of definitions have been introduced, and thus their simplicity is not clear when they are deduced from general theorems.

As in the preceding appendix, the basic probability space is denoted by $(\Omega, \mathcal{C}, P)$. We also introduce the notation L_2 for the set of all random variables on $(\Omega, \mathcal{C}, P)$, whose variance is finite.

Theorem AII.1 (Completeness Theorem). *Let $X_n \in L_2$ for each $n \geq 1$. Let X_n be such that, as n and m tend to infinity, the sequence $E[(X_n - X_m)^2]$ of second moments converges to zero. Then there is a random variable $X \in L_2$ such that $E[(X_n - X)^2] \to 0$ as $n \to +\infty$. Consequently, X_n tends to X in probability.*

Before giving details of the proof, we establish two simple lemmas.

Lemma AII.1. *If U and V belong to L_2, then*

$$E[(U+V)^2] \leq [E^{1/2}(U^2) + E^{1/2}(V^2)]^2. \tag{A14}$$

Proof. We start with the triangle inequality and then apply the well-known inequality of Cauchy and Schwarz. We get

$$E[(U+V)^2] \leq E(|U+V||U|) + E(|U+V||V|)$$

$$\leq E^{1/2}[(U+V)^2] E^{1/2}(U^2) + E^{1/2}[(U+V)^2] E^{1/2}(V^2),$$

which is equivalent to (A14). The proof is complete. ▲

Lemma AII.2 (Fatou's Lemma). *Let $Y_n \in L_2$ for each $n \geq 1$. Then*

$$\int_\Omega \liminf_{n=+\infty} Y_n^2 \, dP \leq \liminf_{n=+\infty} \int_\Omega Y_n^2 \, dP.$$

Proof. Define $V_n = \inf_{k \geq n} Y_k^2$. Then $V_n \leq V_{n+1}$ and V_n tends to $\liminf Y_n^2$ as $n \to +\infty$. Furthermore, $0 \leq V_n \leq Y_n^2$. Thus, by Theorem AI.4,

$$\liminf_{n=+\infty} \int_\Omega Y_n^2 \, dP \geq \liminf_{n=+\infty} \int_\Omega V_n \, dP = \lim_{n=+\infty} \int_\Omega V_n \, dP$$

$$= \int_\Omega \lim_{n=+\infty} V_n \, dP = \int_\Omega \liminf_{n=+\infty} Y_n^2 \, dP,$$

which was to be proved. ▲

Proof of Theorem AII.1. Let X_n satisfy the conditions of the theorem. Let us take $n < m$ such that $E[(X_m - X_n)^2] < \frac{1}{4}$ and label $n = n(1)$ and $m = n(2)$. When $n(k)$ has been fixed, determine $n(k+1) > n(k)$ so that $E[(X_{n(k+1)} - X_{n(k)})^2] < 2^{-2k}$. In this way we select a subsequence $n(k)$, $k \geq 1$, of integers for which

$$\sum_{k=1}^{+\infty} E^{1/2}\left[(X_{n(k+1)} - X_{n(k)})^2\right] < +\infty.$$

Let us define

$$Y_r = |X_{n(1)}| + \sum_{k=1}^{r} |X_{n(k+1)} - X_{n(k)}|.$$

By induction over r, we get from Lemma AII.1 that $Y_r \in L_2$. In fact, we obtain

$$E(Y_r^2) \leq \left\{ E^{1/2}(X_{n(1)}^2) + \sum_{k=1}^{r} E^{1/2}\left[(X_{n(k+1)} - X_{n(k)})^2\right] \right\}^2$$

$$\leq \left\{ E^{1/2}(X_{n(1)}^2) + \sum_{k=1}^{+\infty} E^{1/2}\left[(X_{n(k+1)} - X_{n(k)})^2\right] \right\}^2 < +\infty.$$

Hence,

$$\liminf_{r=+\infty} E(Y_r^2) < +\infty.$$

Thus, using the denotation

$$X^* = \lim_{r=+\infty} Y_r = |X_{n(1)}| + \sum_{k=1}^{+\infty} |X_{n(k+1)} - X_{n(k)}|,$$

Fatou's lemma (Lemma AII.2) yields $X^* \in L_2$. It follows that X^* is finite almost surely and thus the subsequence $X_{n(k)}$ converges almost surely to a finite random variable X. Because $|X_{n(k+1)}| \leq Y_k \leq X^*$, $|X| \leq X^*$ and thus $X \in L_2$. To conclude the proof, we appeal again to Lemma AII.1. We have

$$E^{1/2}\big[(X-X_n)^2\big] \leq E^{1/2}\big[(X-X_{n(k)})^2\big] + E^{1/2}\big[(X_{n(k)}-X_n)^2\big].$$

The last term tends to zero by assumption as $n \to +\infty$ and $n(k) \to +\infty$. On the other hand, the first term on the right hand side converges to zero by the dominated convergence theorem (Theorem AI.5) as $n(k) \to +\infty$. The proof is completed. ▲

For the next theorem, we need the following concept. We say that a subset T of L_2 is a closed linear subset if, for any $X_j \in T$ and real numbers c_j, $c_1 X_1 + \cdots + c_n X_n$ also belongs to T. Furthermore, if $Y \in L_2$ and there is a sequence $X_n \in T$ such that $E[(Y - X_n)^2] \to 0$ as $n \to +\infty$, then Y also belongs to T.

Theorem AII.2. *Let T be a closed linear subset of L_2. Then any random variable $Y \in L_2$ has an almost surely unique representation $Y = X + R$, where $X \in T$ and $E(RV) = 0$ for all $V \in T$.*

Proof. If $Y \in T$, then $X = Y$ and $R \equiv 0$ provides a decomposition as stated. Let therefore Y be a member of L_2 but not of T. Let X_n be a sequence of T such that

$$\lim_{n=+\infty} E\big[(Y-X_n)^2\big] = \inf_{V \in T} E\big[(Y-V)^2\big] = d(Y,T), \text{ say.}$$

The "distance" $d(Y,T) > 0$ because Y is not in T and T is closed. It is easily seen that, as m and n tend to infinity, $E[(X_m - X_n)^2] \to 0$. Hence, by Theorem AII.1, there is a random variable X in L_2 such that $E[(X_n - X)^2] \to 0$ as $n \to +\infty$. Writing $Y = X + R$ with $R = Y - X$, we now show that R indeed has the property of $E(RV) = 0$ for all $V \in T$. Namely, by the minimal property of $d(Y,T)$, for every $V \in T$ and for every real number c,

$$d(Y,T) \leq E\big[(Y-X-cV)^2\big] = E(R^2) - 2cE(RV) + c^2 E(V^2).$$

But

$$E(R^2) = E[(Y-X)^2]$$
$$= E[(Y-X_n+X_n-X)^2]$$
$$= E[(Y-X_n)^2] + E[(X_n-X)^2] + 2E[(Y-X_n)(X_n-X)]$$
$$\to d(Y,T) \quad \text{as } n \to +\infty.$$

Thus, for every real number c,

$$c^2 E(V^2) \geqslant 2cE(RV).$$

This is possible only if $E(RV) = 0$. The possibility of the claimed decomposition of Y is proved. Its uniqueness is immediate by observing that if

$$Y = X_1 + R_1 = X_2 + R_2, \quad X_i \in T, \quad \text{and} \quad E(R_i V) = 0 \quad \text{for } V \in T,$$

then

$$(X_1 - X_2) + (R_1 - R_2) = 0$$

where

$$X_1 - X_2 \in T,$$

and

$$E[(R_1 - R_2)V] = 0 \quad \text{for } V \in T.$$

But then multiplying the first equation by $R_1 - R_2$ and integrating yields $E[(R_1 - R_2)^2] = 0$, which implies $R_1 = R_2$ almost surely. Consequently, $X_1 = X_2$ almost surely, which completes the proof. ▲

Theorem AII.3. *Let $g(x)$ be a Lebesgue integrable function on the finite interval (a,b). Assume that, for all integers $n \geqslant 0$,*

$$\int_a^b g(x) x^n \, dx = 0. \tag{A15}$$

Then $g(x) = 0$ for almost all x.

Notice that if we introduce $g_1(x) = \max(g(x), 0)$ and $g_2(x) = \max(-g(x), 0)$, then (A15) can be reformulated as

$$\int_a^b g_1(x) x^n \, dx = \int_a^b g_2(x) x^n \, dx, \quad n = 0, 1, \ldots, \tag{A16}$$

where $g_1(x) \geq 0$ and $g_2(x) \geq 0$. Therefore, Theorem AII.3 is a special case of the following result.

Theorem AII.4. *Let $g_1(x) \geq 0$ and $g_2(x) \geq 0$ be (Lebesgue) integrable over the finite interval (a,b). Assume that, for all integers $n \geq 0$, (A16) holds. Then, for almost all x, $g_1(x) = g_2(x)$.*

Proof. By replacing x by $x - a$ in (A16), we can achieve that $a = 0$. Furthermore, because (A16) is not affected if we multiply $g_1(x)$ and $g_2(x)$ by the same number, we may assume

$$\int_0^b g_1(x)\,dx = \int_0^b g_2(x)\,dx = 1.$$

Let us define, for $i = 1, 2$,

$$G_i(x) = \int_0^x g_i(t)\,dt, \quad u_i(s) = \int_0^b e^{-st} g_i(t)\,dt, \quad s > 0. \tag{A17}$$

One can easily deduce from the dominated convergence theorem (Theorem AI.5) that $u_i(s)$ is differentiable any number of times and

$$u_i^{(k)}(s) = (-1)^k \int_0^b e^{-st} t^k g_i(t)\,dt. \tag{A18}$$

Hence

$$|u_i^{(k)}(0)| = \int_0^b t^k g_i(t)\,dt \leq b^k$$

and thus the Taylor expansion

$$u_i(s) = \sum_{k=0}^{+\infty} \frac{u_i^{(k)}(0)}{k!} s^k$$

is absolutely convergent for all s. But the coefficients $u_i^{(k)}(0)$ do not depend on i by (A16) and (A18), and thus $u_1(s) = u_2(s)$ for all $s > 0$.

We now show that $u_i(s)$ uniquely determines $G_i(x)$. Indeed, from (A18),

$$\sum_{k=0}^T \frac{(-1)^k s^k}{k!} u_i^{(k)}(s) = \int_0^b \left[e^{-st} \sum_{k=0}^T \frac{(st)^k}{k!} \right] g_i(t)\,dt. \tag{A19}$$

Notice the following probabilistic meaning of the expression in the brackets. Let Y be a Poisson variate with parameter st. Then

$$P(Y \leq T) = e^{-st} \sum_{k=0}^T \frac{(st)^k}{k!}.$$

Thus, by Chebyshev's inequality,

$$\lim_{s=+\infty} P(Y \leqslant sx) = \begin{cases} 1 & \text{if } x > t, \\ 0 & \text{if } x < t. \end{cases}$$

Consequently, the dominated convergence theorem and (A19) yield

$$\lim_{s=+\infty} \sum_{k=0}^{sx} \frac{(-1)^k s^k}{k!} u_i^{(k)}(s) = \int_0^x g_i(t)\,dt = G_i(x).$$

But the left hand side does not depend on i, and thus $G_1(x) = G_2(x)$. Differentiation results in $g_1(x) = g_2(x)$ for almost all x, which was to be proved. ▲

Corollary AII.1. *Let X and Y be random variables (on the same probability space). Assume that there are finite numbers $a < b$ such that $P(a \leqslant X \leqslant b) = P(a \leqslant Y \leqslant b) = 1$. Furthermore, let $E(X^n) = E(Y^n)$, $n \geqslant 1$. Then X and Y are identically distributed.*

Proof. All conditions remain unchanged if we consider $X - a$ and $Y - a$. Hence, we may assume that $a = 0$.

Put $F_1(x) = P(X < x)$ and $F_2(x) = P(Y < x)$. Then, by the method of proof of (A11), we obtain

$$E(X^n) = n \int_0^b [1 - F_1(x)] x^{n-1} dx, \quad E(Y^n) = n \int_0^b [1 - F_2(x)] x^{n-1} dx.$$

Thus, the equality of all moments of X and Y reduces to (A16) with $g_1(x) = 1 - F_1(x)$ and $g_2(x) = 1 - F_2(x)$. Hence, Theorem AII.4 applies, and we get $F_1(x) = F_2(x)$ for almost all x. But distribution functions are identical if they coincide for almost all x, which completes the proof. ▲

We prove at this point a property of sets of distribution functions. This property is very useful for analyzing moment sequences, but we do not exploit this possibility.

Theorem AII.5 (The Compactness of Distribution Functions). *Any infinite set of distribution functions contains a weakly convergent sequence.*

Proof. Let $R = \{r_1, r_2, \ldots\}$ be the set of rational numbers arranged in a sequence. Let $F_n(x)$, $n \geqslant 1$, be a sequence from our set. Consider the sequence $F_n(r_1)$ of numbers. Because $0 \leqslant F_n(r_1) \leqslant 1$ for all $n \geqslant 1$, by an elementary result of calculus, we can select a subsequence $F_{n(k)}(r_1)$ that converges. Let us keep only the terms $F_{n(k)}(x)$ from the original sequence. By the same argument, the numerical sequence $F_{n(k)}(r_2)$ contains a convergent subsequence. Keeping this new subsequence, we select an additional subsequence that converges at r_3, etc. Thus, we can produce inductively an infinite subsequence $F_m^*(x)$, which is convergent for all rational numbers r_k. Let

$$F(x;R) = \lim_{m=+\infty} F_m^*(x), \quad x \text{ rational,}$$

and

$$F(x) = \lim_{r \to x} F(r;R), \quad r < x, \quad r \in R.$$

Both limits exist—the first one by the selection procedure and the second by $F(r;R)$'s being nondecreasing on R. It follows easily that $F(x)$ is nondecreasing, continuous from the left, and $0 \leq F(x) \leq 1$. Let $r_t < x < r_s$, where $r_t, r_s \in R$. Because

$$F_m^*(r_t) \leq F_m^*(x) \leq F_m^*(r_s),$$

letting $m \to +\infty$ results in

$$F(r_t;R) \leq \liminf_{m=+\infty} F_m^*(x) \leq \limsup_{m=+\infty} F_m^*(x) \leq F(r_s;R).$$

Let x be a continuity point of $F(x)$ and let $r_t \to x$, $r_s \to x$. We get

$$F(x) \leq \liminf_{m=+\infty} F_m^*(x) \leq \limsup_{m=+\infty} F_m^*(x) \leq F(x)$$

—that is, for continuity points x of $F(x)$, $F_m^*(x)$ converges to $F(x)$ as $m \to +\infty$. The theorem is established. ▲

APPENDIX III

Slowly Varying Functions

Let $L(x)$ be a real-valued, positive, and Lebesgue measurable function with domain $[A, +\infty)$, where A is some positive number. We say that $L(x)$ is slowly varying if, for each $t > 0$, as $x \to +\infty$,

$$\lim \frac{L(tx)}{L(x)} = 1. \tag{A20}$$

There is a strong relation between slowly varying functions and tails of distribution functions which belong to the domain of attraction of $H_{1,\gamma}(x)$, when the observations are independent and identically distributed. Namely, if in (10) of Chapter 2 we write $L(x) = x^\gamma [1 - F(x)]$, then $L(x)$ satisfies (A20); that is, $L(x)$ is slowly varying. However, the present book is not based on this concept, and thus we do not develop its theory here. We limit this appendix to quoting two special properties of slowly varying functions which are referred to in Chapter 4. Proofs and further details, including a detailed bibliography, can be found in a monograph by E. Seneta (1976); the following theorems are on pp. 2–6.

Theorem AIII.1. *If $L(x)$ is slowly varying, then (A20) holds uniformly in t on any finite closed subinterval of the positive real line.*

Theorem AIII.2. *If $L(x)$ is slowly varying with domain $[A, +\infty)$, $A > 0$, then there is a real number $B \geq A$ such that, for all $x \geq B$,*

$$L(x) = u(x) \exp\left\{ \int_B^x \frac{e(y)}{y} dy \right\}, \tag{A21}$$

where $u(x)$ is bounded on $[B, +\infty)$ and converges to a finite positive value u^* as $x \to +\infty$ and where $e(y)$ is continuous on $[B, +\infty)$ and $e(y) \to 0$ as $y \to +\infty$. Conversely, any such function $L(x)$ is slowly varying.

An immediate consequence of the representation (A21) is that, for any $\varepsilon > 0$,

$$L(x) < x^\varepsilon \qquad (x \to +\infty). \tag{A22}$$

Indeed, choose $d \geq B$ such that $|e(y)| < \varepsilon/2$ for $y \geq d$. Then

$$0 < L(x) \leq M \exp\left\{ \int_B^d \frac{e(y)}{y} dy + \int_d^x \frac{e(y)}{y} dy \right\} \leq M_1 \left(\frac{x}{d}\right)^{\varepsilon/2} < x^\varepsilon$$

for all large x.

The representation (A21) can also be applied to show that, as $x \to +\infty$,

$$\int_0^1 \left(\frac{L(tx)}{L(x)} - 1 \right) dt \to 0, \tag{A23}$$

whenever $L(x)$ is defined for all $x > 0$ in such a way that the integral above is finite (notice that (A21) is applicable for $x \geq B$ only, and thus, for a fixed x, the contribution to the integral as t varies from zero to B/x should remain finite by assumption). For example, in our case in Chapter 4, $L(tx)/L(x)$ is bounded in a neighborhood of $t = 0$. In such a case, if we cut the integral

$$\int_0^1 \cdots = \int_0^{B/x} \cdots + \int_{B/x}^1 \cdots,$$

the first term is of the magnitude $B/x \to 0$ as $x \to +\infty$, while in the second term we can apply (A21) to obtain fine estimates. Since simple applications of Taylor's expansion suffice, we omit the details.

We conclude this appendix by drawing attention to the newly developed theory of slowly varying sequences. It was initiated by J. Galambos and E. Seneta (1973) and further developed by R. Bojanic and E. Seneta. See Seneta's (1976) monograph for details.

References

Afonja, B. (1972). The moments of the maximum of correlated normal and t-variates. *J. Royal Statist. Soc.* B **34**, 251–262.

Anderson, C. W. (1970). Extreme value theory for a class of discrete distributions with applications to some stochastic processes. *J. Appl. Probability* **7**, 99–113.

Anscombe, F. J. (1952). Large sample theory of sequential estimation. *Proc. Cambridge Philos. Soc.* **48**, 600–607.

Antle, C. E., and F. Rademaker (1972). An upper confidence limit on the maximum of m future observations from a type I extreme value distribution. *Biometrika* **59**, 475–477.

Arnold, B.C. (1968). Parameter estimation for a multivariate exponential distribution. *J. Amer. Statist. Assoc.* **63**, 848–852.

Arov, D. Z., and A. A. Bobrov (1960). The extreme terms of a sample and their role in the sum of independent variables. *Teor. verojatnost. i primen.* **5**, 377–396.

Balkema, A. A. (1973). *Monotone transformations and limit laws.* Mathematical Centre Tracts, Amsterdam.

Balkema, A. A., and L. de Haan (1972). On R. von Mises' condition for the domain of attraction of $\exp(-e^{-x})$. *Ann. Math. Statist.* **43**, 1352–1354.

Balkema, A. A. and S. I. Resnick (1977). Max-infinite divisibility. *J. Appl. Probability* **14**, 309–319.

Barlow, R. E. (1971). Averaging time and maxima for air pollution concentrations. *Proc. 38th Session ISI, Washington, D.C.*, pp. 663–676.

Barlow, R. E., and F. Proschan (1975). *Statistical theory of reliability and life testing: Probability models.* Holt, Rinehart and Winston, New York.

Barlow, R. E., and N. D. Singpurwalla (1974). Averaging time and maxima for dependent observations. *Proc. Symposium on Statistical Aspects of Air Quality Data.*

Barlow, R. E., Gupta, S. S., and S. Panchapakesan (1969). On the distribution of the maximum and minimum of ratios of order statistics. *Ann. Math. Statist.* **40**, 918–934.

Barndorff-Nielsen, O. (1961). On the rate of growth of the partial maxima of a sequence of independent identically distributed random variables. *Math. Scand.* **9**, 383–394.

Barndorff-Nielsen, O. (1963). On the limit behavior of extreme order statistics. *Ann. Math. Statist.* **34**, 992–1002.

Barndorff-Nielsen, O. (1964). On the limit distribution of the maximum of a random number of independent random variables. *Acta Math. Acad. Sci. Hungar.* **15**, 399–403.

Barnett, V. (1976). The ordering of multivariate data. *J. Royal Statist. Soc.* **A 139**, 318–354.

Benczur, A. (1968). On sequences of equivalent events and the compound Poisson process. *Studia Sci. Math. Hungar.* **3**, 451–458.

Berkes, I., and W. Philipp (1977). Approximation theorems for independent and weakly dependent random vectors. To appear.

Berman, S. M. (1961). Convergence to bivariate extreme value distributions. *Ann. Inst. Statist. Math.* **13**, 217–223.

Berman, S. M. (1962a). A law of large numbers for the maximum in a stationary Gaussian sequence. *Ann. Math. Statist.* **33**, 93–97.

Berman, S. M. (1962b). Limiting distribution of the maximum term in a sequence of dependent random variables. *Ann. Math. Statist.* **33**, 894–908.

Berman, S. M. (1962c). Equally correlated random variables. *Sankhya* **A 24**, 155–156.

Berman, S. M. (1964). Limit theorems for the maximum term in stationary sequences. *Ann. Math. Statist.* **35**, 502–516.

Bhattacharyya, B. B. (1970). Reverse submartingale and some functions of order statistics. *Ann. Math. Statist.* **41**, 2155–2157.

Bhattacharyya, P. K. (1974). Convergence of sample paths of normalized sums of induced order statistics. *Ann. Statist.* **2**, 1034–1039.

Biondini, R., and M. M. Siddiqui (1975). Record values in Markov sequences. In: *Statistical inference and related topics*, Vol. 2 (Ed.: Puri). Academic Press, New York, pp. 291–352.

Bobrov, A. A. (1954). The growth of the maximal summand in sums of independent random variables. *Mat. Sb. Kievskogo Gosuniv.* **5**, 15–38 (in Russian).

Bofinger, Eve, and V. J. Bofinger (1965). The correlation of maxima in samples drawn from a bivariate normal distribution. *Austr. J. Statist.* **7**, 57–61.

REFERENCES

Bofinger, V. J. (1970). The correlation of maxima in several bivariate nonnormal distributions. *Austr. J. Statist.* **12**, 1–7.

Book, S. A. (1972). Large deviation probabilities for weighted sums. *Ann. Math. Statist.* **43**, 1221–1234.

Bortkiewicz, L. von (1922). Variationsbreite und mittlerer Fehler. *Sitzungsberichte Berliner Math. Ges.* **21**.

Bury, K. V. (1974). Distribution of smallest lognormal and gamma extremes. *Statistische Hefte,* **15**, 105–114.

Cacoullos, T., and H. DeCicco (1967). On the distribution of the bivariate range. *Technometrics,* **9**, 476–480.

Campbell, J. W., and C. P. Tsokos (1973). The asymptotic distribution of maxima in bivariate samples. *J. Amer. Statist. Assoc.* **68**, 734–739.

Canfield, R. V. and L. E. Borgman (1975). Some distributions of time to failure for reliability applications. *Technometrics* **17**, 263–268.

Chan, L. K., and G. A. Jarvis (1970). Convergence of moments of some extreme order statistics. *J. Statist. Res.* **4**, 37–42.

Chandler, K. N. (1952). The distribution and frequency of record values. *J. Royal Statist. Soc.* **B 14**, 220–228.

Chung, K. L., and P. Erdös (1952). On the applications of the Borel-Cantelli lemma. *Trans. Amer. Math. Soc.* **72**, 179–186.

Clough, D. J. (1969). An asymptotic extreme-value sampling theory for estimation of a global maximum. *Canad. Op. Res. Soc. J.* **7**, 102–115.

Clough, D. J., and S. Kotz (1965). Extreme-value distributions with a special queueing model application. *Canad. Op. Res. Soc. J.* **3**, 96–109.

Cohen, J. W. (1967). The distribution of the maximum number of customers present simultaneously during a busy period for the queuing systems $M/G/1$ and $G/M/1$. *J. Appl. Probability* **4**, 162–179.

Cohn, H. and A. G. Pakes (1977). *Convergence rate results for the explosive Galton-Watson process.* Technical Report No. 120, Department of Statistics, Princeton Univ., Princeton, N. J.

Cramér, H. (1946). *Mathematical methods of statistics.* Princeton Univ. Press, Princeton, N. J.

Daniels, H. E. (1945). The statistical theory of the strength of bundles of threads. *Proc. Royal Soc.* **A 183**, 405–435.

Darling, D. A. (1952). The influence of the maximum term in the addition of independent random variables. *Trans. Amer. Math. Soc.* **73**, 95–107.

David, H. A. (1970). *Order statistics.* John Wiley & Sons, New York.

David, H. A. (1973). Concomitants of order statistics. *Bull. Inst. Internat. Statist.* **45**, 295–300.

David, H. A., and J. Galambos (1974). The asymptotic theory of concomitants of order statistics. *J. Appl. Probability* **11**, 762–770.

David, H. A., and M. L. Moeschberger (1978). *The theory of competing risks.* Griffin's Statistical Monographs, London.

David, H. A., O'Connell, M. J., and S. S. Yang (1977). Distribution and expected value of the rank of a concomitant of an order statistic. *Ann. Statist.* **5**, 216–223.

Dawson, D. A., and D. Sankoff (1967). An inequality for probabilities. *Proc. Amer. Math. Soc.* **18**, 504–507.

Deheuvels, P. (1974). Valeurs extrémales d'échantillons croissants d'une variable aléatoire réelle. *Ann. Inst. H. Poincaré* **B 10**, 89–114.

Denzel, G. E., and G. L. O'Brien (1975). Limit theorems for extreme values of chain-dependent processes. *Ann. Probability* **3**, 773–779.

Deo, C. M. (1971). On maxima of Gaussian sequences. *Abstract. Ann. Math. Statist.* **42**, 2176.

Deo, C. M. (1973a). A note on strong mixing Gaussian sequences. *Ann. Probability* **1**, 186–187.

Deo, C. M. (1973b). A weak convergence theorem for Gaussian sequences. *Ann. Probability* **1**, 1061–1064.

Dodd, E. L. (1923). The greatest and the least variate under general laws of error. *Trans. Amer. Math. Soc.* **25**, 525–539.

Downton, F. (1971). Stochastic models for successive failures. *Proc. 38th Session ISI,* Washington, D.C., pp. 677–697.

Dykstra, R. L., Hewett, J. E., and W. A. Thompson, Jr. (1973). Events which are almost independent. *Ann. Statist.* **1**, 674–681.

Dziubdziela, W. (1972). Limit distribution of extreme order statistics in a sequence of random size. *Applicationes Math.* **13**, 199–205 (in Russian).

Dziubdziela, W., and B. Kopocinski (1976a). Limiting properties of the distance random variables. *Przeglad Stat.* **23**, 471–477 (in Polish).

Dziubdziela, W., and B. Kopocinski (1976b). Limiting properties of the kth record values. *Applicationes Math.* **15**, 187–190.

Dwass, M. (1960). Some k-sample rank order tests. *Contributions to probability and statistics.* Stanford, Calif., Stanford Univ. Press, pp. 198–202.

Dwass, M. (1964). Extremal processes. *Ann. Math. Statist.* **35**, 1718–1725.

Dwass, M. (1966). Extremal processes II. *Ill. J. Math.,* **10**, 381–391.

Dwass, M. (1973). Extremal processes III. Discussion Paper 41, Northwestern Univ.

Enns, E. G. (1970). The optimum strategy for choosing the maximum of n independent random variables. *Operations Res.* **14**, 89–96.

Epstein, B. (1948). Applications of the theory of extreme values in fracture problems. *J. Amer. Statist. Assoc.* **43**, 403–412.

Epstein, B. (1960). Elements of the theory of extreme values. *Technometrics,* **2**, 27–41.

Erdös, P. (1942). On the law of the iterated logarithm. *Ann. Math.* **43**, 419–436.
Erdös, P., and A. Rényi (1959). On Cantor's series with convergent $\Sigma 1/q_n$. *Ann. Univ. Sci. Budapest, Sectio Math.* **2**, 93–109.
Fabens, A. J., and M. F. Neuts (1970). The limiting distribution of the maximum term in a sequence of random variables defined on a Markov chain. *J. Appl. Probability* **7**, 754–760.
Feller, W. (1966). *An introduction to probability theory and its applications.* Vol. 2. John Wiley & Sons, New York.
Finetti, B de (1930). Funzione carattenstica di un fenomeno aleatorio. *Atti. Accad. Naz. Lincei Rend. Cl. Sci. Fiz. Mat. Nat.* **4**, 86–133.
Finkelstein, B. V. (1953). On the limiting distributions of the extreme terms of a variational series of a two-dimensional random quantity. *Dokl. Akad. SSSR* (N.S.) **91**, 209–211 (in Russian).
Fisher, R. A., and L. H. C. Tippett (1928). Limiting forms of the frequency distributions of the largest or smallest member of a sample. *Proc. Cambridge Philos. Soc.* **24**, 180–190.
Food and Drug Administration Advisory Committee on Protocols for Safety Evaluation (1971). Panel on carcinogenesis report on cancer testing in the safety evaluation of food additives and pesticides. *Toxicol. Appl. Pharmacol.* **20**, 419–438.
Foster, F. G., and A. Stuart (1954). Distribution free tests in time-series based on the breaking of records. *J. Royal Statist. Soc.* **B 16**, 1–13.
Frank, O. (1966). Generalization of an inequality of Hájek and Rényi. *Skand. Aktuar.* **49**, 85–89.
Fréchet, M. (1927). Sur la loi de probabilité de l'écart maximum. *Ann. de la Soc. polonaise de Math. (Cracow)* **6**, 93.
Fréchet, M. (1940). *Les probabilités associées à un système d'évènements compatibles et dépendants.* Hermann, Paris.
Fréchet, M. (1951). Sur les tableaux de corrélation dont les marges sont données. *Ann. Univ. Lyon,* Section A, Series 3, **14**, 53–77.
Freudenberg, W. and D. Szynal (1976). Limit laws for a random number of record values. *Bull. Acad. Pol. Sci. Ser. Math. Astron. et Phys.* **24**, 193–199.
Galambos, Éva (1965). Discussion of probabilistic inequalities by the method of A. Rényi. Dissertation, L. Eötvös Univ., Budapest (in Hungarian).
Galambos, J. (1966). On the sieve methods in probability theory I. *Studia Sci. Math. Hungar.* **1**, 39–50.
Galambos, J. (1969). Quadratic inequalities among probabilities. *Ann. Univ. Sci. Budapest, Sectio Math.* **12**, 11–16.
Galambos, J. (1970). On the sieve methods in probability theory II. *Ghana J. Sci.* **10**, 11–15.

Galambos, J. (1972). On the distribution of the maximum of random variables. *Ann. Math. Statist.* **43**, 516–521.

Galambos, J. (1973a). A general Poisson limit theorem of probability theory. *Duke Math. J.* **40**, 581–586.

Galambos, J. (1973b). The distribution of the maximum of a random number of random variables with applications. *J. Appl. Probability* **10**, 122–129.

Galambos, J. (1974). A limit theorem with applications in order statistics. *J. Appl. Probability,* **11**, 219–222.

Galambos, J. (1975a). Methods for proving Bonferroni type inequalities. *J. London Math. Soc.* (2) **9**, 561–564.

Galambos, J. (1975b). Characterizations of probability distributions by properties of order statistics I (continuous distributions). In: *Statistical distributions in scientific work,* Vol. 3 (Ed.: G.P. Patil et al.). D. Reidel, Dordrecht, pp. 71–88.

Galambos, J. (1975c). Characterizations of probability distributions by properties of order statistics II (discrete distributions). As above, pp. 89–101.

Galambos, J. (1975d). Order statistics of samples from multivariate distributions. *J. Amer. Statist. Assoc.* **70**, 674–680.

Galambos, J. (1975e). Limit laws for mixtures with applications to asymptotic theory of extremes. *Zeitschrift fur Wahrschein. verw. Geb.* **32**, 197–207.

Galambos, J. (1976). A remark on the asymptotic theory of sums with random size. *Math. Proc. Cambridge Philos. Soc.* **79**, 531–532.

Galambos, J. (1977a). The asymptotic theory of estreme order statistics. In: *The theory and applications of reliability,* Vol. 1 (Ed.: C. Tsokos and I. Shimi). Academic Press, New York, pp. 151–164.

Galambos, J. (1977b). Bonferroni inequalities. *Ann. Probability,* **5**, 577–581.

Galambos, J. (1978). The speed of convergence of the distribution of extremes. To appear.

Galambos, J., and S. Kotz (1978). *Characterizations of probability distributions.* Lecture Notes in Mathematics, Springer Verlag, Heidelberg.

Galambos, J., and R. Mucci (1978). Inequalities for linear combinations of binomial moments. To be published in *Publ. Math. Debrecen.*

Galambos, J., and A. Rényi (1968). On quadratic inequalities in the theory of probability. *Studia Sci. Math. Hungar.* **3**, 351–358.

Galambos, J., and E. Seneta (1973). Regularly varying sequences. *Proc. Amer. Math. Soc.* **41**, 110–116.

Galambos, J., and E. Seneta (1975). Record times. *Proc. Amer. Math. Soc.* **50**, 383–387.

Gallott, S. A. (1966). A bound for the maximum of a number of random variables. *J. Appl. Probability* **3**, 556–558.

REFERENCES

Gaver, D. P. (1976). Random record models. *J. Appl. Probability* **13**, 538–547.
Geffroy, Jean (1958/1959). Contributions à la théorie des valeurs extrèmes. *Publ. Inst. Statist. Univ. Paris,* **7/8**, 37–185.
Ghosh, M., Jogesh Babu, G., and N. Mukhopadhyay (1975). Almost sure convergence of sums of maxima and minima of positive random variables. *Zeitschrift fur Wahrschein. verw. Geb.* **33**, 49–54.
Gnedenko, B. V. (1943). Sur la distribution limite du terme maximum d'une série aléatoire. *Ann. Math.* **44**, 423–453.
Goldstein, N. (1963). Random numbers for the extreme value distribution. *Publ. Inst. Statist. Univ. Paris,* **12**, 137–158.
Govindarajulu, Z. (1966). Characterizations of the exponential and power distributions. *Skand. Aktuar.* **49**, 132–136.
Green, R. F. (1976a). Outlier prone and outlier resistant distributions. *J. Amer. Statist. Assoc.* **71**, 502–505.
Green, R. F. (1976b). Partial attraction of maxima. *J. Appl. Probability* **13**, 159–163.
Greig, Margaret (1967). Extremes in a random assembly. *Biometrika* **54**, 273–282.
Grenander, U. (1965). A limit theorem for sums of minima of stochastic variables. *Ann. Math. Statist.* **36**, 1041–1042.
Grigelionis, B. (1962). On an asymptotic decomposition of the remainder term in the case of convergence to the Poisson law. *Lietuv. Mat. Rink.* **2**, 35–48 (in Russian).
Grigelionis, B. (1970). Limit theorems for the sums of multidimensional stochastic step processes. *Lietuv. Mat. Rink.* **10**, 29–49 (in Russian).
Guess, H. A., and K. S. Crump (1977). Can we use animal data to estimate safe doses for chemical carcinogens? In: *Environmental health: Quantitative methods.* (Ed.: A. S. Whittemore). SIAM, Philadelphia, pp. 13–30.
Guiasu, S. (1971). On the asymptotic distribution of the sequences of random variables with random indices. *Ann. Math. Statist.* **42**, 2018–2028.
Gumbel, E. J. (1958). *Statistics of extremes.* Columbia Univ. Press, New York.
Gumbel, E. J. (1960). Bivariate exponential distributions. *J. Amer. Statist. Assoc.* **55**, 698–707.
Gumbel, E. J. (1962). Statistical theory of extreme values. In: *Contributions to order statistics* (Ed.: A. S. Sarhan and B. G. Greenberg). John Wiley & Sons, New York, pp. 56–93 and 406–431.
Gumbel, E. J., and N. Goldstein (1964). Analysis of empirical bivariate extremal distributions. *J. Amer. Statist. Assoc.* **59**, 794–816.
Gumbel, E. J., and C. K. Mustafi (1967). Some analytical properties of

bivariate extreme value distributions. *J. Amer. Statist. Assoc.* **62**, 569–588.
Gyires, B. (1975). Linear order statistics in the case of samples with non-independent elements. *Publ. Math. Debrecen* **22**, 47–63.
Haan, L. de (1970). *On regular variation and its application to the weak convergence of sample extremes.* Mathematical Centre Tracts, Vol. 32, Amsterdam.
Haan, L. de (1971). A form of regular variation and its application to the domain of attraction of the double exponential distribution. *Zeitschrift fur Wahrschein. verw. Geb.* **17**, 241–258.
Haan, L. de (1974a). Equivalence classes of regularly varying functions. *Stochastic Processes Appl.* **2**, 243–259.
Haan, L. de (1974b). Weak limits of sample range. *J. Appl. Probability* **11**, 836–841.
Haan, L. de, and A. Hordijk (1972). The rate of growth of sample maxima. *Ann. Math. Statist.* **43**, 1185–1196.
Haan, L. de, and S. I. Resnick (1973). Almost sure limit points of record values. *J. Appl. Probability* **10**, 528–542.
Haan, L. de, and S. I. Resnick (1977). Limit theory for multivariate sample extremes. *Zeitschrift fur Wahrschein. verw. Geb.* **40**, 317–337.
Haghighi-Talab, D., and C. Wright (1973). On the distribution of records in a finite sequence of observations, with an application to a road traffic problem. *J. Appl. Probability,* **10**, 556–571.
Harris, R. (1970). An application of extreme value theory to reliability theory. *Ann. Math. Statist.* **41**, 1456–1465.
Heyde, C. C. (1971). On the growth of the maximum queue length in a stable queue. *Operations Res.* **19**, 447–452.
Hoglund, T. (1972). Asymptotic normality of sums of minima of random variables. *Ann. Math. Statist.* **43**, 351–353.
Holmes, P. T., and W. E. Strawderman (1969). A note on the waiting times between record observations. *J. Appl. Probability* **6**, 711–714.
Homma, T. (1951). On the asymptotic independence of order statistics. *Rep. Stat. Appl. Res. JUSE* **1**, 1–3.
Huang, J. S. (1974). Characterizations of the exponential distribution by order statistics. *J. Appl. Probability* **11**, 605–608.
Hunter, D. (1976). An upper bound for the probability of a union. *J. Appl. Probability* **13**, 597–603.
Ibragimov, I. A., and Y. V. Linnik (1971). *Independent and stationary sequences of random variables.* Wolters-Noorhoff, Groningen.
Ibragimov, I. A. and Yu. A Rozanov (1970). *Gaussian stochastic processes.* Izdatelstvo Nauka, Moscow.
Iglehart, D. L. (1972). Extreme values in the $GI/G/1$ queue. *Ann. Math. Statist.* **43**, 627–635.

Iglehart, D. L. (1974). Weak convergence in applied probability. *Stochastic Processes Appl.* **2**, 211–241.

Ikeda, S., and T. Matsunawa (1970). On asymptotic independence of order statistics. *Ann. Inst. Statist. Math. Tokyo* **22**, 435–449.

Ivcenko, G. I. (1971). The limit distributions of order statistics of a polynomial scheme. *Teor. verojatmost. i primen.* **16**, 102–115.

Jensen, A. (1969). A characteristic application of statistics in hydrology. *37th Session ISI, London*.

Johnson, N. L., and S. Kotz (1968/1972). *Distributions in statistics*. 4 vols. John Wiley & Sons, New York.

Jordan, K. (1927). The foundations of the theory of probability. *Mat. Phys. Lapok*, **34**, 109–136 (in Hungarian).

Judickaja, P. I. (1974). On the maximum of a Gaussian sequence. *Theory Probability and Math. Statist. (Kiev)*, **2**, 259–267.

Juncosa, M. L. (1949). On the distribution of the minimum in a sequence of mutually independent random variables. *Duke Math. J.* **16**, 609–618.

Kalinauskaite, N. (1973). The effect of the maximum modulus of a term on the sum of independent random vectors I. *Lietuv. Mat. Rink.* **13**, No. 4, 117–123.

Kalinauskaite, N. (1976). Part II of the preceding paper, same journal, **16**, 41–48.

Kawata, T. (1951). Limit distributions of single order statistics. *Rep. Stat. Appl. Res. JUSE* **1**, 4–9.

Kendall, D. G. (1967). On finite and infinite sequences of exchangeable events. *Studia Sci. Math. Hungar.* **2**, 319–327.

Kolchin, V. F. (1969). On the limiting behavior of extreme order statistics in a polynomial scheme. *Theory Probability Appl.* **14**, 458–469.

Kotz, S. (1973). Normality vs. log normality with applications. *Comm. in Statist.* **1**, 113–132.

Kounias, E. G. (1968). Bounds for the probability of a union, with applications. *Ann. Math. Statist.* **39**, 2154–2158.

Kounias, S., and J. Marin (1976). Best linear Bonferroni bounds. *SIAM J. Appl. Math.* **30**, 307–323.

Kudo, A. (1958). On the distribution of the maximum value of an equally correlated sample from a normal population. *Sankhya* **A20**, 309–316.

Kunio, T., Shimizu, M. Yamada, K., and Y. Kimura (1974). An interpretation of the scatter of fatigue limit on the basis of the theory of extreme value. *Trans. JSME* **40**, 2101–2109 (in Japanese).

Kwerel, S. M. (1975a). Most stringent bounds on aggregated probabilities of partially specified dependent systems. *J. Amer. Statist. Assoc.* **70**, 472–479.

Kwerel, S. M. (1975b). Bounds on the probability of the union and intersection of m events. *Adv. Appl. Probability* **7**, 431–448.

Kwerel, S. M. (1975c). Most stringent bounds on the probability of the union ..., *J. Appl. Probability* **12**, 612–619.

Lai, T. L., and H. Robbins (1976). Maximally dependent random variables. *Proc. Nat. Acad. Sci. USA* **73**, 286–288.

Lai, T. L., and H. Robbins (1977). A class of dependent random variables and their maxima. To appear.

Lamperti, J. (1964). On extreme order statistics. *Ann. Math. Statist.* **35**, 1726–1737.

Larsen, R. I. (1969). A new mathematical model of air pollutant concentration averaging time and frequency. *J. Air Pollution Control Assoc.* **19**, 24–30.

Leadbetter, M. R. (1974). On extreme values in stationary sequences. *Zeitschrift fur Wahrschein. verw. Geb.* **28**, 289–303.

Leadbetter, M. R. (1975). Aspects of extreme value theory for stationary processes—A survey. In: *Stochastic processes and related topics*, Vol. 1 (Ed: Puri). Academic Press, New York.

Loève, M. (1942). Sur les systèmes d'évènements. *Ann. Univ. Lyon*, Sect. A, **5**, 55–74.

Loynes, R. M. (1965). Extreme values in uniformly mixing stationary stochastic processes. *Ann. Math. Statist.*, **36**, 993–999.

Mahmoud, M. W., and A. Ragab (1975). On order statistics in samples drawn from the extreme value distribution. *Math. Operationsforschung u. Stat.* **6**, 800–816.

Mann, N. R., Schaefer, R. E., and N. D. Singpurwalla (1974). *Methods for statistical analysis of reliability and life data*. John Wiley & Sons, New York.

Marcus, M. B., and M. Pinsky (1969). On the domain of attraction of exp $(-e^{-x})$. *J. Math. Anal. Appl.* **28**, 440–449.

Mardia, K. V. (1964a). Some results on the order statistics of the multivariate normal and Pareto type I populations. *Ann. Math. Statist.* **35**, 1815–1818.

Mardia, K. V. (1964b). Asymptotic independence of bivariate extremes. *Calcutta Stat. Assoc. Bull.* **13**, 172–178.

Mardia, K. V. (1967). Correlation of the ranges of correlated samples. *Biometrika* **54**, 529–539.

Mardia, K. V. (1970). *Families of bivariate distributions*. Griffin, London.

Marshall, A. W., and I. Olkin (1967). A generalized bivariate exponential distribution. *J. Appl. Probability* **4**, 291–302.

Marszal, E., and J. Sojka (1974). Limiting distributions and their application. *Wyz. Szkol. Ped. Krakow Rocznik Nauk* **7**, 59–76.

Matsunawa, T. and S. Ikeda (1976). Uniform asymptotic distribution of extreme values. In: *Essays in probability and statistics in honor of J. Ogawa* (Ed.: S. Ikeda). Shinko Tsusho, Tokyo, pp. 419–432.

McCord, J. R. (1964). On asymptotic moments of extreme statistics. *Ann. Math. Statist.* **35**, 1738–1745.

McCormick, W., and Y. Mittal (1976). *On weak convergence of the maximum.* Technical Report 81, Stanford Univ.

Meeker, W. Q., and W. B. Nelson (1975). *Tables for the Weibull and smallest extreme value distribution.* General Electric Co.

Mejzler, D. G. (1949). On a theorem of B. V. Gendenko. *Sb. Trudov. Inst. Mat. Akad. Nauk Ukrain. SSR* **12**, 31–35 (in Russian).

Mejzler, D. G. (1950). On the limit distribution of the maximal term of a variational series. *Dopovidi Akad. Nauk Ukrain. SSR* **1**, 3–10 (in Ukrainian).

Mejzler, D. G. (1953). The study of the limit laws for the variational series. *Trudy Inst. Mat. Akad. Nauk Uzbek. SSR* **10**, 96–105 (in Russian).

Mejzler, D. G. (1956). On the problem of the limit distribution for the maximal term of a variational series. *L'vov Politechn. Inst. Naucn. Zp. (Fiz.-Mat.)* **38**, 90–109 (in Russian).

Meyer, R. M. (1969). Note on a multivariate form of Bonferroni's inequalities. *Ann. Math. Statist.* **40**, 692–693.

Mijnheer, J. L. (1975). *Sample path properties of stable processes.* Mathematical Centre Tracts, Vol. 59, Amsterdam.

Mikhailov, V. G. (1974). Asymptotic independence of vector components of multivariate extreme order statistics. *Teor. verojatnost. i primen.* **19**, 817–821.

Mises, R. von (1923). Über die Variationsbreite einer Beobachtungsreihe. *Sitzungsberichte Berlin. Math. Geo.* **22**, 3.

Mises, R. von (1936). La distribution de la plus grande de n valeurs. Reprinted in *Selected Papers II.* Amer. Math. Soc., Providence, R. I., 1954, pp. 271–294.

Mittal, Y. (1974). Limiting behavior of maxima in stationary Gaussian sequences. *Ann. Probability*, **2**, 231–242.

Mittal, Y. (1976). *Maxima of partial samples in Gaussian sequences.* Technical Report No. 3, Stanford Univ.

Mittal, Y., and D. Ylvisaker (1975). Limit distributions for the maxima of stationary Gaussian processes. *Stochastic Processes Appl.* **3**, 1–18.

Mittal, Y., and D. Ylvisaker (1976). Strong laws for the maxima of stationary Gaussian processes. *Ann. Probability*, **4**, 357–371.

Mogyoródi, J. (1967). On the limit distribution of the largest term in the order statistics of a sample of random size. *Magyar Tud. Akad. Mat. Fiz. Oszt. Kozl.* **17**, 75–83 (in Hungarian).

Morgenstern, D. (1956). Einfache Beispiele zweidimensionaler Verteilungen. *Mitteilungsblatt Math. Stat.* **8**, 234–235.

Mori, T. (1976a). Limit laws for maxima and second maxima for strong mixing processes. *Ann. Probability*, **4**, 122–126.

Mori, T. (1976b). The strong law of large numbers when extreme terms are excluded from sums. *Zeitschrift fur Wahrschein. verw. Geb.* **36**, 189–194.

Mori, T. (1977). Stability for sums of i.i.d. random variables when extreme terms are excluded. *Zeitschrift fur Wahrschein. verw. Geb.* **40**, 159–167.

Mori, T., and H. Oodaira (1976). A functional law of the iterated logarithm for sample sequence. *Yokohama Math. J.* **24**, 35–49.

Morrison, M., and F. Tobias (1965). Some statistical characteristics of a peak to average ratio. *Technometrics* **7**, 379–385.

Mucci, R. (1977). *Limit theorems for extremes.* Thesis for Ph.D., Temple University.

Nagaev, A. V. (1970). On the role of the largest of the order statistics in the formation of a large deviation of a sum of independent random variables. *Sovjet Math. Dokl.* **11**, 972–974.

Nagaev, A. V. (1971). A limit distribution of extreme terms of a variational series under conditions of the type of large deviations that are imposed on the sample mean. *Theor. Prob. Appl.* **16**, 126–140.

Nair, K. A. (1976). Bivariate extreme value distributions. *Comm. in Statist.* **5**, 575–581.

National Bureau of Standards (1953). *Probability tables for the analysis of extreme value data.* Appl. Math. Series, 22.

Neuts, M. F. (1967). Waiting times between record observations. *J. Appl. Probability* **4**, 206–208.

Newell, G. F. (1964). Asymptotic extremes for m-dependent random variables. *Ann. Math. Statist.* **35**, 1322–1325.

Nisio, M. (1967). On the extreme values of Gaussian processes. *Osaka J. Math.* **4**, 313–326.

O'Brien, G. L. (1974a). Limit theorems for the maximum term of a stationary process. *Ann. Probability* **2**, 540–545.

O'Brien, G. L. (1974b). The maximum term of uniformly mixing stationary processes. *Zeitschrift fur Wahrschein. verw. Geb.* **30**, 57–63.

O'Connell, M. J., and H. A. David (1976). Order statistics and their concomitants in some double sampling situations. In: *Essays in probability and statistics in honor of J. Ogawa* (Ed.: S. Ikeda). Shinko Tsusho, Tokyo, pp. 451–466.

Pakes, A. G. (1975). On tails of waiting time distributions. *J. Appl. Probability* **12**, 555–564.

Philipp, W. (1967). Some metrical theorems in number theory. *Pacific J. Math.* **20**, 109–127.
Philipp, W. (1971). Mixing sequences of random variables and probabilistic number theory. *Memoirs Amer. Math. Soc.*, Vol. 114.
Philipp, W. (1976). A conjecture of Erdös on continued fractions. *Acta Arithm.* **28**, 379–386.
Phoenix, S. L., and H. M. Taylor (1973). The asymptotic strength distribution of a general fiber bundle. *Adv. Appl. Probability* **5**, 200–216.
Pickands, J. III (1967a). Sample sequences of maxima. *Ann. Math. Statist.* **38**, 1570–1574.
Pickands, J. III (1967b). Maxima of stationary Gaussian processes. *Zeitschrift fur Wahrschein. verw. Geb.* **7**, 190–233.
Pickands, J. III (1968). Moment convergence of sample extremes. *Ann. Math. Statist.* **39**, 881–889.
Pickands, J. III (1969). An iterated logarithm law for the maximum in a stationary Gaussian sequence. *Zeitschrift fur Wahrschein. verw. Geb.* **12**, 344–353.
Pickands, J. III (1971). The two-dimensional Poisson process and extremal processes. *J. Appl. Probability* **8**, 745–756.
Pickands, J. III (1975). Statistical inference using extreme order statistics. *Ann. Statist.* **3**, 119–131.
Pickands, J. III (1977). Multivariate extreme value distributions. To be published.
Plackett, R. L. (1954). A reduction formula for normal multivariate integrals. *Biometrika* **41**, 351–360.
Pólya, G. (1920). Über der zentralen Grenzwertsatz der Wahrscheinlichkeitsrechnung und das Momentenproblem. *Math. Z.* **8**, 171–181.
Posner, E. C. (1965). The application of extreme value theory to error free communication. *Technometrics*, **7**, 517–529.
Posner, E. C., Rodenich, E. R., Ashlock, J. C., and Lurie, S. (1969). Application of an estimator of high efficiency in bivariate extreme value theory. *J. Amer. Statist. Assoc.* **64**, 1403–1415.
Prabhu, N. U. (1965). *Queues and inventories.* John Wiley & Sons, New York.
Rényi, A. (1953). On the theory of order statistics. *Acta Math. Acad. Sci. Hungar.* **4**, 191–231.
Rényi, A. (1958). Quelques remarques sur les probabilités d'événements dépendants. *J. Math. Pures Appl.* **37**, 393–398.
Rényi, A. (1961). A general method for proving theorems in probability theory and some of its applications. Original in Hungarian, but now translated into English in: *Selected Papers of A. Rényi*, Vol. 2. Akadémiai Kiadó, Budapest, 1976, 581–602.

Rényi, A. (1962). On outstanding values of a sequence of observations (in Hungarian). Translated in: *Selected Papers of A. Rényi*, Vol. 3. Akadémiai Kiadó, Budapest, 1976, 50–65.

Rényi, A. (1963). On stable sequences of events. *Sankhya* **A 25**, 293–302.

Resnick, S. I. (1971a). Tail equivalence and applications. *J. Appl. Probability* **8**, 136–156.

Resnick, S. I. (1971b). Asymptotic location and recurrence properties of maxima of a sequence of random variables defined on a Markov chain. *Zeitschrift fur Wahrschein. verw. Geb.* **18**, 197–217.

Resnick, S. I. (1972a). Products of distribution functions attracted to extreme value laws. *J. Appl. Probability* **8**, 781–793.

Resnick, S. I. (1972b). Stability of maxima of random variables defined on a Markov chain. *Adv. Appl. Probability* **4**, 285–295.

Resnick, S. I. (1973a). Limit laws for record values. *Stochastic Processes Appl.* **1**, 67–82.

Resnick, S. I. (1973b) Record values and maxima. *Ann. Probability* **1**, 650–662.

Resnick, S. I. (1973c). Extremal processes and record value times. *J. Appl. Probability* **10**, 863–868.

Resnick, S. I. (1974). Inverses of extremal processes. *Adv. Appl. Probability* **6**, 392–406.

Resnick, S. I. (1975). Weak convergence to extremal processes. *Ann. Probability*, **3**, 951–960.

Resnick, S. I., and M. F. Neuts (1970). Limit laws for maxima of a sequence of random variables defined on a Markov chain. *Adv. Appl. Probability* **2**, 323–343.

Resnick, S. I., and M. Rubinovitch (1973). The structure of extremal processes. *Adv. Appl. Probability* **5**, 287–307.

Resnick, S. I., and R. J. Tomkins (1973). Almost sure stability of maxima. *J. Appl. Probability* **10**, 387–401.

Richter, W. (1965). Das Null-Eins-Gesetz und ein Grenzwertsatz für zufällige Prozesse mit diskreter zufälliger Zeit. *Wiss. Zeitschrift Techn. Univ. Dresden* **14**, 497–504.

Riddler-Rowe, C. J. (1967). On two problems of exchangeable events. *Studia Sci. Math. Hungar.* **2**, 415–418.

Robbins, H., and D. Siegmund (1972). On the law of the iterated logarithm for maxima and minima. *Proc. Sixth Berkeley Symp. Math. Statist. Probability* **3**, 51–70.

Rootzén, H. (1974). Some properties of convergence in distribution of sums and maxima of dependent random variables. *Zeitschrift fur Wahrschein. verw. Geb.* **29**, 295–307.

Rossberg, H. J. (1960). Über die Verteilungsfunktionen der Differenzen und Quotienten von Ranggrössen. *Math. Nachr.* **21**, 37–79.
Rossberg, H. J. (1965a). Über die stochastische Unabhängigkeit gewisser Funktionen von Ranggrössen. *Math. Nachr.* **28**, 157–167.
Rossberg, H. J. (1965b). Die asymptotische Unabhängigkeit der kleinsten und grössten Werte einer Stichprobe vom Stichprobenmittel. *Math. Nachr.* **28**, 305–318.
Sarhan, A. E., and B. G. Greenberg (1962). *Contributions to order statistics.* John Wiley & Sons, New York.
Sen, P. K. (1961). A note on the large sample behavior of extreme sample values from distributions with finite endpoints. *Bull. Calcutta Statist. Assoc.* **10**, 106–115.
Sen, P. K. (1970). A note on order statistics for heterogeneous distributions. *Ann. Math. Statist.* **41**, 2137–2139.
Sen, P. K. (1972). On weak convergence of extremal processes for random sample sizes. *Ann. Math. Statist.* **43**, 1355–1362.
Sen, P. K. (1973a). On fixed size confidence bands for the bundle strength of filaments. *Ann. Statist.* **1**, 526–537.
Sen, P. K. (1973b). An asymptotically efficient test for the bundle strength of filaments. *J. Appl. Probability* **10**, 586–596.
Sen, P. K. (1976). A note on invariance principles for induced order statistics. *Ann. Probability* **4**, 474–479.
Seneta, E. (1976). *Regularly varying functions.* Lecture Notes in Mathematics, Vol. 508, Springer Verlag, Heidelberg.
Sethuraman, J. (1965). On a characterization of the three limiting types of the extreme. *Sankhya* A **27**, 357–364.
Shorrock, R. W. (1972a). A limit theorem for inter-record times. *J. Appl. Probability* **9**, 219–223.
Shorrock, R. W. (1972b). On record values and record times. *J. Appl. Probability* **9**, 316–326.
Shorrock, R. W. (1973). Record values and interrecord times. *J. Appl. Probability* **10**, 543–555.
Shorrock, R. W. (1974). On discrete time extremal processes. *Adv. Appl. Probability* **6**, 580–592.
Sibuya, M. (1960). Bivariate extreme statistics. *Ann. Inst. Stat. Math.* **11**, 195–210.
Siddiqui, M. M., and R. W. Biondini (1975). The joint distribution of record values and inter-record times. *Ann. Probability*, **3**, 1012–1013.
Singh, C. (1967). On the extreme values and range of samples from non-normal populations. *Biometrika* **54**, 541–550.
Singpurwalla, N. D. (1972). Extreme values from a lognormal law with

applications to air pollution problems. *Technometrics* **14**, 703–711.
Slepian, D. (1962). The one-sided barrier problem for Gaussian noise. *Bell System Tech. J.* **41**, 463–501.
Smid, B. and A. J. Stam (1975). Convergence in distribution of quotients of order statistics. *Stochastic Processes Appl.* **3**, 287–292.
Smirnov, N. V. (1949 and 1952). Limit distributions for the terms of a variational series. Original Russian in *Trudy Mat. Inst. Steklov* **25** (1949), 1–60. Translated in 1952 by *Amer. Math. Soc. Transl.* **67**, 1–64.
Sobel, M., and V. R. R. Uppuluri (1972). On Bonferroni-type inequalities of the same degree for the probability of unions and intersections. *Ann. Math. Statist.* **43**, 1549–1558.
Srivastava, O. P. (1967). Asymptotic independence of certain statistics connected with the extreme order statistics in a bivariate distribution. *Sankhya* **A 29**, 175–182.
Stam, A. J. (1973). Regular variation of the tail of a subordinated probability distribution. *Adv. Appl. Probability* **5**, 308–327.
Steinebach, J. (1976). *Exponentielles Konvergenzverhalten von Wahrscheinlichkeiten grosser Abweichungen.* Thesis for Ph.D., Univ. Düsseldorf.
Stevens, W. L. (1939). Solution to a geometrical problem in probability. *Ann. Eugenics, London* **9**, 315–320.
Strawderman, W. E., and P. T. Holmes (1970). On the law of the iterated logarithm for inter-record times. *J. Appl. Probability* **7**, 432–439.
Suh, M. W., Bhattacharyya, B. B., and A. Grandage (1970). On the distribution and moments of the strength of a bundle of filaments. *J. Appl. Probability* **7**, 712–720.
Sukhatme, P. V. (1937). Tests of significance for samples of the χ^2-population with two degrees of freedom. *Ann. Eugenics, London*, **8**, 52–56.
Takács, L. (1958). On a general probability theorem and its applications in the theory of stochastic processes. *Proc. Cambridge Philos. Soc.* **54**, 219–224.
Takács, L. (1965). A moment problem. *J. Austral. Math. Soc.* **5**, 487–490.
Takács, L. (1967a). *Combinatorial methods in the theory of stochastic processes.* John Wiley & Sons, New York.
Takács, L. (1967b). On the method of inclusion and exclusion. *J. Amer. Statist. Assoc.* **62**, 102–113.
Takács, L. (1975). Combinatorial and analytic methods in the theory of queues. *Adv. Appl. Probability* **7**, 607–635.
Takács, L. (1977). *Theory of random fluctuations.* To be published.
Tata, M. N. (1969). On outstanding values in a sequence of random variables. *Zeitschrift fur Wahrschein. verw. Geb.* **12**, 9–20.
Thomas, D. I. (1972). On limiting distributions of a random number of dependent random variables. *Ann. Math. Statist.* **43**, 1719–1726.

Thompson, W. A., Jr. (1969). *Applied probability*. Holt, Rinehart and Winston, New York
Tiago de Oliveira, J. (1958). Extremal distributions. *Revista da Fac. Ciencias, Univ. Lisboa,* **A 7**, 215–227.
Tiago de Oliveira, J. (1961). The asymptotical independence of the sample mean and the extremes. *Revista da Fac. Ciencias, Univ. Lisboa,* **A 8**, 299–310.
Tiago de Oliveira, J. (1962/1963). Structure theory of bivariate extremes: extensions. *Estudos de Math. Estat. Econom.* **7**, 165–195.
Tiago de Oliveira, J. (1968). Extremal processes: Definition and properties. *Publ. Inst. Stat. Univ. Paris* **17**, 25–36.
Tiago de Oliveira, J. (1970). Biextremal distributions: statistical decision. *Trab. Estat. Invest. Oper.* **21**, 107–117.
Tiago de Oliveira, J. (1971). A new model of bivariate extremes: Statistical decision. In: *Studi Prob. Stat. ricerca oper. in onore di G. Pompilj. Oderisi, Gubbio,* pp. 1–13.
Tiago de Oliveira, J. (1972a). Statistics for Gumbel and Frechet distributions. In: *Structural safety and reliability* (Ed: A. Freudenthal). Pergamon, New York, pp. 91–105.
Tiago de Oliveira, J. (1972b) An extreme markovian stationary sequence; quick statistical decision. *Metron* **30**, 1–11.
Tiago de Oliveira, J. (1973). An extreme markovian stationary process. *Proc. Fourth Conf. Probability Theory.* Acad. Romania, Brasov, pp. 217–225.
Tiago de Oliveira, J. (1974). Regression in the nondifferentiable bivariate extreme models. *J. Amer. Statist. Assoc.* **69**, 816–818.
Tiago de Oliveira, J. (1975). Bivariate extremes. Extensions. *Proc.* 40*th Session ISI, Warsaw.*
Tiago de Oliveira, J. (1976). Asymptotic behavior of maxima with periodic disturbances. *Ann. Inst. Statist. Math.* **28**, 19–23.
Tippett, L. H. C. (1925). On the extreme individuals and the range of samples taken from a normal population. *Biometrika* **17**, 364–387.
Vervaat, W. (1972). *Success epochs in Bernoulli trials with applications in number theory*. Mathematical Centre Tracts, Vol. 42, Amsterdam.
Vervaat, W. (1973a). Limit theorems for records from discrete distributions. *Stochastic Processes Appl.* **1**, 317–334.
Vervaat, W. (1973b). *Limit theorems for partial maxima and records*. Technical Report, Univ. of Washington, Dept. Math., Seattle.
Vervaat, W. (1977). On records, maxima and a stochastic difference equation. *Math. Inst. Katholieke Univ., Nijmegen,* Report 7702.
Villasenor, J. (1976). *On univariate and bivariate extreme value theory*. Thesis for Ph.D., Iowa State University.

Walsh, J. E. (1969a). Asymptotic independence between largest and smallest of a set of independent observations. *Ann. Inst. Statist. Math. Tokyo*, **21**, 287–289.

Walsh, J. E. (1969b). Approximate distributions for largest and for smallest of a set of independent observations. *S. Afr. Statist. J.* **3**, 83–89.

Walsh, J. E. (1970). Sample size for approximate independence of order statistics. *J. Amer. Statist. Assoc.* **65**, 860–863.

Watson, G. S. (1954). Extreme values in samples from m-dependent stationary stochastic processes. *Ann. Math. Statist.* **25**, 798–800.

Weinstein, S. B. (1973). Theory and application of some classical and generalized asymptotic distributions of extreme values. *IEEE Trans. Inf. Theor.* **19**, 148–154.

Weisman, I. (1975a). Extremal processes generated by independent nonidentically distributed random variables. *Ann. Probability* **3**, 172–177.

Weisman, I. (1975b). On location and scale functions of a class of limiting processes with application to extreme value theory. *Ann. Probability* **3**, 178–181.

Weisman, I. (1975c). Multivariate extremal processes generated by independent nonidentically distributed random variables. *J. Appl. Probability* **12**, 477–487.

Weiss, L. (1969). The joint asymptotic distribution of the k-smallest sample spacings. *J. Appl. Probability* **6**, 442–448.

Welsch, R. E. (1971). A weak convergence theorem for order statistics from strong mixing processes. *Ann. Math. Statist.* **42**, 1637–1646.

Welsch, R. E. (1972). Limit laws for extreme order statistics from strong mixing processes. *Ann. Math. Statist.* **43**, 439–446.

Welsch, R. E. (1973). A convergence theorem for extreme values from Gaussian sequences. *Ann. Probability* **1**, 398–404.

Whittle, P. (1959). Sur la distribution du maximum d'un polynome trigonométrique à coefficients aléatoires. *Calc. Prob. Appl. Internat. Centre Nat. Res. Sci.* **87**, 173–184.

Wichura, M. J. (1974). On the functional form of the law of iterated logarithm for the partial maxima of independent identically distributed random variables. *Ann. Probability* **2**, 202–230.

Williams, D. (1973). On Rényi's record problem and Engel's series. *Bull. London Math. Soc.* **5**, 235–237.

Yang, M. C. K. (1975). On the distribution of the inter-record times in an increasing population. *J. Appl. Probability* **12**, 148–154.

Yang, S. S. (1977). General distribution theory of the concomitants of order statistics. *Ann. Statist.* **5**, 996–1002.

Index

For authors, see the sections entitled "Survey of the Literature," which appear at the end of each chapter. For special topics on Extremes, Maxima, and Minima, see the relevant individual entries.

Air pollution, 2, 118, 194, 200
Almost sure results, 213-232, 238, 240
Applications
   engineering, 25-26, 90-91; *see also*
     Failure of equipment
   insurance, 32, 40
   medical, 25, 200
   see also Air pollution, Characterizations, Droughts, Floods
Asymptotic distribution; *see* Limiting distribution
Asymptotic independence
   of extremes in the univariate case: i.i.d. variables, 106, 118; m-dependence and mixing, 203
   of extremes of components of vectors, 257, 272, 276

Bayesian approach, 148, 197
Binomial moments, 17, 42, 132, 143
Bivariate; *see* Multivariate distributions
Borel-Cantelli lemma, 210-211, 239
Bonferroni inequalities; *see* Inequalities

Cauchy distribution, 68, 233
Chain dependence, 199
Characterizations, 31, 33, 38, 41, 43-44, 188
   applications of, 188
   see also Events, Exchangeable events, Limiting distribution
Compactness, of distribution functions, 327
Comparison of maxima
   for two Gaussian sequences, 164, 169, 173-174
   for two uniformly close population distributions, 90
Competing risks, 196, 200
Completeness theorem, 322
Concomitants, of order statistics, 267-270, 273
Convolutions, 118, 123
Corrosion, 2, 4
Criterion of von Mises, for domains of attraction, 117-118

Degenerate limit laws, 205, 238
Dependence function, of multivariate distributions, 250-251; *see also* Limiting distribution
Dependent models; *see* Dependent variables
Dependent variables, 7, 15, 42, 44
   asymptotic distribution of extremes: existence of, *see* Weak convergence; possible forms, *see* Limiting distribution
   distribution of order statistics: exact formulas, 16, 19, 129, 134; examples, 25-30; general estimates, 20-21, 24
   laws of large numbers, for extremes in, 206, 208, 212
   special structures, definition: E_n-sequence, 176; E_n^*-sequence, 177; exchangeable, 127; Gaussian, 163; m-dependent, 156; mixing (in tail), 156; stationary, 156; theoretical model, 143
Discrete populations, extremes in, 118, 120, 122-123, 239, 243
Domains of attraction, 70

349

INDEX

criteria for i.i.d. variables, 71, 93-94, 117, 121-123
multivariate extremes, 252, 258, 263
relations of sums and extremes, 233, 235
the role of moments, 97, 99
Dominated convergence theorem, 317
Droughts, 195

E_n-sequences, 176
applications: failure models, 196; floods, 194; strength of materials, 189-191
limiting distribution of minima, 177, 199
weak convergence, sufficient conditions, 176-177
E_n^*-sequences, 177
limiting distribution of maxima, 177, 199
weak convergence, sufficient conditions, 177
Events, distribution of the number of occurrences: exact formulas, 19, 134; estimates, 20-21, 24, 47-48; relation to exchangeable events, 129, 199
Exchangeable events, 129
characterization of probabilities of intersections of, 129
finite and infinite sequences, 127-128, 201
relation to arbitrary events, 129
relation to sampling without replacement (Kendall's representation), 133-134, 199
see also Exchangeable variables
Exchangeable variables, 127
finite sequences, 127-128
infinite sequences: the deFinetti representation, 148
limiting distribution of extremes, 150-155
role in Bayesian statistics, 148, 197
weak convergence, of normalized extremes, 150-155
see also Exchangeable events
Exponential distribution, 12, 30
a Bayesian example, 154
bivariate: 202-203, 247, 249; Morgenstern's, 247, 250, 266; Gumbel's, 247; Marshall-Olkin, 247, 249; Mardia's, 247, 249, 250
characterizations, 31-33, 38, 43-44, 188
dependent examples, 155, 186
as limiting distribution, 91

range and midrange in, 110
speed of convergence of extremes in, 115, 121
Extremal processes, 304-308
Extremes, definitions, 16, 244, 277; *for special topics, see relevant individual entries*

Failure of equipment, 2, 4, 28, 91, 124, 188, 195
Fatou's lemma, 323
Floods, 1, 3, 5, 194
Fréchet's bounds, for multivariate distributions, 245, 259

Gaussian sequences, 163, 198, 203, 207, 239
Geometric distribution, 120
Goodness-of-fit tests, 90, 200, 265

Hazard rate, 196-197

Independent and identically distributed variables (i.i.d), 7
almost sure results, 213-225, 227, 238
approximation by, 192, 194-195, 197
bounds on the distribution of extremes in, 10, 15
characterizations, 33, 38, 41, 44, 188
joint density of order statistics, 36-37
limiting distributions of extremes in, 51-52, 56-57, 71, 103, 117
midrange, 105-106
range, 105, 108
reduction of inequalities to, 15, 42
relation of extremes and sums, 232-233, 236, 239
speed of convergence, 113, 115, 121
strong laws for extremes, 227, 230
weak convergence of extremes: necessary conditions, 81, 83, 97, 117; necessary and sufficient conditions, 71, 102-103, 117; sufficient conditions, 11, 51-52, 56-57, 93-94, 117; *see also* Domains of attraction, Multivariate distributions, Normalizing constants
weak laws, 208
Independent variables, 135, 180
almost sure results for extremes in, 225-226, 240

INDEX

approximation of dependent models by, 180, 199
limiting distribution of extremes in, 180-181, 199
uniformity assumption, 180
weak convergence of extremes, special cases, 204
see also Independent and identically distributed variables
Inequalities, 8, 9, 47
with binomial moments (Bonferroni inequalities), 20-21, 24, 42-43, 46
for distribution of extremes, 10, 15, 164
methods of proof for, 43-45
for multivariate distributions, 245-246
for multivariate (limiting) distributions of extremes, 255, 259, 267, 277
for multivariate normal distributions, 164, 198
Iterated logarithm theorem, for sums, 320
for extremes; see Lim inf and Lim sup

Lack of memory, 30
Large deviations, 119-120, 123
Laws of large numbers; see Strong law, Weak law
Lim inf, of normalized extremes, 226-227, 238-239, 243
Limiting distribution
of normalized extremes: E_n-sequence, 177; E_n^*-sequence, 177; exchangeable variables, 150-153; Gaussian variables, 166, 169-171; general structure, 144, 185; independent variables, 180-181; i.i.d. variables, 51-52, 56-57, 71, 103, 117; m-dependence, 157, 185; multivariate, 251, 253-255, 258-262, 265; in samples with random size, 282, 284, 289, 308; stationary mixing sequences, 157, 185
the parametric form of von Mises, 52
for records, 302
on a subsequence, of maxima (partial attraction of maxima), 92, 121
Lim sup, of normalized extremes, 225, 227, 238-239, 241-243
Lognormal distribution, 67, 90, 118, 120, 194

m-dependent sequences, 156, 192, 203, 276; see also Dependent variables, Limiting distribution, Weak convergence
Maximum, notations, 3-5, 16, 244; for special topics, see relevant individual entries
Midrange, 105
weak convergence, 108
Minimum, notations, 3-5, 16, 244; for special topics, see relevant individual entries
Mixing, in tails, 156-157, 198
limiting distribution, of normalized extremes, 157, 185
weak convergence, 157, 185, 198
Mixtures, limit theorems for, 137, 141-143
relation to sequences of events, 137
Moments, characterizations by, 33, 327
of extremes: convergence, 118; roles in domains of attraction, 97, 99
Multivariate distributions, 245, 265
asymptotic independence of extremes of the components, 257, 272, 276
dependence function for, 250-254, 258-259, 264
domains of attraction, 252, 258, 263
the Fréchet bounds, 245
limiting distribution of extremes, 251, 253-255, 258-262, 265
nondegenerate, 247
normal, 163; see also Gaussian sequences
the Pickands representation, 265
weak convergence, 247, 248; of extremes, 252, 255, 257, 263, 271-272

Natural disasters, 1, 3-5, 194-195
Normal distribution, 65, 117-118
as limiting distribution, 166, 171, 192, 198, 204
midrange, 109
multivariate: as distribution of sample, see Gaussian sequences; as multivariate population distribution, 257, 265, 267
normalizing constants for extremes, 65, 67
range, 109
speed of convergence, 116
strong laws for extremes, 227-230
uniform closeness to lognormal, 90, 120
weak convergence of extremes, 65, 105
weak law for extremes, 135

Normalizing constants
  in almost sure results, 227
  characteristic to the extremes, 144-145
  choice for extremes, in weak convergence:
    for i.i.d. variables, 51-52, 56-57, 79, 102-103, 123; in multivariate case, 249
  non-uniqueness of, 57, 59
  the structure of, 60-61, 63; with random index, 285
  in weak laws of large numbers, 205-208

Order statistics, definition, 16

Parametric form of von Mises, 52
Pareto distribution, 100
Partial attraction of maxima; see Limiting distribution, on a subsequence
Pickands' representation, 265
Planck distribution, 101
Poisson distribution, 120
Population distribution
  choice of: by asymptotic theory, 189-191, 193-196; by characterizations, 188; by statistical methods, 193-195
  effect on extremes, 90, 200

Queues, 201, 280, 308

Random measurements, 3-4
Random sample size, 282, 308-310
  limiting distribution of extremes with, 282, 284, 289, 308; see also Records
Range, 105
  weak convergence of, 108
Record times, 290-291, 309, 312-313
  distribution of, 291-293, 295-296, 298
Records, 291, 300, 309-310, 312
  limiting distribution of, 302
Reliability theory, 197, 200-201; see also Applications, engineering, Failure of equipment

Safety regulation, 200
Service time, 2
Simplex, 264
Slowly varying functions, 118, 235, 329-330
Speed of convergence, 111, 113, 115, 119, 121
Stationary sequences, 156, 198
Statistical samples, 3, 7, 90
Strength, breaking, 2, 4, 189
  bundles of threads, 191, 200
Strong law of large numbers, 229-230
Sums, with random size, 310-311
  relation to extremes, 232-239

Tail equivalence, 118, 122
Theoretical model, 143-148
Total probability rule, 315
  continuous version, 315
Transformation, of distributions, 6-7, 41, 48, 64, 89, 102, 272
Type of distribution, 70, 193, 195
Types of limiting distribution; see Limiting distribution

Uniform convergence, of distributions, 111
Uniform distribution, 64, 91, 110, 120, 222-223
Uniformity assumption, for independent variables, 180

Weak convergence, 50, 247
  of extremes: E_n-sequence, 176; E_n^*-sequence, 177; exchangeable, 150-153; Gaussian variables, 166, 169-170; general structure, 144-145, 185-186; independent variables, 204; i.i.d. variables, 11, 51-52, 56-57, 71, 81, 83, 97, 102-103, 117; m-dependent variables, 157, 162; multivariate, 252, 255, 257, 263, 271-272; in samples with random size, 282, 289; for stationary mixing sequences, 157, 161, 198, 203
Weak law of large numbers, 206, 238-239
  additive, 207-208
  multiplicative, 207-208
  relation to outliers, 239, 241
Weakest-link principle, 191
Weibull distribution, 91, 189-191, 207

Applied Probability and Statistics (*Continued*)

GROSS and CLARK · Survival Distributions: Reliability Applications in the Biomedical Sciences
GROSS and HARRIS · Fundamentals of Queueing Theory
GUTTMAN, WILKS, and HUNTER · Introductory Engineering Statistics, *Second Edition*
HAHN and SHAPIRO · Statistical Models in Engineering
HALD · Statistical Tables and Formulas
HALD · Statistical Theory with Engineering Applications
HARTIGAN · Clustering Algorithms
HILDEBRAND, LAING, and ROSENTHAL · Prediction Analysis of Cross Classifications
HOEL · Elementary Statistics, *Fourth Edition*
HOLLANDER and WOLFE · Nonparametric Statistical Methods
HUANG · Regression and Econometric Methods
JAGERS · Branching Processes with Biological Applications
JESSEN · Statistical Survey Techniques
JOHNSON and KOTZ · Distributions in Statistics
    Discrete Distributions
    Continuous Univariate Distributions-1
    Continuous Univariate Distributions-2
    Continuous Multivariate Distributions
JOHNSON and KOTZ · Urn Models and Their Application: An Approach to Modern Discrete Probability Theory
JOHNSON and LEONE · Statistics and Experimental Design in Engineering and the Physical Sciences, Volumes I and II, *Second Edition*
KEENEY and RAIFFA · Decisions with Multiple Objectives
LANCASTER · An Introduction to Medical Statistics
LEAMER · Specification Searches: Ad Hoc Inference with Nonexperimental Data
McNEIL · Interactive Data Analysis
MANN, SCHAFER, and SINGPURWALLA · Methods for Statistical Analysis of Reliability and Life Data
MEYER · Data Analysis for Scientists and Engineers
OTNES and ENOCHSON · Digital Time Series Analysis
PRENTER · Splines and Variational Methods
RAO and MITRA · Generalized Inverse of Matrices and Its Applications
SARD and WEINTRAUB · A Book of Splines
SEARLE · Linear Models
THOMAS · An Introduction to Applied Probability and Random Processes
WHITTLE · Optimization under Constraints
WILLIAMS · A Sampler on Sampling
WONNACOTT and WONNACOTT · Econometrics
WONNACOTT and WONNACOTT · Introductory Statistics, *Third Edition*
WONNACOTT and WONNACOTT · Introductory Statistics for Business and Economics, *Second Edition*
YOUDEN · Statistical Methods for Chemists
ZELLNER · An Introduction to Bayesian Inference in Econometrics